The Art and Science of
Analog Circuit Design

The EDN Series for Design Engineers

J. Williams	*The Art and Science of Analog Circuit Design*
J. Lenk	*Simplified Design of Switching Power Supplies*
V. Lakshminarayanan	*Electronic Circuit Design Ideas*
J. Lenk	*Simplified Design of Linear Power Supplies*
M. Brown	*Power Supply Cookbook*
B. Travis and I. Hickman	*EDN Designer's Companion*
J. Dostal	*Operational Amplifiers, Second Edition*
T. Williams	*Circuit Designer's Companion*
R. Marston	*Electronics Circuits Pocket Book: Passive and Discrete Circuits (Vol. 2)*
N. Dye and H. Granberg	*Radio Frequency Transistors: Principles and Practical Applications*
Gates Energy Products	*Rechargeable Batteries: Applications Handbook*
T. Williams	*EMC for Product Designers*
J. Williams	*Analog Circuit Design: Art, Science, and Personalities*
R. Pease	*Troubleshooting Analog Circuits*
I. Hickman	*Electronic Circuits, Systems and Standards*
R. Marston	*Electronic Circuits Pocket Book: Linear ICs (Vol. 1)*
R. Marston	*Integrated Circuit and Waveform Generator Handbook*
I. Sinclair	*Passive Components: A User's Guide*

The Art and Science of Analog Circuit Design

....

Edited by

Jim Williams

Butterworth–Heinemann

An Imprint of Elsevier

Boston   Oxford   Melbourne   Singapore   Toronto   Munich   New Delhi   Tokyo

ISBN-13: 978-0-7506-7062-3   ISBN-10: 0-7506-7062-2

A catalogue record for this book is available from the British Library.

The publisher offers special discounts on bulk orders of this book.
For information, please contact:
Manager of Special Sales
Butterworth-Heinemann
225 Wildwood Avenue
Woburn, MA 01801-2041
Tel: 781-904-2500
Fax: 781-904-2620

For information on all Butterworth–Heinemann publications available, contact our World Wide
Web home page at: http://www.bh.com

Transferred to Digital Printing 20011

MIT building 20 at 3:00 A.M.
Tek. 547, pizza, breadboard.
That's Education.

Contents

....

Preface   ix
Contributors   xi

Part One   Learning How
....

Part Two   Making It Work
....

Preface

This book continues the approach originated in an earlier effort, "Analog Circuit Design—Art, Science, and Personalities." In that book twenty-six authors presented tutorial, historical, and editorial viewpoints on subjects related to analog circuit design. The book encouraged readers to develop their own approach to design. It attempted this by presenting the divergent methods and views of people who had achieved some measure of success in the field. A complete statement of this approach was contained in the first book's preface, which is reprinted here (immediately following) for convenience.

The surprisingly enthusiastic response to the first book has resulted in this second effort. This book is similar in spirit, but some changes have occurred. The most obvious difference is that almost all contributors are new recruits. This seems a reasonable choice: new authors with new things to say, hopefully augmenting the first book's message.

Although accomplished, some of this book's writers are significantly younger and have less experience at analog design than the previous book's authors. This is deliberate, and an attempt to maintain a balanced and divergent forum unencumbered by an aging priesthood.

A final difference is the heavy capitalistic and marketeering influence in many of the chapters. This unplanned emphasis is at center stage in sections by Grant, Williams, Brown, and others, and appears in most chapters. The influence of economics was present in parts of the earlier book, but is much more pronounced here. The pristine pursuit of circuit design is tempered by economic realities, and the role of money as design motivator and modulator is undeniable.

We hope this book is as well received as the earlier effort, even as it broadens the scope of topics and utilizes new authors. As before, it was fun to put together. If we have done our job, it should be rewarding for the reader.

Preface to "Analog Circuit Design—Art, Science, and Personalities"

This is a weird book. When I was asked to write it I refused, because I didn't believe anybody could, or should, try to explain how to do analog design. Later, I decided the book might be possible, but only if it was written by many authors, all with their own styles, topics, and opinions.

There should be an absolute minimum of editing, no subject or style requirements, no planned page count, no outline, no nothing! I wanted the book's construction to reflect its subject. What I asked for was essentially a mandate for chaos. To my utter astonishment the publisher agreed and we lurched hopefully forward.

A meeting at my home in February 1989 was well attended by potential participants. What we concluded went something like this: everyone would go off and write about anything that could remotely be construed as relevant to analog design. Additionally, no author would tell any other author what they were writing about. The hope was that the reader would see many different styles and approaches to analog design, along with some commonalities. Hopefully, this would lend courage to someone seeking to do analog work. There are many very different ways to proceed, and every designer has to find a way that feels right.

This evolution of a style, of getting to know oneself, is critical to doing good design. The single greatest asset a designer has is self-knowledge. Knowing when your thinking feels right, and when you're trying to fool yourself. Recognizing when the design is where you want it to be, and when you're pretending it is because you're only human. Knowing your strengths and weaknesses, prowesses and prejudices. Learning to recognize when to ask questions and when to believe your answers.

Formal training can augment all this, but cannot replace it or obviate its necessity. I think that factor is responsible for some of the mystique associated with analog design. Further, I think that someone approaching the field needs to see that there are lots of ways to do this stuff. They should be made to feel comfortable experimenting and evolving their own methods.

The risk in this book, that it will come across as an exercise in discord, is also its promise. As it went together, I began to feel less nervous. People wrote about all kinds of things in all kinds of ways. They had some very different views of the world. But also detectable were commonalities many found essential. It is our hope that readers will see this somewhat discordant book as a reflection of the analog design process. Take what you like, cook it any way you want to, and leave the rest.

Things wouldn't be complete without a special thanks to Carol Lewis and Harry Helms at High Text Publications, and John Martindale at Butterworth–Heinemann Publishers. They took on a book with an amorphous charter and no rudder and made it work. A midstream change of publishers didn't bother Carol and Harry, and John didn't seem to get nervous over a pretty risky approach to book writing.

I hope this book is as interesting and fun to read as it was to put together. Have a good time.

Contributors

JIM WILLIAMS is the editor-in-chief of this second volume on analog circuit design. As with the first volume, Jim developed the basic concept of the book, identified, contacted, and cajoled potential contributors, and edited the contributions. Jim was at the Massachusetts Institute of Technology from 1968 to 1979, concentrating exclusively on analog circuit design. His teaching and research interests involved application of analog circuit techniques to biochemical and biomedical problems. Concurrently, he consulted U.S. and foreign concerns and governments, specializing in analog circuits. In 1979, he moved to National Semiconductor Corporation, continuing his work in the analog area with the Linear Integrated Circuits Group. In 1982 he joined Linear Technology Corporation as staff scientist, where he is presently employed. Interests include product definition, development, and support. Jim has authored over 250 publications relating to analog circuit design. He received the 1992 Innovator of the Year Award from *EDN* Magazine for work in high-speed circuits. His spare time interests include sports cars, collecting antique scientific instruments, art, and restoring and using old Tektronix oscilloscopes. He lives in Palo Alto, California with his son Michael, a dog named Bonillas, and 28 Tektronix oscilloscopes.

CARL BATTJES has worked in the analog design of systems with a focus on detailed design at the bipolar transistor device and bipolar IC level. He has been involved in the design of Tektronix, Inc. oscilloscopes and their components, such as delay lines, filters, attenuators, and amplifiers. For the Grass Valley Group, he developed a precision analog multiplier for video effects. Carl has been a consultant for over ten years and has done major detailed designs for the Tektronix 11A72 pre-amp IC, Seiko message watch receiver IC, and IC for King Radio (Allied Signal) receiver. A registered Professional Engineer in Oregon who holds seven patents, he has a BSEE from the University of Michigan and an MSEE from Stanford University.

JAMES BRYANT is head of European applications at Analog Devices. He lives in England and is a Eur. Ing. and MIEE and has degrees in philosophy and physics from the University of Leeds. He has over twenty years' experience as an analog and RF applications engineer and is well known as a lecturer and author. His other interests include archery, cooking, ham radio (G4CLF), hypnotism, literature, music, and travel.

ART DELAGRANGE, when he was young, took his electric train apart and reassembled it by himself. Since that day, it has not run. He attended MIT, where he studied digital circuitry, receiving a BS/MS in electrical engineering in 1961/62. During his graduate year he worked on a hybrid digital/analog computer. It did not revolutionize the industry. Beginning as a co-op student, he worked for 33 years for the Naval Surface Warfare Center in Silver Spring, Maryland. Among his other achievements are a PhD in electrical engineering from the University of Maryland, ten patents, and 23 articles in the open literature. Retired from the government, he works for Applied Technology and Research in Burtonsville, Maryland. Art lives in Mt. Airy, Maryland, with his wife, Janice, and his cat, Clumsy. His hobbies are cars, boats, sports, music, and opening packages from the wrong end.

RICHARD P. FEYNMAN was professor of physics at the California Institute of Technology. He was educated at MIT and Princeton, and worked on the Manhattan Project during World War II. He received the 1965 Nobel Prize in Physics for work in quantum electrodynamics. His life and style have been the subject of numerous biographies. He was an uncommonly good problem solver, with notable ability to reduce seemingly complex issues to relatively simple terms. His *Feynman Lectures on Physics,* published in the 60s, are considered authoritative classics. He died in 1988.

BARRIE GILBERT has spent most of his life designing analog circuits, beginning with four-pin vacuum tubes in the late 1940s. Work on speech encoding and synthesis at the Signals Research and Development Establishment in Britain began a love affair with the bipolar transistor that shows no signs of cooling off. Barrie joined Analog Devices in 1972, where he is now a Division Fellow working on a wide variety of IC products and processes while managing the Northwest Labs in Beaverton, Oregon. He has published over 40 technical papers and been awarded 20 patents. Barrie received The IEEE Outstanding Achievement Award in 1970, was named an IEEE Fellow in 1984, and received the IEEE Solid-State Circuits Council Outstanding Development Award in 1986. For recreation, Barrie used to climb mountains, but nowadays stays home and tries to write music in a classical style for performance on a cluster of eight computer-controlled synthesizers and other toys.

DOUG GRANT received a BSEE degree from the Lowell Technological Institute (now University of Massachusetts–Lowell) in 1975. He joined Analog Devices in 1976 as a design engineer and has held several positions in engineering and marketing prior to his current position as marketing manager for RF products. He has authored numerous papers and articles on mixed-signal and linear circuits, as well as his amateur radio hobby.

BILL GROSS is a design manager for Linear Technology Corporation, heading a team of design engineers developing references, precision amplifiers, high-speed amplifiers, comparators, and other high-speed products. Mr. Gross has been designing integrated circuits for the semiconductor industry for 20 years, first at National Semiconductor, including three years living and working in Japan, and later at Elantec. He has a BSEE from California State Polytechnic University at Pomona and an MSEE from the University of Arizona at Tucson. He is married and the father of two teenage sons, whose sports activities keep him quite busy.

BARRY HARVEY is a designer of bipolar analog integrated circuits at Elantec, Inc. His first electronic projects were dismantling vacuum tube television sets as a child and later in life rebuilding them. These days he tortures silicon under a microscope.

GREGORY T.A. KOVACS received a BASc degree in electrical engineering from the University of British Columbia, Vancouver, British Columbia, in 1984; an MS degree in bioengineering from the University of California, Berkeley, in 1985; a PhD degree in electrical engineering from Stanford University in 1990; and an MD degree from Stanford University in 1992. His industry experience includes the design of a wide variety of analog and mixed-signal circuits for industrial and commercial applications, patent law consulting, and the co-founding of three electronics companies. In 1991, he joined Stanford University as Assistant Professor of Electronic Engineering, where he teaches analog circuit design and micromachined transducer technologies. He holds the Robert N. Noyce Family Faculty Scholar Chair, received an NSF Young Investigator Award in 1993, and was appointed a Terman Fellow in 1994. His present research areas include neural/electronic interfaces, solid-state sensors and actuators, micromachining, analog circuits, integrated circuit fabrications, medical instruments, and biotechnology.

CARL NELSON is Linear Technology's Bipolar Design Manager. He has 25 years in the semiconductor IC industry. Carl joined Linear Technology shortly after the company was founded. He came from National Semiconductor and before that worked for Teledyne Semiconductor. He has a BSEE from the Northrup Institute of Technology. He is the designer of the first temperature-sensor IC and is the father of the LT1070/1270 family of easy-to-use switching regulators. He holds more than 30 patents on a wide range of analog integrated circuits.

ROBERT REAY became an analog designer after discovering as a teenager that the manual for his Radio Shack electronics kit didn't describe how any of the circuits really worked. His scientific curiosity and realization that he wasn't going to make any money as a pianist led him to Stanford University, where he earned his BSEE and MSEE in 1984. He worked for Intersil, designing data conversion products, for four years before Maxim hired away most of the design team. He is currently managing a group of designers at Linear Technology Corporation, doing interface circuits, battery chargers, DACs, references, comparators, regulators, temperature sensors, and anything else that looks interesting. He regularly plays roller blade hockey with the kids in the neighborhood and is helping his children discover the beauty of a Chopin waltz and a well-designed circuit.

STEVE ROACH received his BS in engineering physics from the University of Colorado in 1984 and his MS in electrical engineering from Ohio State University in 1988. He worked from 1984 to 1986 as a software engineer for Burroughs Corporation and from 1988 to 1992 at Hewlett-Packard Company, designing digital oscilloscopes. From 1992 to 1994, Stephen designed industrial sensors at Kaman Instrumentation Company. He is currently designing digital oscilloscopes for Hewlett-Packard. His hobbies include backpacking, hunting, off-road motorcycling, and tutoring kids at the Boys' and Girls' Club.

KEITARO SEKINE received his BE, ME, and Dr. Eng. degrees in electronics from Waseda University in 1960, 1962, and 1968, respectively. Since 1969, he has been with the Faculty of Science and Technology, Science University of Tokyo, where he is now a professor in the Department of Electrical Engineering. His main research interests are in analog integrated circuits and their application systems. His interests in the physical aspects of analog circuits, such as implementation, mutual electro-magnetic couple within the circuits, and EMC, originated from the experiments at his own amateur radio station, which he has had since 1957. He has been chair of the Committee for Investigative Research and Committee on Analog Circuit Design Technologies at the Institute of Electrical Engineers of Japan (IEEJ) and also a member of the Editorial Committee for the Transactions of IEICE Section J-C. He is now president of the Society for Electronics, Information, and System at the IEEJ, as well as a member of the Board of Directors at the Japan Institute of Printed Circuit (JIPC). Dr. Sekine is a member of the Institute of Electrical and Electronics Engineers, the IEEJ, and the JIPC.

ERIC SWANSON received his BSEE from Michigan State University in 1977 and his MSEE from Cal Tech in 1980. From 1980 to 1985 he worked on a variety of analog LSI circuits at AT&T-Bell Laboratories in Reading, Pennsylvania. In 1985 he joined Crystal Semiconductor in Austin, Texas, where he is currently Vice President of Technology. His development experience includes millions of CMOS transistors, a few dozen bipolar transistors, and nary a vacuum tube. Eric holds 20 patents, evenly divided between the analog and digital domains, and continues to design high-performance data converters. He enjoys swimming and biking with his wife Carol and four children.

JOHN WILLISON is the founder of Stanford Research Systems and the Director of R&D. Considered a renegade for having left "pure research" after completing a PhD in atomic physics, he continues to enjoy designing electronic instruments in northern California. Married with four children, he's in about as deep as you can get.

Learning How

The book's initial chapters present various methods for learning how to do analog design. Jim Williams describes the most efficient educational mechanism he has encountered in "The Importance of Fixing." A pair of chapters from Barry Harvey emphasize the importance of realistic experience and just how to train analog designers. Keitaro Sekine looks at where future Japanese analog designers will come from. He has particularly pungent commentary on the effects of "computer-based" design on today's students. Similar concerns come from Stanford University professor Greg Kovacs, who adds colorful descriptions of the nature of analog design and its practitioners. Finally, Nobel prize-winning physicist Richard P. Feynman's 1974 Cal Tech commencement address is presented. Although Feynman wasn't an analog circuit designer, his observations are exceptionally pertinent to anyone trying to think clearly about anything.

1. The Importance of Fixing

Fall 1968 found me at MIT preparing courses, negotiating thesis topics with students, and getting my laboratory together. This was fairly unremarkable behavior for this locale, but for a 20 year old college dropout the circumstances were charged; the one chance at any sort of career. For reasons I'll never understand, my education, from kindergarten to college, had been a nightmare, perhaps the greatest impedance mismatch in history. I got hot. The Detroit Board of Education didn't. Leaving Wayne State University after a dismal year and a half seemed to close the casket on my circuit design dreams.

All this history conspired to give me an outlook blended of terror and excitement. But mostly terror. Here I was, back in school, but on the other side of the lectern. Worse yet, my research project, while of my own choosing, seemed open ended and unattainable. I was so scared I couldn't breathe out. The capper was my social situation. I was younger than some of my students, and my colleagues were at least 10 years past me. To call things awkward is the gentlest of verbiage.

The architect of this odd brew of affairs was Jerrold R. Zacharias, eminent physicist, Manhattan Project and Radiation Lab alumnus, and father of atomic time. It was Jerrold who waved a magic wand and got me an MIT appointment, and Jerrold who handed me carte blanche a lab and operating money. It was also Jerrold who made it quite clear that he expected results. Jerrold was not the sort to tolerate looking foolish, and to fail him promised a far worse fate than dropping out of school.

Against this background I received my laboratory budget request back from review. The utter, untrammeled freedom he permitted me was maintained. There were no quibbles. Everything I requested, even very costly items, was approved, without comment or question. The sole deviation from this I found annoying. He threw out my allocation for instrument repair and calibration. His hand written comment: "You fix everything."

It didn't make sense. Here I was, under pressure for results, scared to pieces, and I was supposed to waste time screwing around fixing lab equipment? I went to see Jerrold. I asked. I negotiated. I pleaded, I ranted, and I lost. The last thing I heard chasing me out of his office was, "You fix everything."

I couldn't know it, but this was my introduction to the next ten years. An unruly mix of airy freedom and tough intellectual discipline that would seemingly be unremittingly pounded into me. No apprenticeship was ever more necessary, better delivered, or, years later, as appreciated.

I cooled off, and the issue seemed irrelevant, because nothing broke for a while. The first thing to finally die was a high sensitivity, differential 'scope plug-in, a Tektronix 1A7. Life would never be the same.

The problem wasn't particularly difficult to find once I took the time to understand how the thing worked. The manual's level of detail and writing tone were notable; communication was *the* priority. This seemed a significant variance from academic publications, and I was impressed. The instrument more than justified the manual's efforts. It was gorgeous. The integration of mechanicals, layout, and electronics was like nothing I had ever seen. Hours after the thing was fixed I continued to probe and puzzle through its subtleties. A common mode bootstrap scheme was particularly interesting; it had direct applicability to my lab work. Similarly, I resolved to wholesale steal the techniques used for reducing input current and noise.

Over the next month I found myself continually drifting away from my research project, taking apart test equipment to see how it worked. This was interesting in itself, but what I really wanted was to test my

Figure 1–1. Oh boy, it's broken! Life doesn't get any better than this.

understanding by having to fix it. Unfortunately, Tektronix, Hewlett-Packard, Fluke, and the rest of that ilk had done their work well; the stuff didn't break. I offered free repair services to other labs who would bring me instruments to fix. Not too many takers. People had repair budgets . . . and were unwilling to risk their equipment to my unproven care. Finally, in desperation, I paid people (in standard MIT currency—Coke and pizza) to deliberately disable my test equipment so I could fix it. Now, their only possible risk was indigestion. This offer worked well.

A few of my students became similarly hooked and we engaged in all forms of contesting. After a while the "breakers" developed an armada of incredibly arcane diseases to visit on the instruments. The "fixers" countered with ever more sophisticated analysis capabilities. Various games took points off for every test connection made to an instrument's innards, the emphasis being on how close you could get utilizing panel controls and connectors. Fixing without a schematic was highly regarded, and a consummately macho test of analytical skill and circuit sense. Still other versions rewarded pure speed of repair, irrespective of method.[1] It really was great fun. It was also highly efficient, serious education.

The inside of a broken, but well-designed piece of test equipment is an extraordinarily effective classroom. The age or purpose of the instrument is a minor concern. Its instructive value derives from several perspectives.

It is always worthwhile to look at how the designer(s) dealt with problems, utilizing available technology, and within the constraints of cost, size, power, and other realities. Whether the instrument is three months or thirty years old has no bearing on the quality of the thinking that went into it. Good design is independent of technology and basically timeless. The clever, elegant, and often interdisciplinary approaches found in many instruments are eye-opening, and frequently directly applicable to your own design work. More importantly, they force self-examination, hopefully preventing rote approaches to problem solving, with their attendant mediocre results. The specific circuit tricks you see are certainly adaptable and useful, but not nearly as valuable as studying the thought process that produced them.

The fact that the instrument is broken provides a unique opportunity. A broken instrument (or anything else) is a capsulized mystery, a puzzle with a definite and very singular "right" answer. The one true reason why that instrument doesn't work as it was intended to is really there. You are forced to measure your performance against an absolute, non-negotiable standard; the thing either works or it doesn't when you're finished.

---

1. A more recent development is "phone fixing." This team exercise, derived by Len Sherman (the most adept fixer I know) and the author, places a telephone-equipped person at the bench with the broken instrument. The partner, somewhere else, has the schematic and a telephone. The two work together to make the fix. A surprise is that the time-to-fix seems to be less than if both parties are physically together. This may be due to dilution of ego factors. Both partners simply must speak and listen with exquisite care to get the thing fixed.

The reason all this is so valuable is that it brutally tests your thinking process. Fast judgments, glitzy explanations, and specious, hand-waving arguments cannot be costumed as "creative" activity or true understanding of the problem. After each ego-inspired lunge or jumped conclusion, you confront the uncompromising reality that the damn thing still doesn't work. The utter closedness of the intellectual system prevents you from fooling yourself. When it's finally over, and the box works, and you know why, then the real work begins. You get to try and fix you. The bad conclusions, poor technique, failed explanations, and crummy arguments all demand review. It's an embarrassing process, but quite valuable. You learn to dance with problems, instead of trying to mug them.

It's scary to wonder how much of this sort of sloppy thinking slips into your own design work. In that arena, the system is not closed. There is no arbitrarily right answer, only choices. Things can work, but not as well as they might if your thinking had been better. In the worst case, things work, but for different reasons than you think. That's a disaster, and more common than might be supposed. For me, the most dangerous point in a design comes when it "works." This ostensibly "proves" that my thinking is correct, which is certainly not necessarily true. The luxury the broken instrument's closed intellectual system provides is no longer available. In design work, results are open to interpretation and explanation and that's a very dangerous time. When a design "works" is a very delicate stage; you are psychologically ready for the kill and less inclined to continue testing your results and thinking. That's a precarious place to be, and you have to be so careful not to get into trouble. The very humanness that drives you to solve the problem can betray you near the finish line.

What all this means is that fixing things is excellent exercise for doing design work. A sort of bicycle with training wheels that prevent you from getting into too much trouble. In design work you have to mix a willingness to try anything with what you hope is critical thinking. This seemingly immiscible combination can lead you to a lot of nowheres. The broken instrument's narrow, insistent test of your thinking isn't there, and you can get in a lot deeper before you realize you blew it. The embarrassing lessons you're forced to learn when fixing instruments hopefully prevent this. This is the major reason I've been addicted to fixing since 1968. I'm fairly sure it was also Jerrold's reason for bouncing my instrument repair allocation.

There are, of course, less lofty adjunct benefits to fixing. You can often buy broken equipment at absurdly low cost. I once paid ten bucks for a dead Tektronix 454A 150MHz portable oscilloscope. It had clearly been systematically sabotaged by some weekend-bound calibration technician and tagged "Beyond Repair." This machine required thirty hours to uncover the various nasty tricks played in its bowels to ensure that it was scrapped.

This kind of devotion highlights another, secondary benefit of fixing. There is a certain satisfaction, a kind of service to a moral imperative, that comes from restoring a high-quality instrument. This is unquestionably a gooey, hand-over-the-heart judgment, and I confess a long-term love affair with instrumentation. It just seems sacrilege to let a good piece of equipment die. Finally, fixing is simply a lot of fun. I may be the only person at an electronics flea market who will pay more for the busted stuff!

2. How to Grow Strong, Healthy Engineers

Graduating engineering students have a rough time of it lately. Used to be, most grads were employable and could be hired for many jobs. Ten years ago and earlier, there were a lot of jobs. Now, there aren't so many and employers demand relevant course work for the myriad of esoteric pursuits in electrical engineering. Of those grads that do get hired, the majority fail in their first professional placement.

We should wonder, is this an unhealthy industry for young engineers? Well, I guess so. Although I am productive and comfortable now, I was not successful in my first three jobs, encompassing nine years of professional waste. Although I designed several analog ICs that worked in this period, none made it to market.

Let me define what I call professional success:

The successful engineer delivers to his or her employer at least 2½ times the yearly salary in directly attributable sales or efficiency. It may take years to assess this.

For many positions, it's easy to take this measure. For others, such as in quality assurance, one assays the damage done to the company for not executing one's duties. This is more nebulous and requires a wider business acumen to make the measure. At this point, let me pose what I think is the central function of the engineer:

Engineers create, support, and sell machines.

That's our purpose. A microprocessor is a machine; so is a hammer or a glove. I'll call anything which extends human ability a machine.

It doesn't stop with the designer: the manufacturing workers and engineers really make the machines, long-term. There's lots of engineering support, and all for making the machines and encouraging our beloved customers to buy them. Some people don't understand or savor this definition, but it's been the role of engineers since the beginning of the industrial revolution. I personally like it. I like the structure of business, the creation of products, the manufacture of them, and the publicizing of them. Our products are like our children, maybe more like our pets. They have lives, some healthy and some sickly. Four of my ICs have healthy, popular lives; ten are doing just OK; and six are just not popular in the market. Others have died.

A young engineering student won't ever hear of this in school. Our colleges' faculties are uneasy with the engineers' charter. The students don't know that they will be held to standards of productivity. They are taught that engineering is like science, sort of. But science need not provide economic virtue; engineering pursuits must.

So what is the state of engineering for the new grad? Mixed. Hopefully, the grad will initially be given procedural tasks that will be successful and lead to more independent projects. At worst, as in my experience, the young engineer will be assigned to projects better left to seasoned engineers. These projects generally veer off on some strange trajectory, and those involved suffer. Oddly enough, the young engineer receives the same raises per year for each possibility. After all, the young engineer is nothing but "potential" in the company's view.

What, then, is the initial value of a young engineer? The ability to support ongoing duties in a company? Not usually; sustaining engineering requires specific training not available in college, and possibly not transferable between similar companies. Design ability due to new topics available in academia? Probably not, for two reasons. First, colleges typically follow rather than lead progress in industry. Second, new grads can't seem to design their way out of a paper bag, in terms of bringing a design through a company to successful customer acceptance. Not just my opinion, it's history.

This is what's wrong with grads, with respect to the electronics industry:

They are not ready to make money for their new employer.

They don't know they're not scientists; that engineers make and sell things. They don't appreciate the economic foundation we all operate with.

They don't know just how under-prepared they are. They are sophomores—from the ancient Greek, suggesting "those who think they know." They try to change that which they don't really understand. They have *hubris*, the unearned egotistical satisfaction of the young and the matriculated.

They see that many of their superiors are jerks, idiots, incompetents, or lazy. Well, sure. Not in all companies, but too often true enough. Our grads often proclaim this truth loudly and invite unnecessary trouble.

They willingly accept tasks they are ill-suited for. They don't know they'll be slaughtered for their failures. Marketing positions come to mind.

Not all grads actually like engineering. They might have taken the career for monetary reward alone. These folks may never be good at the trade.

So, should we never hire young engineers? Should we declare them useless and damn them to eternal disgrace? Should we never party with them? Well, probably not. I can see that at Elantec, a relatively young and growing company, we need them now and will especially need them when we old farts get more lethargic. It's simple economics; as companies grow they need more people to get more work done. Anyway, young people really do add vitality to our aging industry.

It behooves us all, then, to create a professional growth path where the company can get the most out of its investment, and the new grad can also get the most lifelong result from his or her college investment. I have a practical plan. I didn't invent it; the Renaissance tradespeople did. It's called "apprenticeship."

The "crafts" were developed in the 1400s, mostly in Italy. The work was the production of household art. This might be devotional paintings, could be wondrous inlaid marble tables, might be gorgeous hand-woven tapestries to insulate the walls. In most cases, the artistic was combined with the practical. Let me amplify: the art was profitable. There was no cynicism about it; beauty and commerce were both considered good.

We have similar attitudes today, but perhaps we've lost some of the artistic content. Too bad: our industrial management has very little imagination, and seldom recognizes the value of beauty in the marketplace. At Elantec, we've made our reputation on being the analog boutique of high-speed circuits. We couldn't compete on pure price as a younger company, but our willingness to make elegant circuits gave us a lot of customer loyalty. We let the big companies offer cheap but ugly circuits; we try to give customers their ideal integrated solutions. We truly like our customers and want to please them. We are finally competitive in pricing, but we still offer a lot of value in the cheaper circuits.

Do college grads figure into this market approach? Not at all. You can't expect the grad to immediately understand the marketplace, the management of reliable manufacturing, or even effective design right out of college. Just ain't taught. The Renaissance concept of the "shop" will work, however. The shop was a training place, a place where ability was measured rather than assumed, where each employee was assigned tasks aimed for success. Professional growth was managed.

An example: the Renaissance portrait shop. The frame was constructed by the lowliest of apprentices. This frame was carved wood, and the apprentice spent much of his or her time practicing carving on junk wood in anticipation of real product. The frame apprentice also was taught how to suspend the canvas properly. Much of the area of the canvas was painted by other apprentices or journeyman painters. They were allowed to paint only cherubs or buildings or clouds. The young painters were encouraged to form such small specialties, for they support deeper abilities later. So many fine old paintings were done by gangs; it's surprising. Raphael, Tintoretto, and even Michelangelo had such shops. The masters, of course, directed the design and support effort, but made the dominant images we attribute to them alone. Most of the master painters had been apprentices in someone else's shops. We get our phrase "state of the art" from these people.

Today's engineers do practice an art form. Our management would probably prefer that we not recognize the art content, for it derails traditional business management based on power. We engineers have to ensure that artistic and practical training be given to our novices.

So, how does one train the engineering grad? I can only speak for my own field, analog IC design. I'll give some suggestions that will have equivalents in other areas of engineering. The reader can create a program for his or her own work.

1. The grad will initially be given applications engineering duty. Applications is the company's technical link with the buying public. This group answers phone calls of technical inquiries and helps customers with specific problems with the circuits in the lab, when published or designer information is unavailable. Phone duty is only half of applications; they develop applications circuits utilizing products and get the write-ups published, typically through trade magazines such as *EDN*. They produce application notes, which serve as practical and educational reading for customers. A well-developed department will also create data sheets, lifting the burden from the designers but also enforcing a level of quality and similarity in the company's literature. My first two years in the industry were in this job. In one instance, I forced a redesign of a circuit I was preparing the data sheet for because it simply did not function adequately for the end application. Of course, designers always think their circuits are good enough. A truly seasoned applications engineer can be involved in new product selection.

The point of this assignment is to teach future designers what to design, what customers need (as opposed to what they want), how to interact with the factory, and general market information. I wouldn't let new grads speak to customers immediately; first they would make data sheets for new products and be required to play with circuits in the lab to become familiar with the product line. Making application notes would be required, guided by senior applications engineers. I believe that developing good engineering writing skills is important for the designer.

After a couple of months, the engineer would start phone duty. I think the first few calls should be handled with a senior apps engineer listening, to coach the young engineer after the calls. It's important that the engineer be optimally professional and helpful to the customer so as to represent the company best. Most of us have called other companies for help with some product problem, only to reach some useless clone.

This stint in applications would last full-time for six months, then be continued another six months half-time, say mornings for us West Coast folks.

2. Device modeling would be the next part-time assignment. In analog IC circuit design, it's very important to use accurate and extensive model parameters for the circuit simulators. Not having good models has caused extensive redesign exercises in our early days, and most designers in the industry never have adequate models. As circuits get faster and faster, this becomes even more critical. Larger companies have modeling groups, or require the process development engineers to create models. I have found these groups' data inaccurate in the previous companies where I've worked. We recently checked for accuracy between some device samples and the models created by a modeling group at a well-known simulator vendor, and the data was pure garbage. We modeled the devices correctly ourselves.

This being a general design need, I would have the young engineer create model parameters from process samples, guided by a senior engineer with a knack for the subject. This would also be an opportunity to steep the engineer in the simulation procedures of the department, since the models are verified and adjusted by using them in the circuit simulator to play back the initial measurements. It's a pretty tedious task, involving lots of careful measurements and extrapolations, and would probably take three months, part-time, to re-characterize a process. Modeling does give the engineer truly fundamental knowledge about device limitations in circuits and geometries appropriate to different circuit applications, some really arcane and useful laboratory techniques, and the appreciation for accuracy and detail needed in design.

Because of the tedium of modeling, few companies have accurate ongoing process data.

3. A couple of layouts would then be appropriate. Most of our designers at Elantec have done the mask design for some of their circuits, but this is rare in the industry. The usual approach is to give inadequate design packages to professional mask designers and waste much of their time badgering them through the layout. The designer often does an inadequate check of the finished layout, occasionally insisting on changes in areas that should have been edited earlier. When the project runs late, the engineer can blame the mask designer. You see it all the time.

I would have the young engineer take the job of mask designer for one easy layout in the second three months of half-time. He would lay out another designer's circuit and observe all the inefficiencies heaped upon him, hopefully with an eye to preventing them in the future. Actually, we designers have found it very enlightening to draw our own circuits here; you get a feel for what kind of circuitry packs well on a die and what is good packing, and you confront issues of component matching and current/power densities. The designer also gains the ability to predict the die size of circuits before layout. The ultimate gain is in improving engineers' ability to manage a project involving other people.

4. The first real design can be started at the beginning of the second year. This should be a design with success guaranteed, such as splicing the existing circuit A with the existing circuit B; no creativity desired but economy required. This is a trend in modern analog IC design: elaborating functions around proven working circuitry. The engineer will be overseen by a senior engineer, possibly the designer of the existing circuitry to be retrofitted. The senior engineer should be given management power over the young engineer, and should be held responsible for the project results. We should not invest project leadership too early in young engineers; it's not fair to them. The engineer will also lay the circuit out, characterize it, and make the data sheet. Each step should be overseen by an appropriate senior engineer. This phase is a full-time effort for about five months for design, is in abeyance while waiting for silicon, and full-time again for about two months during characterization.

5. The first solo design can now begin. The engineer now has been led through each of the steps in a design, except for product development. Here the designer (we'll call the young engineer a designer only when the first product is delivered to production) takes the project details from the marketing department and reforms them to a more producible definition of silicon. At the end of the initial product planning, the designer can report to the company what the expected specifications, functionality, and die size are. There are always difficulties and trade-offs that modify marketing's initial request. This should be overseen by the design manager. The project will presumably continue through the now-familiar sequence. The designer should be allowed to utilize a mask designer at this point, but should probably characterize the silicon and write the data sheet one last time.

This regimen takes a little over two years, but is valuable to the company right from the start. In the long run, the company gains a seasoned designer in about three years, not the usual seven years minimum. It's also an opportunity to see where a prospective designer will have difficulties without incurring devastating emotional and project damage. The grad can decide for himself or herself if the design path is really correct, and the apprenticeship gives opportunities to jump into other career paths.

I like the concepts of apprentice, journeyman, and master levels of the art. If you hang around in the industry long enough, you'll get the title "senior" or "staff." It's title inflation. I have met very few masters at our craft; most of us fall into the journeyman category. I put no union connotation on the terms; I just like the emphasis on craftsmanship.

There are a few engineers who graduate ready to make a company some money, but very few. Most grads are fresh engineering meat, and need to be developed into real engineers. It's time for companies to train their people and eliminate the undeserved failures. I worked for five years at a well-known IC company that was fond of bragging that it rolled 20% of its income into research and development. The fact is, it was so poorly organized that the majority of development projects failed. The projects were poorly managed, and the company was fond of "throwing a designer and a project against the wall and seeing which ones stick." Most of the designers thrown were recent graduates.

We should guide grads through this kind of apprenticeship to preserve their enthusiasm and energy, ensuring a better profession for us all.

When I read the first Williams compendium (the precursor to this book), I was shocked by the travelogs and editorials and downright personal writings. Myself, I specialize in purely technical writing. But after Jim gave me the opportunity to offer something for the second book, the first book seemed more right and I couldn't resist this chance for blatant editorialization. I'm mad, see, mad about the waste of young engineers. Waste is bad.

3. We Used to Get Burned a Lot, and We Liked It

I'm a fortunate engineer. My employer sponsors the hobby I've had for thirty of my forty-year life. We don't disagree much; I like most of the aspects of my job, even the tedious ones. However, I'm no lackey. I don't really listen to many people, although I try to appear to. There's no cynicism here; all my associates agree with me that we will produce nifty new ICs and make money. That's the job.

This entry of Jim's compendium is offered to relate what an earlier generation of engineers experienced in preparation for a career in electronics. Many of my associates were quite functional in electronics when they entered college. We were apparently different from most of the students today. We were self-directed and motivated, and liked the subject. I have detected a gradual decrease in proficiency and enthusiasm in college graduates over the last fifteen years; perhaps this writing will explain some of the attitudes of their seniors. I've included some photographs of lovely old tube equipment as background.

My experiences with electronics started with construction projects involving vacuum tubes, then transistors, eventually analog ICs, raw microprocessor boards, and finally the design of high-frequency analog ICs. Through all the years, I've tried to keep the hobby attitude alive. I'm not patient enough to grind through a job for years on end if I don't really enjoy it. I recommend that anyone who finds his or her job boring decide what they do like to do, quit the current job, and do the more enjoyable thing.

My first memory of vacuum tubes is a hot Las Vegas, Nevada morning around 1 A.M. I was young, about ten years old. It was too hot to sleep and the AM radio was gushing out Johnny Cash, Beach Boys, Beatles, and the House of the Rising Sun, as well as cowboy music. It was pretty psychedelic stuff for the time, and with a temperature of 100°F at night, the low humidity and the rarefied air, I spent a lot of late nights awake with the radio.

As I lay listening to the music I noticed that the tubes of the radio projected more blue light on the ceiling than the expected yellow-red filament glow. It's hard to imagine that simple, beautiful, blue projection upon your wall which comes from the miniature inferno within the tubes. It comes from argon gas which leaks into the tube and fluoresces in the electric fields within. Occasionally, you can see the music modulate the light of the output tubes.

My radio, which sat next to my bed so that I could run it quietly without waking the parents, was a humble GE table model. It was built in the mid-50s, so it was made of cheap pine with ash (or maple?) veneer. Typical of the times, it had sweeping rounded corners between the top and front, and inlaid edging. They never did figure out how to make a true accurate corner with cheap wood processes. This radio was B-grade, though; it had a magic-eye tube and included the "MW" band-low MHz AM reception. Allegedly, you could hear ships and commercial service on MW, but in Las Vegas all I heard were ham radio 1.8MHz "rag chewer" conversations. At length.

Radios were magic then. TV wasn't nearly as entrancing as now, being black-and white in most homes and generally inane (the good adult stuff was on too late for me to see). On radio you heard world news, pretty much the only up-to-the-minute news. You heard radio stations that didn't know from anything but variety in music. They didn't go for demographics or intense advertising; they just tried to be amusing. When I was that young, the people who called into the talk shows were trying to be intelligent. Shows what an old fart I am.

The electronic product market of the time was mostly TV and radios. Interestingly, the quality living-room TV of that time cost around $600, just like now. Then you also got a big console, radio, speakers, and

Figure 3–1.
A lovely TRF radio from the 1920s and '30s. This was before superheterodyne reception; you had to tune all three dials to get your station. More or less gain was dialed in with the rheostats in series with the input tubes' filaments. A lot of farm as well as city dwellers used these. The coils were hand-wound, and every component was available for scrutiny. This set will be usable after a nuclear attack. From the John Eckland Collection, Palo Alto, California. Photo by Caleb Brown.

record player for the price (it even played a stack of records in sequence). It worked poorly, but it was a HOME ENTERTAINMENT SYSTEM. We pay only a little more for similar but better today. Lab equipment was really rotten then compared to today. There was no digital anything. Want to measure a voltage? You get a meter, and if you're lucky it has a vacuum-tube amplifier to improve its range, versatility, and resistance to burnout. I couldn't afford one; I had a 20KΩ/V multimeter. I eventually did wreck it, using it on a wrong range.

In the vacuum-tube days, things burned out. The tubes might only last a year, or they might last 20 years. Early 2-watt resistors had wax in them, and always burned out. The later carbon resistors could still burn out. When I say burn out, I mean exactly that: they went up in smoke or even flame. That's where the term came from. Where we have cute switching power supplies today, then the tubes ran from what we call "linear" supplies that included power transformers which in quality gear weighed a dozen pounds or more. The rectifiers might be massive tubes, or they could be selenium rectifiers that also burned up, and they were poisonous when they did. The bypass capacitors were a joke. They would eventually fail and spew out a caustic goop on the rest of the innocent electronics. Let's face it, this stuff was dangerous.

I almost forgot to mention the heat. A typical vacuum tube ran hot; the glass would burn you if you touched it. The wood cabinets needed to be regularly oiled or waxed because the heat inside discolored and cooked them. A power tube ran really hot, hot enough to make the plate glow cherry-red in normal operation. You could get an infrared sunburn from a few inches' proximity to a serious power tube. From a couple of feet away your face would feel the heat from an operating transmitter.

But it wasn't burnout or heat that was the most dangerous thing to an electronics enthusiast; it was the voltage. The very wimpiest tube ran from 45V plate potential, but the usual voltage was more like 200V for a low-power circuit. I made a beautiful supply for my ham transmitter that provided 750V for the output amplifier. Naturally, it knocked me across the room one day when I touched the wrong thing; a kind of coming-of-age ritual. This event relieved me of all fear of electricity, and it gave me an inclination to think before acting. Nowadays, I sneer at bare electrodes connected to semiconductors. I routinely touch nodes to monitor the effect of body capacitance and damping on circuit behavior. I have often amazed gullible peasants by curing oscillations or fixing bypasses with only my touch. Of course, the off-line power supplies command my respect. For them, I submit and use an isolation transformer.

At this point, I think we can explain the lack of females attracted to electronics at the time. In the 50s and 60s, society protected women but offered men up to danger. The same is true for the earlier industrial revolution: women were huddled into protective work environments and men were fodder for the dangerous jobs. I think this attitude was prevalent with respect to vacuum tube electronics. Women (girls, in particular) were not encouraged to enjoy the shock hazards, the burns, the excessive weights of the equipment, or the dirtiness of the surfaces.

Boys, of course, found all this attractive. I suppose this is the historical basis of the male domination of the field. The duress of dealing with this kind of electronics really appealed to young men's macho, just like working on cars appealed to the gearhead set. The difference between the groups was that electronics required a lot more education and intellect than cars, and so appealed to more bookish types. The girls never caught on to how cool electronics was, probably because a radio can't get you out of the house. The electronics hobbyists (creators of today's nerd stereotype) simply found another way to get away from the parents. It worked; the old folks really did keep out of the garage, the rightful dominion of hobby electronics.

A social difference between then and now is how much more prevalent hobbies were. As I mentioned, TV did not occupy as much of people's time. Kids got as bored as now, so they turned to hobbies. When boys got together, they needed something to do, and they could share cars or electronics. This led to a much more capable young workforce, and getting a job after high school seemed easier than now. Furthermore, you probably had strong interests that could guide you through college. Changing majors or not having a major was unusual. Now, kids are generally far less self-directed. They haven't had to resolve boredom; there's too much en-

Figure 3–2.
An original breadboard. The components are on the board, and hopefully Ma has another. This is a phonograph pre-amp and power amplifier, just like 1930-to-1960 home project assemblies. You can really see your solder joints in this construction style. From the John Eckland Collection, Palo Alto, California. Photo by Caleb Brown.

tertainment easily available to them today. Further, drugs destroy hobbies. As a result, the college students I've interviewed over the years have gradually lost pre-college experience with their field. Twenty years ago college grads had typically been working with electronics for two to seven years before college, and the new grad could perform well in industry. Regrettably, it now takes up to three years of professional experience to build a junior engineer, titles notwithstanding.

Perhaps worse is the attitude change over the years. The new grad was considered an amateur; "amateur" from the Latin, meaning "one who loves a field": motivated but inexperienced. Increasingly, the grads are in electronics for the bucks, and seldom play in the art for their own amusement. Present company excepted; I know the readers of this book are not in that category. To be fair, present electronics focuses on computers and massive systems that are hard to comprehend or create in youth. Construction of projects or repairing home electronics is mostly out of the realm of kids not encouraged by a technical adult.

I think this places an obligation on families and schools to support electronics projects for kids, if we are to generate really capable and wise engineers in the future. By the time a present grad has had enough years of experience to become an expert in some area, the technology is liable to change. Breadth of technical experience is the only professional answer

Figure 3–3.
A really beautiful radio from the 1950s. A so-called Tombstone radio; the fins are wood decoration. This is electronics as furniture; the radio is good but the cabinet is exquisite. The dial is artistic and several frequency bands await the curious. Not fully visible is the same radio flanked by different cabinets made by competitive groups within Zenith. From the John Eckland Collection, Palo Alto, California. Photo by Caleb Brown.

to this problem. Employers do not encourage nor support the engineer's development outside his narrow field, so breadth seems something best developed by hobbies before college, and a more varied engineering training during college.

But we digress. Somewhere around 1964 I saw the first transistor radios. They were kind of a novelty; they didn't work too well and were notoriously unreliable. They replaced portable tube radios, which were just smaller than a child's lunch box. They weighed about seven pounds, and used a 45V or 67V battery and a couple of "D" cells for the filaments. The tubes were initially normal-sized but had low-power filaments in the portables, but the latest were socketless and had cases only 1½" long and ½" diameter. These tubes were also used in satellites and were quite good. Even so, the transistor radios were instant winners. They were cheaper than any tube radio, were truly portable, and could be hidden in classrooms. The miniature earphone really made it big.

The transistor radio easily doubled the audience for musicians and advertisers. Perhaps it was the portable transistor radio that accounted for the explosive growth of rock music. . . . While it's true that rock-and-roll was popular as hell in the late 50s and early 60s, the sales of records and the number of radio stations just didn't compare with the activity at the end of the 60s.

As I said, the transistor radios were unreliable. I made spending money repairing radios when I was in grade school. Attempting to repair them; my hit ratio was only 50%. These repairs were on bad hand-soldered joints, on broken circuit boards (they were made of so-called Bakelite—a mixture of sawdust and resin), and unreliable volume controls. Replacement parts were grudgingly sold by TV repair shops; they'd rather do the servicing, thank you. The garbage line of 2SK-prefix transistors was offered. These Japanese part numbers had nothing to do with the American types and surprisingly few cross-references were available. I had no equipment, but most of the failures were due to gross construction or device quality problems.

Only a few years after the transistor radios emerged they became too cheap to repair. They made for a poor hobby anyway, so I turned to ham radio. This was the world-wide society of folks who like to talk to each other. The farther away the better; it's more fun to talk to a fellow in Panama than one in Indiana. People were more sociable then, anyway. The world community seemed comfortably far off and "foreign" had an attraction.

I didn't have enough money to buy real commercial ham gear. Luckily for me, many hams had the same inclinations as I and a dynamic home-construction craze was ongoing. Hams would build any part of a radio station: receivers, transmitters, or antennas. They were quite a game group (of mostly guys), actually; grounded in physics and algebra, they used little calibrated equipment but actually furthered the state of radio art. Congress gave them wide expanses of spectrum to support this renaissance of American engineering. We got a generation of proficient

Figure 3–4.
Here's the chassis of a first-rate radio. The base metal is chrome-plated for longevity. All coils are shielded in plated housings, and string tuning indicator mechanisms are replaced with steel wire. These components are as uncorrupted as they were when they were made in 1960. The designers gave extra attention to the quality of everything the customer would see and feel (the knobs play very well). From the John Eckland Collection, Palo Alto, California. Photo by Caleb Brown.

engineers from radio. Hams performed feats of moon bounce communications and even made a series of Oscar repeater satellites. Imagine that, a group of civilians building satellites that NASA launched into space for free. I myself have heard aurora skip signals on the 6-meter band—the bouncing of signals off the northern lights. All this in the days of early space travel and Star Trek. Some fun.

Soon after transistor radios were common, industrial transistors became cheap and available in volume. The hobby books were out with good circuit ideas in them, so I finally started making transistor projects about 1966. I was a bit reluctant at first, because the bipolars were delicate, physically and electrically, and had poor gain and frequency response. Tubes were still superior for the hobbyist because of their availability. You could salvage parts from radios and TVs found at the dump, or discarded sets awaiting the trashman. Because the circuits were relatively simple, we would dismantle old sets right down to separated components and chassis, which would be reassembled into the next hobby project. I began to tap the surplus parts suppliers, and the added supply of tube and related parts delayed my interest in solid-state circuits.

The first commercial transistors were germanium PNP, and they sucked. They just wouldn't work correctly at high temperatures, and their

Figure 3–5.
A medium-quality table radio of the 1950s. Being decorative, the cabinet and dial are of good quality. In the upper-right corner is a magic-eye tube, an oscilloscope-like gizmo that gives an analog indication of tuning accuracy. From the John Eckland Collection, Palo Alto, California. Photo by Caleb Brown.

leakage currents skyrocketed past 100°C to the extent of debiasing circuits. Their Vbe went to zero at 200°C; that is, the whole transistor became intrinsic and was a short-circuit. Furthermore, you couldn't find two devices that halfway matched with respect to Vbe and beta and output impedance. You didn't bother making instrumentation circuits with those devices; there just weren't any matched pairs to be found. The Vbe's also suffered from terrible long-term drift, I think because germanium could never be alloyed adequately for a solid contact. It didn't matter; chopper-stabilized tube op amps were common and worked well. I still have one of the best VTVMs ever made, a Hewlett-Packard chopper-stabilized model that has sensitive DC ranges and a 700MHz active AC probe.

What really made my decision to use transistors was the advent of the silicon NPN device. Silicon could tolerate temperature, and was insensitive to excessive soldering. It never went intrinsic, and beta control allowed for matched pairs. The high-quality differential input stage made the industry of hybrid op amps possible, and some of them could handle the same signal voltages as the tube op amps. Silicon transistors even gave decent frequency responses, although the faster devices were still electrically delicate. Silicon made TVs and radios work better too.

Circuit design changed overnight. The threshold voltage of tubes (analogous to the threshold of JFETs) would vary over a 3:1 range. Because of the poor bias point accuracies, most circuits were AC coupled. This precluded them from many industrial applications. Although

Figure 3–6.
The electronics of the previous radio. Because this set was not of the highest caliber, the electronics are humble and have no precious elements. From the John Eckland Collection, Palo Alto, California. Photo by Caleb Brown.

the chopper-stabilized op amp was very accurate, it was expensive and the chopper could wear out, being a mechanical vibrator. The uncertainty of transistor Vbe was really negligible, relative to supply voltages, and biasing transistors was a snap, although not widely understood then. Transistors could seemingly do anything that didn't involve too much power. But until perhaps 1966, if you had to handle power with a transistor, you used a cow of a germanium device.

But between 1961 and 1967, the choice of transistor or tube was often made by the prejudice of the designer. Some applications demanded one device or the other, but in the case of audio amplifiers, there was free choice.

Construction of electronics changed radically in this time. Tubes were mounted in sockets whose lugs served as the supports for components, and a solid steel chassis supported the circuits. Steel was necessary, since the tubes couldn't tolerate mechanical vibration and the massive power supplies needed support. The most elegant construction was found in Tektronics oscilloscopes. They used molded ceramic terminal strips to support components, and only about eight components could be soldered into a pair of terminal strips. Cheaper products used Bakelite strips. These were all rather three-dimensional soldered assemblies: point-to-point wiring literally meant a carpet of components connected to each other and to tubes in space. The assemblies were also very three dimensional; the tubes sprouted vertically above the chassis by three to five inches and the other components sprawled in a two-inch mat below the chassis.

Transistors made construction more two dimensional. The transistors weren't tall, generally the size of our TO-39 package of today, and circuit boards were practical since they didn't have to support heavy or hot components. All passive components became short too. A layer of transistor circuitry thinned to one inch or less. There was a volume reduction of about 20:1 over equivalent tube circuits. For industrial electronics, however, transistors afforded only a 2:1 overall product cost reduction.

In the 1960s, the quality of cabinets really degraded. Transistor equipment was considered cheap, relative to tube gear, and only received cheesy plastic cases. The paint and decals on the plastic rubbed or flaked off, and impact could shatter it altogether. Tube equipment, on the other hand, had enjoyed quality wood casings for decades. Since the tube chassis were so large and heavy, furniture-quality cabinets were needed simply to transport the electronics. The radios and TVs were so obtrusive in tube form that manufacturers really made the cabinets fine furniture to comply with home decor.

Quality in the tube years came to mean both mass and the use of precious materials. Greater mass meant you could transport or physically abuse the equipment with no damage. It also meant that the components would suffer less from thermal changes and microphonics (electrical sensitivity to mechanical vibrations). A really sturdy chassis would not need alignment of the tuned circuits as often as a flimsy frame. Precious materials included quality platings—such as chrome or vanadium—of the chassis, to avoid corrosion and extend useful life. Heavier transformers allowed more power for better bass response and greater volume. A heavier power transformer would burn out less frequently, as would oversize power tubes. Components came in quality levels from cheap organic-based resistors and capacitors that cockroaches could eat to more expensive and long-lived sealed components. The general attitude about electronics construction was akin to furniture: the more mass and the more precious the material, the better.

Since the transistor circuits had no thermal nor microphonic problems, the poorest of cases were given to them. They weighed next to nothing, and a hard fall wouldn't cause too much damage. Since the products had no mass nor special materials in their construction, people thought of transistor products as low-quality. The manufacturers made sure this was true by using the poorest materials available. The circuit boards did indeed tarnish and warp, and the copper could crack and cause opens. The wires soldered to the boards seemed always stressed from assembly and often broke. Even the solder had corrosive rosin.

Because the transistor circuits were small, the traditional soldering guns and irons were far too hot and large to use; we now had to buy new small irons. We even had to get more delicate probes for oscilloscopes and voltmeters. These problems were moot; you couldn't effectively repair transistor stuff then anyway. Even if you could troubleshoot a bad

Figure 3–7.
Electronics for the masses: the 1960 Knight-Kit audio amplifier. For $70, you get a kit of parts and a chassis which can become a stereo 50W audio power amplifier. This was a good deal; since labor was expensive, building the thing at home saved money, and the experience was somewhat educational. More than 100,000 were sold. From the John Eckland Collection, Palo Alto, California. Photo by Caleb Brown.

board, you had only a 50-50 chance of not damaging it when you tried to replace a component. You could not make a profit repairing transistor products.

It got harder to make hobby circuits too. In the mid-60s, printed circuit boards were so bad you might as well try to make your own. So I bought a bottle of ferric chloride and tried it myself. For masking, I tried direct painting (house exterior paint wasn't bad) and resist ink pens. This sort of worked; I had to blob-solder across many splits in the copper of my homemade boards. "Hobby boards" were the solution. These are the pre-etched general-purpose breadboards in printed circuit form. They had DIP package regions and general 0.1" spacing solder holes. Analog hobbyists would obediently solder interconnect wires between pads, but the digital hobbyists had too many connections to make and adopted wire-wrap construction.

Suddenly construction projects lost their artistic appeal. Tubes arrayed on a chassis with custom wiring are very attractive, but the scrambled wire masses of transistor projects are about as pretty as a Brillo pad. You could hardly see the connections of transistor circuits, and this only got worse as ICs displaced groups of transistors. I knew a couple of old codgers who gave up hobby electronics due to failing eyesight. They wouldn't have had trouble with tube projects. Funny thing was, semiconductor projects still cost as much as tube equivalents but were uglier, more difficult to build, and harder to debug and tune.

Professional breadboards were similar to the hobbyboards until perhaps the early '80s. At work you built circuits on higher-quality breadboards. But within only a few years, critical ICs were available in surface-mount packages, or more expensive and clumsy socketed alternatives. The pin count of the packages just skyrocketed. The sockets are expensive and fragile. A transition began which is almost complete today: breadboards are simply not attempted to develop each subsystem of a board; the first tentative schematic will be laid out on a full-fledged circuit board. Any corrections are simply implemented as board revisions. These boards contain mostly surface-mount components. This technique is not practical for the hobbyist.

God, what a nightmare it is to troubleshoot these boards. They are generally multilayer and the individual traces can't be seen, so finding interconnects is impossible. The only connections that can be probed or modified are the IC's leads themselves. You generally can't read the markings on resistors or capacitors, because they are so small. Development work is accomplished with stereo microscopes.

So hobby electronics has taken a major beating in the last twenty years. It's become intellectually difficult to build a really significant project, to say nothing of increased expense and construction difficulty. This portends a generation of relatively green engineers who have only college experience with electronics. God help us. I suppose there still are some handy people, as demonstrated by the continuing component sales of Radio Shack. Too bad that they have diminished the component content of their stores over the years, and traditional hobby suppliers like Lafayette and Heathkit have altogether disappeared. There is no substitute for pre-college electronics experience.

Gone too is the magic people used to see in electronics. As a kid, I saw that other kids and their parents were amazed that radios and TVs worked at all. Our folks used to think of installing a TV antenna as an electronics project. Parents gave their kids science toys. These were great; we had chemistry sets, metal construction kits, build-your-own-radio-from-household-junk sets, model rockets, crystal-growing kits, all sorts of great science projects. The television stations even kept Mr. Wizard alive, the weekly science experiment program.

It seems now that people assume they can't understand science or technology, and accept this ignorance. Kind of like religious belief. People seem to enjoy technology less, and expect more. We even predict future advancements when we have no idea how to accomplish them. We don't give our young children these science toys, even though the kids would find them wondrous. Parents are imposing jaded attitudes on kids.

This would be all right, except that electronics has grown in scope beyond the ability of college to teach it well. Students graduating today have insufficient breadth of knowledge of the field, and not enough depth to really take on a professional project. I don't blame them; it's probably impossible to be the master of anything with a college diploma but no real experience.

I don't know all of the answers, just the problem. As long as our society considers engineering unglamorous and nerdy, kids won't be attracted to it. Industry will wonder why young engineers are not highly productive. Companies never really train people; they just give them opportunities. We'll see a general malaise in design productivity, just as we now see a problem with software production. I could be getting carried away with all this, but we should promote science and technology as suitable hobbies for our kids.

4. Analog Design Productivity and the Challenge of Creating Future Generations of Analog Engineers

Introduction

Recently, digital techniques are very commonly used in the fields of electronics. According to the statistics taken by MITI (Figure 4–1), Japanese integrated circuits industry has shown a growth of 5.5 times in the last one decade (from 1980 to 1991). While digital ICs (MOS and bipolar digital) grew 6.24 times in this period, analog ICs did only 3.57 times. This reflects to a analog vs. digital percentage ratio, showing that analog decreases from 25.9% on 1980 to 16.7% on 1991 (Figure 4–2). From these facts, many people in the electronics fields might think that the age of analog has been finished.

	'80	'85	'90	'91
MOS Digital	100	346	650	691
Bipolar Digital	100	352	340	336
Total of Digital	100	348	591	624
Linear	100	261	309	357
Grand Total	100	325	518	555

Figure 4–1. Percentage of Japanese IC production.

Institute of Electronics, Information and Communication Engineers (IEICE), one of the largest academic societies in electronics fields in Japan, held special sessions to discuss many problems with respect to the analog technologies in Japan at the IEICE National Convention in 1989 and again in 1992 chaired by the author. Both sessions attracted much more participants than expected and proved that many serious engineers were still recognizing the importance of analog technology. We discussed the present status of analog technologies, how to create new analog technologies, how to hand them down to the next generation engineers and how to use CAD in design of analog circuits to enhance productivity. This paper is based on several discussions in these sessions and author would like to acknowledge to those who discussed on the problems.

	'80	'85	'90	'91
MOS Digital	60.0	63.9	75.3	74.8
Bipolar Digital	14.1	15.3	9.2	8.5
Total of Digital	74.1	79.2	84.5	83.3
Linear	25.9	20.8	15.5	16.7
Grand Total	100.0	100.0	100.0	100.0

Figure 4–2.
Digital–Analog
Percentage Ratio
(MITI).

To summarize those discussions, we could categorized the problems in to the following three major classes;[1]

First, because of many people cannot understand that analog circuits technologies are not out of date but they really a key to develop digital technologies, the number of students who want to learn analog circuits technologies are has been decreasing year by year. Even student who willingly study analog circuits tends to prefer computer simulation rather than experiments, so they lose a sensitivity to the real world. Accordingly this lead the results that only a very few number of universities in Japan still publish technical papers in the field of analog circuits.

Secondly, in the industries, although the importance of the analog circuits technologies are aware, two things make the number of analog circuits engineer decreased: increasing production of digital hardware system need to increase digital circuits engineers, and analog engineers easily understand digital technologies.

Third, while CAD makes design of digital system very popular, design of analog circuits are still difficult, it requires still expert's skill. It has very insufficient productivity. Besides it takes a long time to educate engineers to be an analog circuits expert. Finally many factories tend to change their main productions from analog to digital systems.

Analog circuits, however, have many advantages over digital technologies: very high functional densities for the same chip size, high speed abilities and high potentials.

So we must make a effort to increase the number of analog engineer and to hand analog circuits technologies down to next generations.

Analog Design Productivity

CAD (Computer Aided Design, but some peoples think it as Computer Automated Design) has been widely adopted in the design of digital integrated circuits. Computers can do everything from logic synthesis to mask pattern generation, taking the place of average design engineers, only if they got functional specification of the system written in some high level descriptive language. Meanwhile analog circuits CAD also become in great request according to the rise of several novel technologies such as personal communication system, multimedia and so on, because we have insufficient number of analog circuits design engineers to cope with this situations. (The reason why they have been decreased shall be mentioned in later section of this paper.) But unfortunately it is believed that there should be no such a powerful analog CAD system like a digital for a while.

Analog circuits design technologies have following features which prevent us from realizing unified approach schemes:

1. While digital systems can be described with a couple of logic equations in principle, specifications of analog circuits are too much complicated to describe in a clear format. For instance, it sometimes is requested to design "excellent sound quality HiFi amplifier." We have no definition for "excellent sound quality" at all. It depends on individual judgment, some feels good the others feels no good, listening to the same amplifier. Besides a feeling judgment, amplifier has many characteristic items such as gain, frequency characteristics, dynamic range, distortion, temperature characteristics, input and output impedance, power consumption and so on. And normally we could not find evident correspondence between these characteristic items and the total performance.

2. Several specifications on a single circuit usually conflict each other, so many trade off should be indispensable during the design procedure, taking restrictions such as performance of devices available, cost, deadline etc. into account. As these compromises could be done with the designer's personal experience and knowledge, there was no straightforward scheme to do them. There were many papers with respect to the optimization of electronic circuits, but difficulties are not in how to do it but in where one should place the goal.

3. To design a good analog circuits, a step by step method is quite insufficient and a breakthrough should be mandatory. Only man of talents can do that. But perhaps he cannot explain how he comes to the breakthrough.

4. There are many circuit topologies and their combinations to realize the same specification. It should be so difficult for CAD to get a unique solution.

Above mentioned features of analog circuits design are based on very essential characteristics of analog. We can not write any program without the knowledge about how it works. We think "computer-automated-design" of analog circuits are still one of challenging problems for us.

We have, however, powerful tools for analog circuit design, a circuit simulator. Among them "SPICE" and its derivatives are widely used by the design engineers. It is very useful as far as he use as literally "computer-aided-design" tools. Circuit simulator requires good understanding of circuits from the design engineer. We discussed about merits/demerits of using circuit simulator in the National Convention of IEICE in 1992 to find the following problems:

1. Simulator could be very useful only for design engineers who really understand how the circuit works.
2. It is very difficult to simulate such a circuit as having more than two widely spread time constants, for instance PLL, AM/FM detector, crystal oscillator.
3. It is also difficult to derive device parameters, and installed model does not reflect many parasitic elements such as substrate current, parasitic transistors, thermal coupling etc. Some of them can be avoided by adding some appropriate circuits, however this is not so easy for the average engineers.
4. It cannot cope with a variation of circuit topology. We need to rewrite net lists and restart program whenever we change a circuit topology.

These show that circuit simulators are indeed user dependent program therefore it is very important to teach beginners how to use it.

Although the author mentioned about the shadow of circuit simulator, it is still very powerful tool. Dr. Minoru Nagata, Director of Central Research Laboratory Hitachi Ltd., showed the following evidence as an example.

In the past 2 years, analog LSI has been developing, number of transistors per chip increases twice while available time for design decreases two thirds. But design engineers have 20% decreased in their fail rate at the first cut. Dr. Nagata also said that layout productivity increased 10 times and design correction decreased one tenth during this period. He stressed that these result could not be got without circuit simulators.

The author pointed out how Japanese engineers thinking about analog circuit design productivity and circuit simulator. However analog circuit design still strongly depends on the designer of talent. Comparing the design of logic system to analog circuit, we would find that an one of apparent difference between them is that analog circuits has usually more than one complex function while one logic circuit element has only one function. Most digital system designers think their design in logic element or logic gate level, while analog designs are carried out in circuit element level such as transistors, resistor etc. A resistor in collector circuit works as a voltage dropper and same time it governs gain and frequency characteristics of that circuit. Analog circuits design engineer should always pay his attention to trade-off between these complex functions. Professional analog circuit designer is a man who knows these trade-off technology and who success to realize compact and high performance circuits.

As demands for analog circuit rising, we should solve this design productivity problem. How could we make beginner or computer designed analog circuits? Professor Nobuo Fujii at Tokyo Institute of Technologies and other members in the Technical Committee for Analog Circuit Design at Institute of Electrical Engineers of Japan (IEEJ), chaired by the author, has been discussed about these problem. We thought at first use of "Expert System" which installed many knowledge of experienced professional designers as a element functional circuit. We tried to categorize analog circuits by their function. However this idea did not work. Because of above mentioned reason, each circuit has complex functions, it was very difficult to find functional element circuit in a database format.

Analog systems can be described with a couple of differential equations and "analog computer" is a tool to solve differential equation. Analog computer consist of some operational element such as integrator, adder, multiplier, limiter etc. Recently we come to the conclusion that by taking this operational circuit as an element we could compose any analog circuit using them in principle, although the circuit compactness should be lost. Several case studies in the committee show that this idea works[6]. There needs further investigation before this idea would be real.

Analog Circuit Engineers in Japanese Industry

It is thought that rising digital technologies has been taking over analog circuits technologies. A number of laboratories in Japanese universities whose activities are in analog circuits fields, has been decreased recently. Dr. Minoru Nagata at Hitachi Ltd. questionnaired managers in several electronics factories to investigate what leading electronics engineers thinking about[2].

The followings are the results of Dr. Nagata's questionnaires.

QUESTIONNAIRE 1
Q. How do you think about an ability of newcome electronics engineers at your company? Please choice from the followings.

a) Newcomers know neither digital circuits nor analog circuit. Nothing about circuits technology.
b) Newcomers know about digital circuit very well but nothing about analog circuits.
c) Newcomers have average knowledge about either analog or digital circuits.
d) Newcomers know about analog circuit very well but nothing about digital circuits.
e) Newcomers know about computer software very well but nothing about hardware technologies.

RESULTS:

a).... 24   b) 16
c).... 11   d) 0
e).... 26

QUESTIONNAIRE 2

Q. We have two professional circuit engineers, one is in digital and the other in analog, available to add to your project troop. Which do you prefer, analog or digital?

RESULTS:

Analog .. 32(62%)   Digital .. 20(38%)

QUESTIONNAIRE 3

Q. To support your urgent project, you can add ten more circuit engineers to your troop. What ratio of engineers, analog to digital, do you like?

RESULTS:

10 digital engineers 1
1 analog, 9 digital 1
2 analog, 8 digital 14
3 analog, 7 digital 16
4 analog, 6 digital 9
5 analog, 5 digital 2
6 analog, 4 digital 3
7 analog, 3 digital 3
8 analog, 2 digital 2
9 analog, 1 digital 0
10 analog engineers 1

Results of Questionnaire 1 confirm that a few universities are interested in analog circuit technology and most student are fond of computer software rather than hardware technology. This shows at the same time that most general people's interests are in digital field. It is, however, very interesting that industries need a lot of analog circuit engineers. Dr. Nagata said "Analog technology is a Key technology, while digital is a Main technology." It means that what governs the final performance of digital system such as speed and reliability is an analog circuit technology. Digital circuits are analog circuits in topological sense, they use only two states of the circuits. Therefore faster the digital LSI, more troubles arise which analog technologies are mandatory to solve.

As mentioned at the beginnings main productions of Japanese IC industries are digital LSI, they need much digital circuit engineer to hold their production. It is difficult for a digital circuit engineer to understand rather complicated analog circuit, but to the contrary analog circuit engineer can easily design digital circuits. By this reason analog engineers are tend to be thrown into digital project, it forms one way flow (diode) of engineers from analog to digital, making the number of analog circuit engineers in the industry decreased year by year. Nevertheless many leading project managers become aware of importance of analog technologies. Results of questionnaire 2 and 3 seem to show this situation.

Recent high speed digital LSI such as memory and CPU requests much more analog circuit technology and digital signal processing system (DSP) need AD/DA converter at their interface most of which are analog circuits. Furthermore raising new system such as VHF/UHF communication, HDTV, multimedia etc. should request much analog circuit engineers.

From historical view, in the field of high speed and high frequency, systems are implemented with analog technology at first, then according process technologies developing, they are took over by digital. For example in communication digital system are implemented in 9.6 kbit/s, while coaxial 400 Mbit/s and light 1.6 Gbit/s use analog technology. Another very interesting difference between two technologies are the number of transistors to realize the same function. Digital systems use a lot of transistors while analog use only one hundreds or less transistors. (Unfortunately this does not mean that design of analog system needs less human resources including designer's skill.)

To summarize, our industries become aware of importance of analog technologies and look for newcome analog engineer from university, but insufficient number of analog circuit engineers are supplied by universities.

Creation and Education of Next-Generation Engineers at the University

It is said recently that the number of Japanese high school students who want to take entrance examination for science or technology course of university has been decreasing year by year. Meanwhile the number of graduating students in technology course of university who want to get job at non-industrial company such as securities company and bank. For 30 years ago most student in department of electronics selected their course because they wanted to be an electronics engineer. But at present time, more than two thirds of them came with other reasons. In other words, many students in electronics course do not have their interest in electronics and study their curriculum only with a sense of duty. Instead, many students are fond of hitting a keyboard. They tend to play not in real world but in computer created virtual world. As a result, they think what circuit simulator outputs as a real circuit itself. Even young researcher in the doctor course sometimes write a paper using simulator only without simple experiment.

This seems an origin of why young analog circuits engineers disappear. Our discussion at the National Convention came to the conclusion that it is because of disappearance of "Radio boy." Radio boy means such a boy who likes assembling parts to make a radio receiver, HiFi reproducer or transmitter as his hobby. We think many of them grew up to be analog engineers and play an important role in the development of Japanese electronics industries. Professor Yanagisawa at Tokyo Institute

Technology (now moves to Shibaura Institute Technology) pointed out that the criminal of disappearance of radio boy is spread of LSI into electronics. LSI is quite a "black box" and to look into a package of LSI can never stimulate his curiosity! Therefore, in most university, professors are gradually increasing a percentage of basic experiments in their curriculum such as assembling a simple transistor circuits using a solder iron after designing it himself with a SPICE simulator. The author's experience shows that most student are attracted by these type of experiments.

The author believes that to increase "radio boy" is one of the most efficient means to increase good analog circuit engineers and it is an urgent matter for creating next generation analog engineer. Therefore it is very important to create system which inspire young people to be interesting in real electronics world. We must pay our effort to looking for such a system.

Conclusion

The author describes several problems with respect to the analog circuits technologies in Japan, design productivities, challenge to creation and how hand them down to the next generations. Potential analog circuits engineer are decreasing here. But it should be stressed that analog circuit technologies are always necessary in the wave front region of electronics technologies, therefore the key technologies to develop much higher performance digital system and much high frequency circuits. So we must make as many younger peoples as possible to be interesting in learning analog technologies.

Acknowledgment

The author acknowledges Dr. Minoru Nagata at the Central Research Laboratory of Hitachi Ltd. for his encouragement and valuable advice. Thanks are also due to Professor Ken Yanagisawa, Professor Nobuo Fujii and Dr. Nagata for the permission to cite their opinions and discussions.

References

1. K. Sekine, "Creation and Hand down of Analog Circuit Technology Part 1," *Journal of IEICE* 73 (Sep. 1990): 1009–1010.

2. K. Yanagisawa, "Creation and Hand down of Analog Circuit Technology Part 2," *Journal of IEICE* 73 (Sep. 1990): 1010–1011.

3. M. Nagata, "Creation and Hand down of Analog Circuit Technology Part 3," *Journal of IEICE* 73 (Sep. 1990): 1012–1015.

4. H. Yamada, "Creation and Hand down of Analog Circuit Technology Part 4," *Journal of IEICE* 73 (Sep. 1990): 1015–1017.

5. N. Fujii, "Why Must You Study Analog Circuits in These Digital Technology Days?" *Journal of IEICE* 72 (Aug. 1989): 925–928.

6. K. Sekine, "Analysis and Design of Analog Circuits," IEEJ Papers of Technical Meeting on Electronic Circuits ECT92-14 (March 1992): 57–62.

5. Thoughts on Becoming and Being an Analog Circuit Designer

Special commentary by Laurel Beth Joyce, Greg's wife

"My favorite programming language is solder."
—Todd K. Whitehurst
Stanford University, 1988

Well, here I am, finally writing this book chapter! Instead of trying to tell the reader how to design analog circuits (I'll leave it to the folks with circuits named after them to do that, unless you take my courses), I will discuss several aspects of becoming and being an analog circuit designer. I will try to cover a few areas that I think are important, particularly to someone considering a career in this field. My wife's comments near the end of this chapter will also be of considerable interest to the significant other (S.O.) of anyone considering this career choice.

Analog Circuit Designers

What type of person becomes an analog circuit designer? Perhaps the best way to address that question is to start by describing the types of people who do *not* become analog circuit designers! Examples are folks whose second career choice would have been accounting, people who say "dude" a lot, people who have time to sit around wondering why their belly-button lint is gray,[1] people who wear Birkenstock sandals and eat alfalfa, people who are frustrated by devices more complex than a paper clip, and people who are repeatedly abducted by space aliens.

In other words, analog circuit designers tend to be a creative, practical, and curious bunch of folks who are rarely abducted by space aliens. The typical analog designer doesn't worry too much about shaving on weekends (especially the female ones), drinks beer and eats pizza, owns an oscilloscope (see "Things You Need to Survive as a 'Real' Analog Designer" below), thinks modern art consisting of blank white canvases is a bunch of crap, occasionally uses "swear words," and may be considered a bit "eccentric" by his or her friends and colleagues. Over the years, knowing a fair number of analog designers, I have only encountered one notable exception: Jim Williams.[2]

---

1. Actually, my friends at the Office of Navel Research in Washington, DC, have studied this issue extensively. They have found that belly-button lint color is a complex function of clothing color, belly-button humidity, and the amount of cheese consumed.
2. He doesn't drink beer.

41

Why should anyone want to become an analog designer? Aside from the large amounts of money you earn, the hordes of attractive members of the opposite sex that are drawn to you by the heady smell of solder, the ability to simulate circuits in your head, and the undying respect of all other engineers, there is one really important advantage to this line of work: it's fun!

In fact, designing circuits can be absolutely wonderful. You create, from scratch, a complete working[3] circuit that accomplishes a function you (or your boss) desire. Once you get some experience, you can visualize how the circuit building blocks you know can be combined to get what you want. Sometimes you realize that you need to invent something really new to do a particular function. Creativity and a bit of insanity really helps with that.

You don't need big power tools, a yard full of old cars up on blocks, or a trip to the Himalayas to build analog circuits. Actually, what you do need are small power tools, a garage full of old oscilloscopes up on blocks, and a trip to some surplus stores in Mountain View. In any case, once you reach some level of "analog enlightenment," it is really addictive. This is good, because the majority of engineers have gotten so seduced by digital circuits and software that some very big electronics companies exist that do not have a single decent analog circuit designer in house. In other words, if you learn analog circuit design, you can get a job!

"I've heard enough! Sign me up!" If that's what you are thinking,[4] you may want to know how you can become an analog designer. One way is to learn "on the street" ("Hey buddy, wanna pick up some transistors cheap? . . . They've got high betas and they're clean!"). That works eventually (the word "eventually" is key), but most people go to a university and learn there. If you are remotely interested in the latter option, please read on . . .

Analog Boot Camp: One Way to Become an Analog Designer

I teach analog circuit design at Stanford,[5] along with my colleagues in the Department of Electrical Engineering. In recent years, we have taken great pains to upgrade the electronics courses to include more practical, design-oriented material. My own courses are considered "analog boot camp" for undergraduates who think of transistors only in

---

3. (eventually)
4. (if not, please put this book down and read that biography of Bill Gates over there to the left)
5. The opinions and/or other crap in this chapter are completely the fault of the author and do not reflect the opinions and/or other crap of Stanford University in any way.

terms of band diagrams. I'll share with you some of our "indoctrination" techniques . . . [6]

First, we administer an exam to weed out the people who should really be learning about French history or something like that. Here are a few sample questions:

Choose the single best answer.

1) The best all-around programming language is:

a) C

b) C++

c) BASIC

d) Fortran

e) solder

2) A "GUI" is:

a) a productivity-enhancing graphical user interface for modern computers

b) useful for opening beer bottles

c) a voltage regulation circuit invented by famous Dutch EE Cornelius von Fritzenfratz

d) who gives a crap, this test is about analog circuits!

3) Analog circuits are:

a) circuits involving only resistors and capacitors, like in first-year electronics, dude

b) circuits built with digital logic and no more than two discrete transistors that you debug by reprogramming EPROMS until they work

c) not needed now that we have the "Newton"

d) really cool

4) SPICE is:

a) stuff like salt and pepper you put on your food

b) the reason nobody needs to build real circuits at all

c) a program designed to see how quickly your computer bogs down when doing floating-point operations

d) the only reason we need computers, other than Tetris.™

5) "Solder suckers" are:

a) PG-rated, but can occasionally be seen on National Geographic specials

b) the black holes of circuits, often seen running around with current sources invented by Mr. Wilson (from "Dennis the Menace")

c) people who are lured into analog circuit design by evil professors

d) plastic pumps used to remove solder from component leads where those uneducated about analog design have made mistakes

---

6. These techniques have been developed over several decades by carefully selected teams of scientists from all over the world.

That sort of thing helps weed out the sick, the feeble-minded, and the history majors. Then we begin analog "basic training," which involves learning the following song for drill practice and considerable healthful marching and shouting.

Analog Boot Camp Drill Routine

by G. Kovacs

(The words are first barked out by the professor, then shouted back by
    students marching in formation.)
Analog circuits sure are fine,
Just can't get 'em off my mind.

Digital circuits ain't my kind,
Zeros and ones for simple minds.

I guess NAND gates aren't all that bad,
'Cause I need them for circuit CAD.

One, two, three, four,
Gain and bandwidth, we want more.

Five, six, seven, eight,
We don't want to oscillate.

Widlar, Wilson, Brokaw too,
They've got circuits, how 'bout you?

(*repeat*)

I also ask a few random questions and have been known to order a few push-ups here and there if, for example, a student cannot correctly distinguish between the Miller and Budweiser Effects. Now the students are ready for their plunge into the world of analog . . .

At this point, they are taught theory in one class and hands-on aspects in another. Essentially, the idea is to progress from the basic idea of an operational amplifier (op amp) through the necessary circuit building blocks that are required to design one. Finally, we reach the point where the students know enough to do that, and then we get into feedback and stability. Meanwhile, in the laboratory part of the class, the students are learning how to destroy most of the circuits covered in lecture. It is in the lab that we teach them the all-important "smoke principle" of solid-state devices. This is the formerly very closely guarded industrial secret that each discrete or integrated circuit is manufactured with a certain amount of smoke pre-packaged inside. If, through an inadvertent wiring error, conditions arise through which the smoke is permitted to escape, the device ceases to function. We also train the students to recognize and distinguish the smells of different burning components ("Ah yes, a carbon resistor seems to have burned up in this circuit . . . smells like 220KΩ.").

I am not kidding about this, but not more than ⅓ of the EE students at this level have ever used a soldering iron before! In contrast, nearly all of them have driven a BMW and can explain leveraged buyouts in great detail (I presume this is a phenomenon more common at schools where yuppy pupae are present in large numbers). After a little trial and error, most of them learn which end of the soldering iron is hot (I am told that those who never really figure this out generally transfer to a local state-run university where they can just write software, but I have no concrete evidence of this). Pretty soon, they not only know how to solder, but also how to use a wide range of up-to-date test equipment. (I worry about the ones who keep looking for an "auto setup" button on a voltmeter, though! . . . more on this below.)

At this point, we get the students into the guts of Boot Camp: design it, SPICE it, make it work, and examine the differences between the SPICE model and the real thing. The idea is to teach simulation as "virtual instruments" and then introduce the real ones (the type with knobs). We provide SPICE decks[7] for each circuit that are already on the student computers. We leave out critical component values for the students to choose. They have to come to lab with a running simulation and then build the circuit. This can be fun to watch the first time, as the students look around the lab for 10,000 amp current sources, diodes with forward voltages of exactly 0.700V, and 13.4567E3 ohm resistors. Eventually, they figure things out and get things working.[8]

We ask them to simulate and build a lot of discrete circuits, including power supplies, basic op amp circuits, single-transistor amplifiers, a simple op amp built from discretes, and power amplifiers. After that they build a project of their own choosing, demonstrating their analog design skills. This exercise gives them a chance to construct a complete circuit from scratch and write an instruction manual, specification sheet, and marketing sheet for whatever it is. Some students have built really amazing things, such as a waveform synthesizer, a heterodyne spectrum analyzer, an infrared remote control system, an acoustic rangefinder, etc. Some have built devices that are also humorous, including a fake leopard

---

7. "Gee, Dad, why do they call them SPICE decks?"
   "Well, son, way back before they found a practical use for the 'Newton' in 2027, computers used punched paper cards as a way to enter data and programs. We called a stack of those cards a 'deck'."
8. Our current sources only go to 9,000 amps, we keep the 0.700-V diodes in another room, and they need to specify resistor values to a few more decimals or our component supplier doesn't know which value to provide.

fur-covered[9] laser/galvanometer system for a light show, a guitar amplifier that "goes to eleven," and a contraption that the student proudly described as a "large vibrator" (he meant "multivibrator," but it was terribly funny at the time).

Does it work? Are we able to turn out decent analog designers? Well, it seems to be working, and feedback from companies who have hired our students is positive.[10] For me, success can be measured by the number of students who actually learn to love analog circuit design despite the fact that they are growing up in a world devoid of Heathkits and basements full of surplus electronics to hack circuits with.

To illustrate the transformations that occur, I have reproduced a letter home from one of the students on his first and last days in Boot Camp (the names have been changed to protect the student's identity):

Day 1 of Boot Camp:

Dear Mom,

Things are going fine here at Stanford! Today we learned about "operational amplifiers." They are triangle-shaped things that can do basically anything. The textbook says they have an "ideal voltage source" inside. Tell Pop that this means I can hook one up to power the whole farm when I get home this summer! I can't wait!

Love,
Billy

Last day of Boot Camp:

Dear Mom,

I just finished my analog circuit training at Stanford! I now know I was wrong about operational amplifiers being able to power the whole farm! That was totally silly, because they are simply integrated circuits, and thus require external power. Also, their non-zero output resistance and short-circuit protection circuitry means that they can only supply a few milliamps of current.

Do you know why smoke comes out of transistors when they get too hot? I will explain it all to you, Pop, and the farmhands when I get back there in a few weeks.

I think we should consider turning the barn into a circuit design laboratory. Bossie could stay in my room, since I will probably spend most of my time out there. Please let me know if this is OK, because I would rather do this than take a job doing software-

---

9. Of course, we use only fake leopard fur because it is an endangered species, and we are very politically correct. The only type of skin that is still OK to use for decorative purposes is that of Caucasian heterosexual males, but we were out of it at the time.

10. We all know that positive feedback can lead to oscillations, so we will have to keep an eye on this situation. Raising tuition seems to provide the necessary negative feedback to keep the system stable.

simulated power consumption validation of a subset of indirect-jump instructions of the new Valium computer chip at Interola's new lab in Lumbago, Oregon.

Love,
Billy

What Should Aspiring Analog Designers Read?

There is good stuff on analog circuits to read out there, and generally it is reasonably easy to locate. I am not going to go into the large number of books available other than to point out that you really need to have Horowitz and Hill, *The Art of Electronics* (Cambridge Press) and Gray and Meyer, *Analysis and Design of Analog Integrated Circuits* (John Wiley and Sons). Those two books are simply the essentials;[11] it's easy to supplement them from the droves of texts out there.

As far as journals go, there are several good ones out there. Of course, the IEEE has a few. Then there's *Wireless World*, put out by a bunch of hackers in the United Kingdom, with real depth mixed right in there with fun projects. Another good foreign offering is *Elektor*, which is put out by a bunch of hackers in Holland (the closed-loop cheese fondue controller project last year was awesome). The *Computer Applications Journal* (alias *Circuit Sewer*) is worth reading, but is aimed at those who think debugging a piece of hardware involves mainly fixing software (it is 90% digital subject matter, with occasional forays into scary things like op amps). What about those old standards like *Popular Electronics*? Well, they are OK for the occasional project idea, but as for technical content, I generally say, "Later!" (especially to ones with names like *Electronics Now!*).

One of the richest sources of information, and probably the least obvious to beginners, is the application notes written by the manufacturers of integrated circuits. Just think about it . . . they are trying to sell their wares by getting you excited about their uses.[12] They are absolutely packed with interesting circuits! Usually, you can get them for free, as well as sets of data books, just by calling the manufacturers. Saying you are a student usually helps, and will often get you free samples too. In case you don't know, the best ones are from National Semiconductor, Linear Technology, Maxim, Analog Devices, and Burr Brown.

---

11. Did I mention that this book is also one of the essentials? In any case, you are already clever enough to be reading it, so why bother!

12. They have to accomplish this by showing you cool circuits you can build, as opposed to traditional marketing approaches, such as those used to sell beer. I am still waiting for the Swedish Bipolar Bikini Team, though!

Things You Need to Survive as a "Real" Analog Designer

I am occasionally asked what you need to survive as a "real" analog designer. Well, this is a highly personal matter, but I can at least give my standard answer, which is the things I need (in order of importance):

1. An understanding significant other (S.O.)
2. A laboratory dog to keep my feet warm
3. A basic supply of discrete and integrated components
4. A decent oscilloscope
5. A power supply
6. A soldering iron
7. Basic hand tools
8. Cheap beer
9. A pad and pencil

An understanding S.O. is critical, because when you start coming home with large chunks of blue-colored equipment and go misty-eyed when you see an old Tektronix catalog, it takes a special kind of person to understand! Analog designers tend to build up huge collections of old oscilloscopes, circuit boards, random metal boxes, and all sorts of "precious" items that will come in handy some day. I think meeting an analog designer who isn't a packrat is about as likely as meeting the Swedish Bipolar Bikini Team.

A typical workbench for analog circuit design is shown in Figure 5–1. In addition, the "analog workstation," where most of the really good circuit ideas are developed, is shown in Figure 5–2. The very useful labora-

Figure 5–1.
A typical workbench used for analog circuit design.

Figure 5–2.
An analog work station. This is the place many great circuit designs are developed.

tory dog (black Labrador called Rosie) is shown in Figure 5–3. She is better with a soldering iron than most engineers I know!

Comments on Test Instruments

Good test instruments are critical to a person's success as an analog circuit designer! They are the equivalents of musical instruments to a musician . . . you never share your Stradivarius (i.e., Tektronix 7904A oscilloscope) and need to be intimately familiar with its nuances to get the best performance out of it. Bottom lines here: 1) don't buy cheesy foreign test gear unless you absolutely have to, and 2) when you find

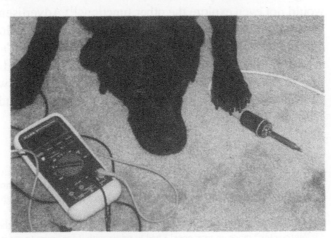

Figure 5–3.
Rosie, the laboratory dog in our house. She will debug any circuit for a piece of beef jerky.

your beautiful oscilloscope, spot-weld it to some part of your body so that it is not borrowed without your knowledge.

I am an absolute hard-core fan of Tektronix test equipment. Tektronix oscilloscopes (the most important item) are available with a wonderful user interface and provide extremely high performance plus real versatility. The only problem is that they don't make that kind any more.

In recent years, there has been a trend toward computer-controlled, menu-driven test instruments, rather than instruments that use a dedicated switch or knob for each function (so-called "knob-driven" instruments). In most cases, the push for menu-driven test instruments has an economic basis—they are simply cheaper to build or provide more features for the same price. However, there are practical drawbacks to that approach in many cases. A common example, familiar to anyone who has ever used an oscilloscope, is the frequent need to ground the input of a vertical channel to establish a "zero" reference. With a knob-driven instrument, a simple movement of the index finger and thumb will suffice. With a menu-driven instrument, one often has to fumble through several nested menus. This really sucks, and I think it is because they are starting to let MBAs design oscilloscopes. (I suppose one possible benefit of this is that soon 'scopes will have a built-in mode that tells you when to refinance your mortgage!)

Grounding a vertical channel's input is something you need to do often, and it is quite analogous to something familiar even to digital engineers, like going to the bathroom. You simply wouldn't want to scroll through a bunch of menus during your mad dash to the bathroom after the consumption of a bad burrito! There are several similar annoyances that can crop up when using menu-driven instruments (how about ten keystrokes to get a simple sine wave out of a signal generator?!).

To be fair, menu-driven instruments do have advantages. However, since I am not a big fan of them, I'll conveniently omit them here.[13] It always pisses me off to watch students hitting the "auto setup" button on the digital 'scopes in our teaching lab and assuming it is doing the right thing for them every time (not!). If we didn't force them to, most of them would not even explore the other functions![14] Advertisements for these new instruments often brag that they have a more "analog-like feel" (as opposed to what, a "primordial slime ooze feel"?). Let's get real here . . . at least in part, this is just another incarnation of the old engineering saying, "If you can't fix it, make it a feature." Since when was a "more chocolate-like taste" a real key reason to buy brown sludge instead of chocolate?

---

13. One of the key advantages is that they can help us lure would-be engineers into the lab. The type of EE student who doesn't like hands-on hardware engineering (you know, the ones who end up working for Microsloth) can be attracted by the nice menus long enough to actually see how much fun electronics can be.
14. At this point, I will admit that our VCR does blink "12:00," but I hear there will be an "auto-setup" mode on new ones! I had to fiddle with it for hours to get it to blink "12:00."

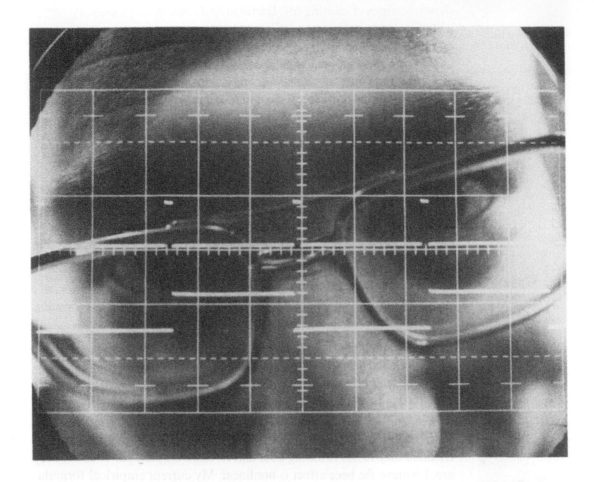

Figure 5–4.
What you look like to your oscilloscope (yuk!). Actually, this is what Jim Williams looks like to his oscilloscope. You probably won't look that silly.

I am sad to report that knob-driven analog test instruments are becoming more difficult to get. I also have to admit that performance is improving while relative prices are dropping, so "user-friendly" instruments aren't all that bad. Students take note: at least try to check out instruments with knobs, in between pressing "auto-setup" and "help" keys! A great place to find this stuff is at your friendly neighborhood university (we'll never surrender!), local "ham radio" swap meets, and companies that specialize in used test equipment. Also, remember to be nice to your oscilloscope! What you look like to that faithful piece of test gear is shown in Figure 5–4.

What Does My Wife Think about All of This?

This section was written by my wife, Laurel Beth Joyce, the pride of Mars, PA.[15] It is added to provide an extra sense of realism and to prepare

---

15. I am not making this up. This is because I don't need to. Western PA has tons of great names of towns, like Beaver, Moon, etc., as well as great names for public utilities, like "Peoples' Natural Gas." Naturally, nobody from there thinks any of this is funny.

a would-be analog circuit designer for the impact this career choice has on one's home life.[16]

If your S.O. is an analog designer, your relationship will be much happier once you come to understand and accept some of the basic differences between analog circuit designers and normal people.

1. Analog circuit designers consider beer one of the major food groups and an essential hacking tool. (See "Things You Need to Survive as a 'Real' Analog Designer.") To avoid major altercations, be sure there's always beer in the house.

Fortunately, my husband's students signed him up for a Beer-of-the-Month club. Each month the UPS lady drops a big box of beer on our doorstep, putting him in hacker heaven and saving me many trips to the beer store.

2. Circuit designers don't tell time in the same way that the rest of us do. Unfortunately, I still haven't figured out the exact formula for converting circuit design time into regular time.

For example, let's say my husband is in the middle of a hacking project at work and he calls to tell me that he's going to head home in about half an hour. If he's alone and I know he's working on a project that doesn't require an oscilloscope, I simply multiply the time by two. If there is an oscilloscope involved, I multiply by three. If he's got any circuit design friends with him, I generally add at least 40 minutes per friend if they're not drinking beer and an extra 2 hours per friend if they are. I believe the beer effect is nonlinear. My current empirical formula for computing circuit design time in minutes is thus:

$$t_{cd} = (2 + N_{scopes})\, t + (40 + 120\, k_{brewski})\, N_{friends}$$

where N_{scopes} is the number of oscilloscopes present, $k_{brewski}$ is the linear approximation for the nonlinear beer effect (taken to be one, but can be replaced by a suitable time-dependent nonlinearity) and $N_{friends}$ is the number of circuit design friends present.

My calculations are rarely perfect, so I'm pretty sure there are some other variables involved. It may have something to do with the number of op amps in the project, but since I'm still trying to figure out what an op amp is, I haven't quite determined how that should factor into the formula.

My suspicion is that this formula varies slightly among hackers, but you're probably safe to use this as a starting point for deriving your own formula.

3. Circuit designers have an interesting concept of economics. Last weekend we wandered down the breakfast cereal aisle of our local

---

16. The opinions and/or other crap written by my wife are completely her fault and do not reflect the opinions and/or other crap of Stanford University or myself in any way.

grocery and my husband was astounded that the big box of Cap'n Crunch cost $4.58. He considered it so expensive, he wanted to put it back on the shelf.

In contrast, he tells me that $2,000 is a bargain for a 20-year-old, used oscilloscope that only smokes a little bit and will only require one or two weekends to fix up. And $1,000 is a great deal on a 'scope that doesn't work at all, because it can be cannibalized for parts to repair the 'scopes that smoke comes out of (assuming that it has enough parts left that never smoked).

4. When an analog circuit designer brings home a new piece of equipment, the S.O. becomes invisible for several hours.

I used to get jealous every time a new 'scope or signal generator came into the house. He'd burst in the door all breathless and say, "Hi, Laurel, look what I found today. Isn't she beautiful? I'm just going to take her upstairs for a few minutes." The two would disappear into the lab and I'd hear lots of cooing and giddy chatter that went on until daybreak. It was as if my S.O. was bringing home his mistress and dashing up to our bedroom right under my nose.

If the dog or I went into the room, he wouldn't even notice us. I could tell him that beer had just been outlawed in the United States or the dog could vomit on his shoes. He'd just say, "I'll be with you in a minute," and go back to grinning and twiddling the knobs of his new toy.

When you realize it's no use being jealous and that you'll never be able to compete with these machines (unless you want to turn to the folks at Tektronix for fashion advice and get some clothes in that particular shade of blue, some 'scope knob earrings and some WD-40 cologne), you can actually have some fun when your S.O. is in this condition. If you like to watch TV, you've got the remote control to yourself for a few hours. If you have friends that your S.O. can't stand, invite them over for a party. If you're angry with your S.O. you can stand there and say nasty things ("You solder-sucking slimeball!"), get all the anger out of your system, and he'll remain totally oblivious. Be creative!

I was miserable before I learned that these basic differences and quirks are characteristic of *most* analog circuit designers, not just my husband. When I finally understood that they're simply a different species, my bills for psychoanalysis decreased significantly.

There are a couple of other things that help, too. First, ask all of your relatives to move to towns where there are used test equipment shops or frequent swap meets. If you don't, you may never see them again. It took six years for my husband to meet my Aunt Gertrude, but as soon as he found out that Crazy Egbert's World of 'Scopes was only 12 miles from her house, we were on an airplane—"Because I feel terrible that it has taken me so long to meet your aunt"—within 24 hours.

And, when all else fails, you may have to resort to the spouse alignment unit (SAU). Mine is a wooden rolling pin (shown in Figure 5–5),

Figure 5–5.
The pride of Mars, PA, with her spouse alignment unit (SAU).

but I hear a baseball bat or cast-iron skillet works just as well. The SAU comes in handy, for example, when you're hosting a large dinner party, all the guests have arrived and are waiting for their meal, and your analog circuit designer has said he'll join the party "in just a minute" for the past two hours. In this situation you should quietly hide the SAU up your sleeve, excuse yourself while flashing a charming smile at your guests, waltz into the lab, yank the plug on the soldering iron and strike a threatening pose with the SAU.

It's kind of like training a dog with a rolled-up newspaper—you only have to use it once. After that, the sight of the unit or the threat that you're in the mood to do some baking will yield the desired response.

Conclusion

I hope this chapter has given you some sense of what you need to learn and obtain to become an analog circuit designer, as well as some of the emotional challenges in store for you. It would be great if you considered it as an alternative to the digital- or software-based engineering drudgery that you are statistically likely to end up doing. There may yet be some burnt resistors and oscillations in your future!

6. Cargo Cult Science[*]

During the Middle Ages there were all kinds of crazy ideas, such as that a piece of rhinoceros horn would increase potency. Then a method was discovered for separating the ideas—which was to try one to see if it worked, and if it didn't work, to eliminate it. This method became organized, of course, into science. And it developed very well, so that we are now in the scientific age. It is such a scientific age, in fact, that we have difficulty in understanding how witch doctors could ever have existed, when nothing that they proposed ever really worked—or very little of it did.

But even today I meet lots of people who sooner or later get me into a conversation about UFOs, or astrology, or some form of mysticism, expanded consciousness, new type of awareness, ESP, and so forth. And I've concluded that it's not a scientific world.

Most people believe so many wonderful things that I decided to investigate why they did. And what has been referred to as my curiosity for investigation has landed me in a difficulty where I found so much junk that I'm overwhelmed. First I started out by investigating various ideas of mysticism, and mystic experiences. I went into isolation tanks and got many hours of hallucinations, so I know something about that. Then I went to Esalen, which is a hotbed of this kind of thought (it's a wonderful place; you should go visit there). Then I became overwhelmed. I didn't realize how much there was.

At Esalen there are some large baths fed by hot springs situated on a ledge about thirty feet above the ocean. One of my most pleasurable experiences has been to sit in one of those baths and watch the waves crashing onto the rocky shore below, to gaze into the clear blue sky above, and to study a beautiful nude as she quietly appears and settles into the bath with me.

One time I sat down in a bath where there was a beautiful girl sitting with a guy who didn't seem to know her. Right away I began thinking, "Gee! How am I gonna get started talking to this beautiful nude babe?"

I'm trying to figure out what to say, when the guy says to her, "I'm, uh, studying massage. Could I practice on you?"

---

[*]Adapted from the Cal Tech commencement address given in 1974.

"Sure," she says. They get out of the bath and she lies down on a massage table nearby.

I think to myself, "What a nifty line! I can never think of anything like that!" He starts to rub her big toe. "I think I feel it," he says. "I feel a kind of dent—is that the pituitary?"

I blurt out, "You're a helluva long way from the pituitary, man!"

They looked at me, horrified—I had blown my cover—and said, "It's reflexology!"

I quickly closed my eyes and appeared to be meditating.

That's just an example of the kind of things that overwhelm me. I also looked into extrasensory perception and PSI phenomena, and the latest craze there was Uri Geller, a man who is supposed to be able to bend keys by rubbing them with his finger. So I went to his hotel room, on his invitation, to see a demonstration of both mindreading and bending keys. He didn't do any mindreading that succeeded; nobody can read my mind, I guess. And my boy held a key and Geller rubbed it, and nothing happened. Then he told us it works better under water, and so you can picture all of us standing in the bathroom with the water turned on and the key under it, and him rubbing the key with his finger. Nothing happened. So I was unable to investigate that phenomenon.

But then I began to think, what else is there that we believe? (And I thought then about the witch doctors, and how easy it would have been to check on them by noticing that nothing really worked.) So I found things that even more people believe, such as that we have some knowledge of how to educate. There are big schools of reading methods and mathematics methods, and so forth, but if you notice, you'll see the reading scores keep going down—or hardly going up—in spite of the fact that we continually use these same people to improve the methods. There's a witch doctor remedy that doesn't work. It ought to be looked into; how do they know that their method should work? Another example is how to treat criminals. We obviously have made no progress—lots of theory, but no progress—in decreasing the amount of crime by the method that we use to handle criminals.

Yet these things are said to be scientific. We study them. And I think ordinary people with commonsense ideas are intimidated by this pseudoscience. A teacher who has some good idea of how to teach her children to read is forced by the school system to do it some other way—or is even fooled by the school system into thinking that her method is not necessarily a good one. Or a parent of bad boys, after disciplining them in one way or another, feels guilty for the rest of her life because she didn't do "the right thing," according to the experts.

So we really ought to look into theories that don't work, and science that isn't science.

I think the educational and psychological studies I mentioned are examples of what I would like to call cargo cult science. In the South Seas there is a cargo cult of people. During the war they saw airplanes land with lots of good materials, and they want the same thing to happen now. So they've arranged to make things like runways, to put fires along the sides of the runways, to make a wooden hut for a man to sit in, with two wooden pieces on his head like headphones and bars of bamboo sticking out like antennas—he's the controller—and they wait for the airplanes to land. They're doing everything right. The form is perfect. It looks exactly the way it looked before. But it doesn't work. No airplanes land. So I call these things cargo cult science, because they follow all the apparent precepts and forms of scientific investigation, but they're missing something essential, because the planes don't land.

Now it behooves me, of course, to tell you what they're missing. But it would be just about as difficult to explain to the South Sea Islanders how they have to arrange things so that they get some wealth in their system. It is not something simple like telling them how to improve the shapes of the earphones. But there is one feature I notice that is generally missing in cargo cult science. That is the idea that we all hope you have learned in studying science in school—we never explicitly say what this is, but just hope that you catch on by all the examples of scientific investigation. It is interesting, therefore, to bring it out now and speak of it explicitly. It's a kind of scientific integrity, a principle of scientific thought that corresponds to a kind of utter honesty—a kind of leaning over backwards. For example, if you're doing an experiment, you should report everything that you think might make it invalid—not only what you think is right about it: other causes that could possibly explain your results; and things you thought of that you've eliminated by some other experiment, and how they worked—to make sure the other fellow can tell they have been eliminated.

Details that could throw doubt on your interpretation must be given, if you know them. You must do the best you can—if you know anything at all wrong, or possibly wrong—to explain it. If you make a theory, for example, and advertise it, or put it out, then you must also put down all the facts that disagree with it, as well as those that agree with it. There is also a more subtle problem. When you have put a lot of ideas together to make an elaborate theory, you want to make sure, when explaining what it fits, that those things it fits are not just the things that gave you the idea for the theory; but that the finished theory makes something else come out right, in addition.

In summary, the idea is to try to give all of the information to help others to judge the value of your contribution; not just the information that leads to judgment in one particular direction or another.

The easiest way to explain this idea is to contrast it, for example, with advertising. Last night I heard that Wesson oil doesn't soak through food. Well, that's true. It's not dishonest; but the thing I'm talking about is not just a matter of not being dishonest, it's a matter of scientific integrity, which is another level. The fact that should be added to that advertising statement is that no oils soak through food, if operated at a certain temperature. If operated at another temperature, they all will—including Wesson oil. So it's the implication which has been conveyed, not the fact, which is true, and the difference is what we have to deal with.

We've learned from experience that the truth will come out. Other experimenters will repeat your experiment and find out whether you were wrong or right. Nature's phenomena will agree or they'll disagree with your theory. And, although you may gain some temporary fame and excitement, you will not gain a good reputation as a scientist if you haven't tried to be very careful in this kind of work. And it's this type of integrity, this kind of care not to fool yourself, that is missing to a large extent in much of the research in cargo cult science.

A great deal of their difficulty is, of course, the difficulty of the subject and the inapplicability of the scientific method to the subject. Nevertheless, it should be remarked that this is not the only difficulty. That's why the planes don't land—but they don't land.

We have learned a lot from experience about how to handle some of the ways we fool ourselves. One example: Millikan measured the charge on an electron by an experiment with falling oil drops, and got an answer which we now know not to be quite right. It's a little bit off, because he had the incorrect value for the viscosity of air. It's interesting to look at the history of measurements of the charge of the electron, after Millikan. If you plot them as a function of time, you find that one is a little bigger than Millikan's, and the next one's a little bit bigger than that, and the next one's a little bit bigger than that, until finally they settle down to a number which is higher.

Why didn't they discover that the new number was higher right away? It's a thing that scientists are ashamed of—this history—because it's apparent that people did things like this: When they got a number that was too high above Millikan's, they thought something must be wrong—and they would look for and find a reason why something might be wrong. When they got a number closer to Millikan's value they didn't look so hard. And so they eliminated the numbers that were too far off, and did other things like that. We've learned those tricks nowadays, and now we don't have that kind of a disease.

But this long history of learning how to not fool ourselves—of having utter scientific integrity—is, I'm sorry to say, something that we haven't specifically included in any particular course that I know of. We just hope you've caught on by osmosis.

The first principle is that you must not fool yourself—and you are the easiest person to fool. So you have to be very careful about that. After you've not fooled yourself, it's easy not to fool other scientists. You just have to be honest in a conventional way after that.

I would like to add something that's not essential to the science, but something I kind of believe, which is that you should not fool the layman when you're talking as a scientist. I am not trying to tell you what to do about cheating on your wife, or fooling your girlfriend, or something like that, when you're not trying to be a scientist, but just trying to be an ordinary human being. We'll leave those problems up to you and your rabbi. I'm talking about a specific, extra type of integrity that is not lying, but bending over backwards to show how you're maybe wrong, that you ought to have when acting as a scientist. And this is our responsibility as scientists, certainly to other scientists, and I think to laymen.

For example, I was a little surprised when I was talking to a friend who was going to go on the radio. He does work on cosmology and astronomy, and he wondered how he would explain what the applications of this work were. "Well," I said, "there aren't any." He said, "Yes, but then we won't get support for more research of this kind." I think that's kind of dishonest. If you're representing yourself as a scientist, then you should explain to the layman what you're doing—and if they don't want to support you under those circumstances, then that's their decision.

One example of the principle is this: If you've made up your mind to test a theory, or you want to explain some idea, you should always decide to publish it whichever way it comes out. If we only publish results of a certain kind, we can make the argument look good. We must publish both kinds of results.

I say that's also important in giving certain types of government advice. Supposing a senator asked you for advice about whether drilling a hole should be done in his state; and you decide it would be better in some other state. If you don't publish such a result, it seems to me you're not giving scientific advice. You're being used. If your answer happens to come out in the direction the government or the politicians like, they can use it as an argument in their favor; if it comes out the other way, they don't publish it at all. That's not giving scientific advice.

Other kinds of errors are more characteristic of poor science. When I was at Cornell, I often talked to the people in the psychology department. One of the students told me she wanted to do an experiment that went something like this—it had been found by others that under certain circumstances, X, rats did something, A. She was curious as to whether, if she changed the circumstances to Y, they would still do A. So her proposal was to do the experiment under circumstances Y and see if they still did A.

I explained to her that it was necessary first to repeat in her laboratory the experiment of the other person—to do it under condition X to see if she could also get result A, and then change to Y and see if A changed. Then she would know that the real difference was the thing she thought she had under control.

She was very delighted with this new idea, and went to her professor. And his reply was, no, you cannot do that, because the experiment has already been done and you would be wasting time. This was in about 1947 or so, and it seems to have been the general policy then to not try to repeat psychological experiments, but only to change the conditions and see what happens.

Nowadays there's a certain danger of the same thing happening, even in the famous field of physics. I was shocked to hear of an experiment done at the big accelerator at the National Accelerator Laboratory, where a person used deuterium. In order to compare his heavy hydrogen results to what might happen with light hydrogen, he had to use data from someone else's experiment on light hydrogen, which was done on different apparatus. When asked why, he said it was because he couldn't get time on the program (because there's so little time and it's such expensive apparatus) to do the experiment with light hydrogen on this apparatus because there wouldn't be any new result. And so the men in charge of programs at NAL are so anxious for new results, in order to get more money to keep the thing going for public relations purposes, they are destroying—possibly—the value of the experiments themselves, which is the whole purpose of the thing. It is often hard for the experimenters there to complete their work as their scientific integrity demands.

All experiments in psychology are not of this type, however. For example, there have been many experiments running rats through all kinds of mazes, and so on—with little clear result. But in 1937 a man named Young did a very interesting one. He had a long corridor with doors all along one side where the rats came in, and doors along the other side where the food was. He wanted to see if he could train the rats to go in at the third door down from where he started them off. No. The rats went immediately to the door where the food had been the time before.

The question was, how did the rats know because the corridor was so beautifully built and so uniform that this was the same door as before? Obviously there was something about the door that was different from the other doors. So he painted the doors very carefully, arranging the textures on the faces of the doors exactly the same. Still the rats could tell. Then he thought maybe the rats were smelling the food, so he used chemicals to change the smell after each run. Still the rats could tell. Then he realized the rats might be able to tell by seeing the lights and the arrangement in the laboratory like any commonsense person. So he covered the corridor, and still the rats could tell.

He finally found that they could tell by the way the floor sounded when they ran over it. And he could only fix that by putting his corridor in sand. So he covered one after another of all possible clues and finally was able to fool the rats so that they had to learn to go in the third door. If he relaxed any of his conditions, the rats could tell.

Now, from a scientific standpoint, that is an A-number-one experiment. That is the experiment that makes rat-running experiments sensible, because it uncovers the clues that the rat is really using—not what you think it's using. And that is the experiment that tells exactly what conditions you have to use in order to be careful and control everything in an experiment with rat-running.

I looked into the subsequent history of this research. The next experiment, and the one after that, never referred to Mr. Young. They never used any of his criteria of putting the corridor on sand, or being very careful. They just went right on running rats in the same old way, and paid no attention to the great discoveries of Mr. Young, and his papers are not referred to, because he didn't discover anything about the rats. In fact, he discovered all the things you have to do to discover something about rats. But not paying attention to experiments like that is a characteristic of cargo cult science.

Another example is the ESP experiments of Mr. Rhine, and other people. As various people have made criticisms—and they themselves have made criticisms of their own experiments—they improve the techniques so that the effects are smaller, and smaller, and smaller until they gradually disappear. All the parapsychologists are looking for some experiment that can be repeated—that you can do again and get the same effect—statistically, even. They run a million rats—no, it's people this time—they do a lot of things and get a certain statistical effect. Next time they try it they don't get it any more. And now you find a man saying that it is an irrelevant demand to expect a repeatable experiment. This is science?

This man also speaks about a new institution, in a talk in which he was resigning as Director of the Institute of Parapsychology. And, in telling people what to do next, he says that one of the things they have to do is be sure they only train students who have shown their ability to get PSI results to an acceptable extent—not to waste their time on those ambitious and interested students who get only chance results. It is very dangerous to have such a policy in teaching—to teach students only how to get certain results, rather than how to do an experiment with scientific integrity.

So I have just one wish for you—the good luck to be somewhere where you are free to maintain the kind of integrity I have described, and where you do not feel forced by a need to maintain your position in the organization, or financial support, or so on, to lose your integrity. May you have that freedom.

Making It Work

Five authors in this section give guided tours into what it takes to go from concept to a completed, functional circuit. Steve Roach shows how monstrously complex a "simple" voltage divider can become when it's an oscilloscope input attenuator. Bill Gross gives an eye-opening trip through the development process of an analog integrated circuit, with special emphasis on how tradeoffs must be dealt with. James Bryant explores a fast, flexible way to breadboard analog circuits which is usable from DC to high frequency. A true pioneer in wideband oscilloscope design, Carl Battjes, details the intricacies of T-coil design, an enabling technology for wideband oscilloscopes. In the section's finale, Jim Williams writes about how hard it can be to get your arms around just what the problem is. Imagine taking almost a year to find the right way to turn on a light bulb!

Making It Work

7. Signal Conditioning in Oscilloscopes and the Spirit of Invention

....

The Spirit of Invention

When I was a child my grandfather routinely asked me if I was going to be an engineer when I grew up. Since some of my great-uncles worked on the railroads, I sincerely thought he wanted me to follow in their footsteps. My grandfather died before I clarified exactly what kind of engineer he hoped I would become, but I think he would approve of my interpretation.

I still wasn't sure what an engineer was when I discovered I wanted to be an inventor. I truly pictured myself alone in my basement toiling on the important but neglected problems of humanity. Seeking help, I joined the Rocky Mountain Inventors' Congress. They held a conference on invention where I met men carrying whole suitcases filled with clever little mechanical devices. Many of these guys were disgruntled and cranky because the world didn't appreciate their contributions. One of the speakers, a very successful independent inventor, told of a bankrupt widow whose husband had worked twenty years in isolation and secrecy inventing a mechanical tomato peeler. The tomato peeler had consumed the family savings, and the widow had asked the speaker to salvage the device. With sadness the speaker related the necessity of informing her that tomatoes were peeled in industrial quantities with sulfuric acid. Apparently the inventor had been too narrowly focused to realize that in some cases molecules are more powerful than machines.

I didn't want to become disgruntled, cranky, or isolated and I didn't even own a basement. So I went to engineering school and adopted a much easier approach to inventing. I now design products for companies with such basic comforts as R&D budgets, support staff, and manufacturing operations. Along the way I have discovered many ways of nurturing inventiveness. Here are some techniques that seem to work:

Give yourself time to invent. If necessary, steal this time from the unending rote tasks that your employer so readily recognizes and rewards. I try to work on things that have nothing to do with a particular product, have no schedule, and have no one expecting results. I spend time on highly tangential ideas that have little hope for success. I can fail again and again in this daydream domain with no sense of loss.

Get excited. Enjoy the thrilling early hours of a new idea. Stay up all night, lose sleep, and neglect your responsibilities. Freely explore tangents to your new idea. Digress fearlessly and entertain the absurd. Invent in the morning or whenever you are most energetic. Save your "real" work for when you are tired.

Master the fundamentals of your field. The most original and creative engineers I have known have an astonishing command of undergraduate-level engineering. Invention in technology almost always stems from the novel application of elementary principles. Mastery of fundamentals allows you to consider, discard, and develop numerous ideas quickly, accurately, and fairly. I believe so much in this concept that I have begun taking undergraduate classes over again and paying very careful attention.

Honestly evaluate the utility of your new idea at the right time: late enough not to cut off explorations of alternatives and wild notions, but early enough that your creativity doesn't go stale. In this stage you must ask the hardest questions: "Is this new thing useful to anyone else? Exactly where and how is it useful? Is it really a better solution or just a clever configuration of parts?" Even if you discover that your creation has no apparent utility, savor the fun you had exploring it and be thankful that you don't have the very hard work of developing it.

Creativity is not a competitive process. It is sad that we engineers are so inculcated with the competitive approach that we use it even privately. You must suspend this internal competition because almost all of your new ideas will fail. This is a fact, but it doesn't detract a bit from the fun of inventing.

Now it's time to get on to a very old and interesting analog design problem where there is still a great deal of room for invention.

Requirements for Signal Conditioning in Oscilloscopes

Most of my tenure as an electrical engineer has been spent designing analog subsystems of digital oscilloscopes. A digital oscilloscope is a rather pure and wholesome microcosm of signal processing and measurement, but at the signal inputs the instrument meets the inhospitable real world. The input signal-conditioning electronics, sometimes referred to as the "front-end" of the instrument, includes the attenuators, high-impedance buffer, and pre-amplifier. Figure 7–1 depicts a typical front-end and is annotated with some of the performance requirements.

The combination of requirements makes the design of an oscilloscope front-end very difficult. The front-end of a 500MHz oscilloscope develops nearly 1GHz of bandwidth and must have a very clean step response. It operates at this bandwidth with a 1MΩ input resistance! No significant resonances are allowed out to 5GHz or so (where everything wants to resonate). Because we must maintain high input resistance and low capacitance, transmission lines (the usual method of handling microwave

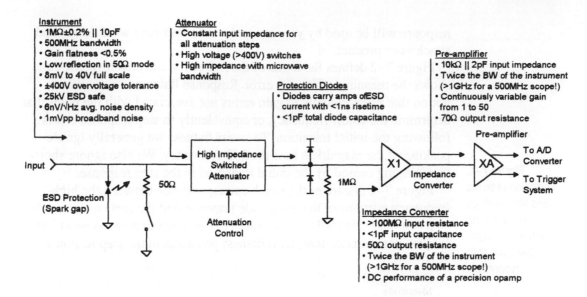

Instrument
- 1MΩ±0.2% || 10pF
- 500MHz bandwidth
- Gain flatness <0.5%
- Low reflection in 50Ω mode
- 8mV to 40V full scale
- ±400V overvoltage tolerance
- 25kV ESD safe
- 6nV/√Hz avg. noise density
- 1mVpp broadband noise

Attenuator
- Constant input impedance for all attenuation steps
- High voltage (>400V) switches
- High impedance with microwave bandwidth

Protection Diodes
- Diodes carry amps of ESD current with <1ns risetime
- <1pF total diode capacitance

Pre-amplifier
- 10kΩ || 2pF input impedance
- Twice the BW of the instrument (>1GHz for a 500MHz scope!)
- Continuously variable gain from 1 to 50
- 70Ω output resistance

ESD Protection (Spark gap)

Input

50Ω

High Impedance Switched Attenuator

Attenuation Control

1MΩ

X1

Impedance Converter

XA

Pre-amplifier

To A/D Converter

To Trigger System

Impedance Converter
- >100MΩ input resistance
- <1pF input capacitance
- 50Ω output resistance
- Twice the BW of the instrument (>1GHz for a 500MHz scope!)
- DC performance of a precision opamp

Figure 7–1.
Annotated diagram of an oscilloscope front-end, showing specifications and requirements at each stage.

signals) are not allowed! The designer's only defense is to keep the physical dimensions of the circuit very small. To obtain the 1GHz bandwidth we must use microwave components. Microwave transistors and diodes are typically very delicate, yet the front-end has to withstand ±400V excursions and high-voltage electrostatic discharges. Perhaps the most difficult requirement is high gain flatness from DC to a significant fraction of full bandwidth.

A solid grasp of the relationships between the frequency and time domains is essential for the mastery of these design challenges. In the following I will present several examples illustrating the intuitive connections between the frequency magnitude and step responses.

The Frequency and Time Domains

Oscilloscopes are specified at only two frequencies: DC and the -3dB point. Worse, the manufacturers usually state the vertical accuracy at DC only, as if an oscilloscope were a voltmeter! Why is a time domain measuring device specified in the frequency domain? The reason is that bandwidth measurements are traceable to international standards, whereas it is extremely difficult to generate an impulse or step waveform with known properties (Andrews 1983, Rush 1990).

Regardless of how oscilloscopes are specified, in actual practice oscilloscope designers concern themselves almost exclusively with the step response. There are several reasons for focusing on the step response: (1) a good step response is what the users really need in a time domain instrument, (2) the step response conveys at a glance information about a very wide band of frequencies, (3) with practice you can learn to intuitively relate the step response to the frequency response, and (4) the step response will be used by your competitors to find your weaknesses and attack your product.

Figure 7–2 defines the terms of the frequency and step responses and shows the meaning of flatness error. Response flatness is a qualitative notion that refers roughly to gain errors not associated with the poles that determine the cutoff frequency, or equivalently to step response errors following the initial transition. To assess flatness we generally ignore peaking of the magnitude near the 3dB frequency. We also ignore short-term ringing caused by the initial transition in the step response.

Figure 7–2.
Definition of terms and relationships between the frequency magnitude and step responses.

Figure 7–2 illustrates the rough correspondence between the high-frequency portions of the magnitude response and the early events in the step response. Similarly, disturbances in the magnitude response at low frequencies generate long-term flatness problems in the step response

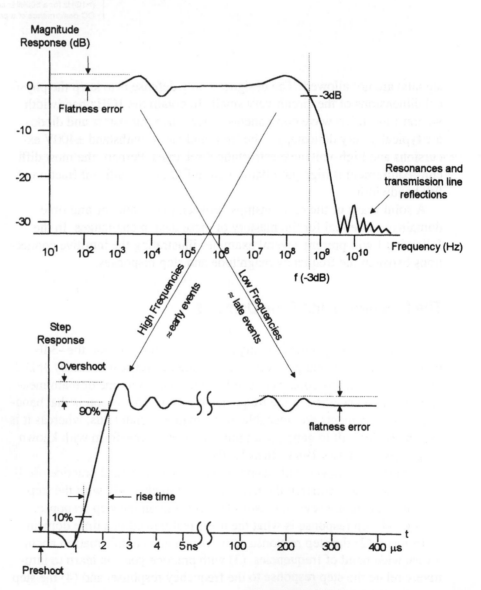

(Kamath 1974). Thus the step response contains information about a very wide band of frequencies, when observed over a long enough time period. For example, looking at the first ten nanoseconds (ns) of the step conveys frequency domain information from the upper bandwidth of the instrument down to approximately 1/(10ns) or 100MHz.

Figure 7–3 shows an RC circuit that effectively models most sources of flatness errors. Even unusual sources of flatness errors, such as dielectric absorption and thermal transients in transistors, can be understood with similar RC circuit models. The attenuator and impedance converter generally behave like series and parallel combinations of simple RC circuits. Circuits of this form often create flatness problems at low frequencies because of the high resistances in an oscilloscope front-end. In contrast, the high-frequency problems are frequently the result of the innumerable tiny inductors and inadvertent transmission lines introduced in the physical construction of the circuit. Notice how in Figure 7–3 the reciprocal nature of the frequency and step responses is well represented.

High Impedance at High Frequency: The Impedance Converter

Oscilloscopes by convention and tradition have $1M\Omega$ inputs with just a few picofarads of input capacitance. The $1M\Omega$ input resistance largely determines the attenuation factor of passive probes, and therefore must be accurate and stable. To maintain the accuracy of the input resistance, the oscilloscope incorporates a very high input impedance unity gain buffer (Figure 7–1). This buffer, sometimes called an "impedance converter," presents more than $100M\Omega$ at its input while providing a low-impedance, approximately 50Ω output to drive the pre-amp. In a 500MHz oscilloscope the impedance converter may have 1GHz of bandwidth and very carefully controlled time domain response. This section

Figure 7–3. A simple circuit that models most sources of flatness errors.

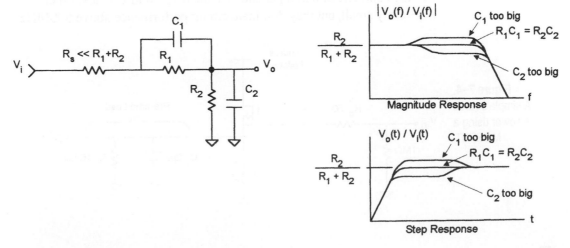

shows one way in which these and the many additional requirements of Figure 7–1 can be met (Rush 1986).

A silicon field effect transistor (FET) acting as a source follower is the only type of commercially available device suitable for implementing the impedance converter. For 500MHz instruments, we need a source follower with the highest possible transconductance combined with the lowest gate-drain capacitance. These parameters are so important in a 500MHz instrument that oscilloscope designers resort to the use of short-channel MOSFETs in spite of their many shortcomings. MOSFETs with short channel lengths and thin gate oxide layers develop very high transconductance relative to their terminal capacitances. However, they suffer from channel length modulation effects which give them undesirably high source-to-drain or output conductance. MOSFETs are surface conduction devices, and the interface states at the gate-to-channel interface trap charge, generating large amounts of 1/f noise. The 1/f noise can contribute as much noise between DC and 1MHz as thermal noise between DC and 500MHz. Finally, the thin oxide layer of the gate gives up very easily in the face of electrostatic discharge. As source followers, JFETs outperform MOSFETs in every area but raw speed. In summary, short-channel MOSFETs make poor but very fast source followers, and we must use a battery of auxiliary circuits to make them function acceptably in the impedance converter.

Figure 7–4 shows a very basic source follower with the required 1MΩ input resistance. The resistor in the gate stabilizes the FET. Figure 7–5 shows a linear model of a typical high-frequency, short-channel MOSFET. I prefer this model over the familiar hybrid-π model because it shows at a glance that the output resistance of the source is $1/g_m$. Figure 7–6 shows the FET with a surface-mount package model. The tiny capacitors and inductors model the geometric effects of the package and the surrounding environment. These tiny components are called "parasitics" in honor of their very undesirable presence. Figure 7–7 depicts the parasitics of the very common "0805" surface-mount resistor. This type of resistor is often used in front-end circuits built on printed circuit boards. Package and circuit board parasitics at the 0.1pF and 1nH level seem negligibly small, but they dominate circuit performance above 500MHz.

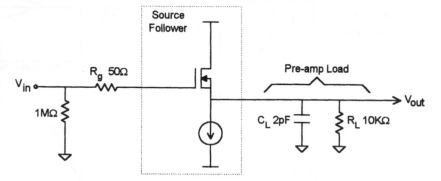

Figure 7–4.
A simple source follower using a MOSFET.

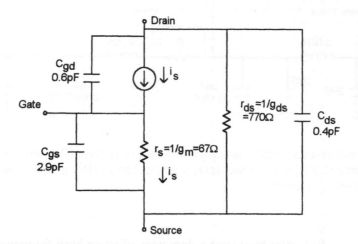

Figure 7–5.
A linear model of a BSD22, a typical high-frequency, short-channel MOSFET. The gate current is zero at DC because the controlled current source keeps the drain current equal to the source current.

In oscilloscope circuits I often remove the ground plane in small patches beneath the components to reduce the capacitances. One must be extremely careful when removing the ground plane beneath a high-speed circuit, because it always increases parasitic inductance. I once turned a beautiful 2GHz amplifier into a 400MHz bookend by deleting the ground plane and thereby effectively placing large inductors in the circuit.

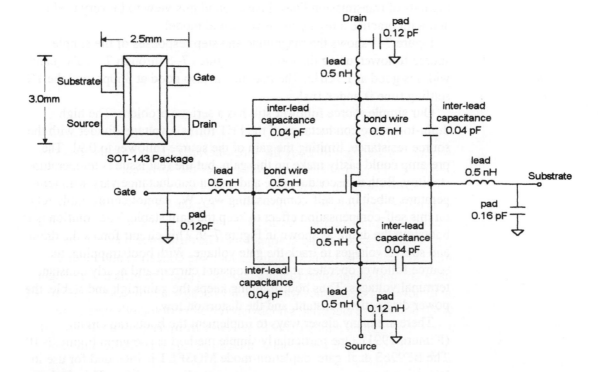

Figure 7–6.
A MOSFET with SOT-143 surface-mount package parasitics. The model includes the effects of mounting on a 1.6mm (0.063") thick, six-layer epoxy glass circuit board with a ground plane on the fourth layer from the component side of the board.

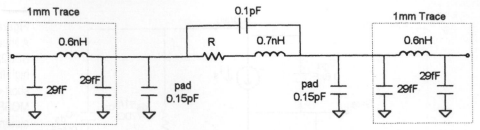

Figure 7–7.
A model of an 0805 surface-mount resistor, including a 1mm trace on each end. The model includes the effects of mounting on a 1.6mm (0.063") thick, six-layer epoxy glass circuit board with a ground plane on the fourth layer from the component side of the board.

Parasitics have such a dominant effect on high-frequency performance that 500MHz oscilloscope front-ends are usually built as chip-and-wire hybrids, which have considerably lower parasitics than standard printed circuit construction. Whether on circuit boards or hybrids, the bond wires, each with about 0.5 to 1.0nH inductance, present one of the greatest difficulties for high-frequency performance. In the course of designing high-frequency circuits, one eventually comes to view the circuits and layouts as a collection of transmission lines or the lumped approximations of transmission lines. I have found this view to be very useful and with practice a highly intuitive mental model.

Figure 7–8 shows the magnitude and step responses of the simple source follower, using the models of Figures 7–5 through 7–7. The bandwidth is good at 1.1GHz. The rise time is also good at 360ps, and the 1% settling time is under 1ns!

Our simple source follower still has a serious problem. The high drain-to-source conductance of the FET forms a voltage divider with the source resistance, limiting the gain of the source follower to 0.91. The pre-amp could easily make up this gain, but the real issue is temperature stability. Both transconductance and output conductance vary with temperature, albeit in a self-compensating way. We cannot comfortably rely on this self-compensation effect to keep the gain stable. The solution is to bootstrap the drain, as shown in Figure 7–9. This circuit forces the drain and source voltages to track the gate voltage. With bootstrapping, the source follower operates at nearly constant current and nearly constant terminal voltages. Thus bootstrapping keeps the gain high and stable, the power dissipation constant, and the distortion low.

There are many clever ways to implement the bootstrap circuit (Kimura 1991). One particularly simple method is shown in Figure 7–10. The BF996S dual-gate, depletion-mode MOSFET is intended for use in television tuners as an automatic gain controlled amplifier. This device acts like two MOSFETs stacked source-to-drain in series. The current source shown in Figure 7–10 is typically a straightforward bipolar transistor current source implemented with a microwave transistor. An ap-

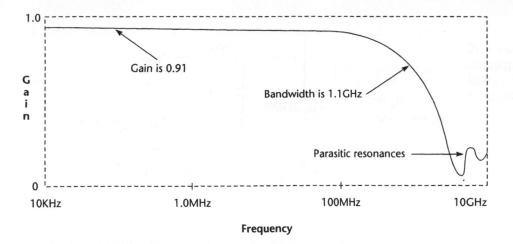

Gain is 0.91

Bandwidth is 1.1GHz

Parasitic resonances

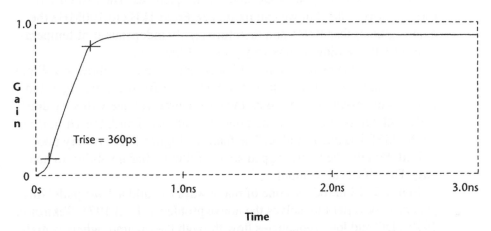

Trise = 360ps

proximate linear model of the BF996S is shown in Figure 7–11. The BF996S comes in a SOT-143 surface-mount package, with parasitics, as shown in Figure 7–6.

Figure 7–12 shows the frequency and step responses of the bootstrapped source follower. The bootstrapping network is AC coupled, so

Figure 7–8.
The magnitude and step responses of the simple source follower.

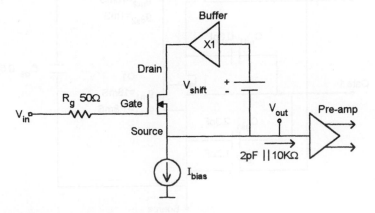

Figure 7–9.
The bootstrapped source follower. Driving the drain with the source voltage increases and stabilizes the gain.

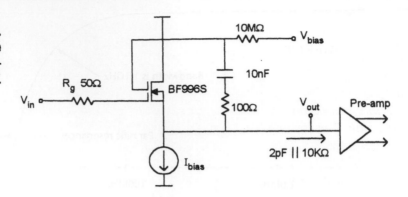

Figure 7–10.
Bootstrapping the drain with a dual-gate MOSFET.

it does not boost the gain at DC and low frequencies. The response therefore is not very flat, but we can fix it later. From 1kHz to 100MHz the gain is greater than 0.985 and therefore highly independent of temperature. The 1% settling time is very good at 1.0ns.

Several problems remain in the bootstrapped source follower of Figure 7–10. First, the gate has no protection whatever from overvoltages and electrostatic discharges. Second, the gate-source voltage will vary drastically with temperature, causing poor DC stability. Third, the 1/f noise of the MOSFET is uncontrolled. The flatness (Figure 7–12) is very poor indeed. Finally, the bootstrapped source follower has no ability to handle large DC offsets in its input.

Figure 7–13 introduces one of many ways to build a "two-path" impedance converter that solves the above problems (Evel 1971, Tektronix 1972). DC and low frequencies flow through the op amp, whereas high frequencies bypass the op amp via C1. At DC and low frequencies, feed-

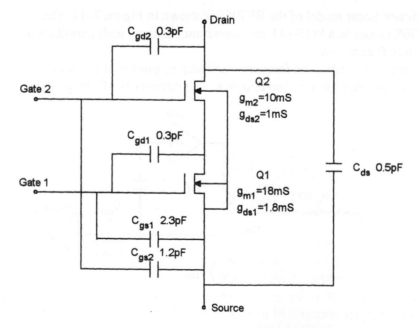

Figure 7–11.
Linear model of the BF996S dual-gate, depletion MOSFET.

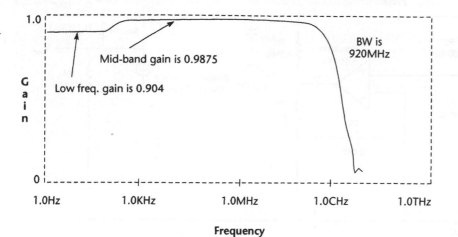

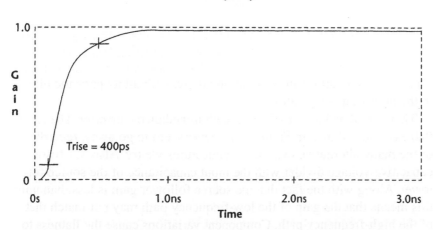

Figure 7–12.
The magnitude and
step responses of
the bootstrapped
source follower.

back gives the two-path source follower the accuracy of a precision op amp. At high frequencies, the signal feeding through C1 dominates control of gate 1, and the source follower operates open loop. The FET is protected by the diodes and the current limiting effects of C1. The 1/f noise of the FET is partially controlled by the op amp, and the circuit can offset large DC levels at the input with the offset control point shown in Figure 7–13.

Figure 7–14 shows the flatness details of the two-path impedance converter. Feedback around the op amp has taken care of the low-frequency gain error exhibited by the bootstrapped source follower (Figure 7–12). The gain is flat from DC to 80MHz to less than 0.1%. The "wiggle" in the magnitude response occurs where the low- and high-frequency paths cross over.

There are additional benefits to the two-path approach. It allows us to design the high-frequency path through C1 and the MOSFET without regard to DC accuracy. The DC level of the impedance converter output is independent of the input and can be tailored to the needs of the pre-amplifier. Although it is not shown in the figures, AC coupling is easily implemented by blocking DC to the non-inverting input of the op amp.

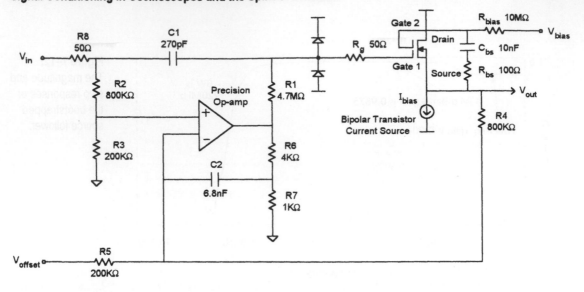

Figure 7–13.
A two-path impedance converter.

Thus we avoid putting an AC coupling relay, with all its parasitic effects, in the high-frequency path.

There are drawbacks to the two-path impedance converter. The small flatness errors shown in Figure 7–14 never seem to go away, regardless of the many alternative two-path architectures we try. Also, C1 forms a capacitive voltage divider with the input capacitance of the source follower. Along with the fact that the source follower gain is less than unity, this means that the gain of the low-frequency path may not match that of the high-frequency path. Component variations cause the flatness to vary further. Since the impedance converter is driven by a precision high-impedance attenuator, it must have a very well-behaved input impedance that closely resembles a simple RC parallel circuit. In this regard the most common problem occurs when the op amp has insufficient speed and fails to bootstrap R1 in Figure 7–13 to high enough frequencies.

Figure 7–14.
Flatness details of the two-path impedance converter.

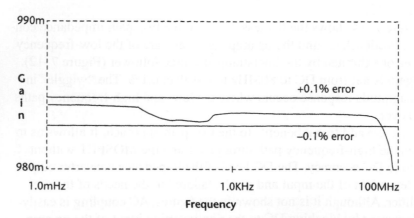

The overdrive recovery performance of a two-path amplifier can be abysmal. There are two ways in which overdrive problems occur. If a signal is large enough to turn on one of the protection diodes, C1 charges very quickly through the low impedance of the diode (Figure 7–13). As if it were not bad enough that the input impedance in overdrive looks like 270pF, recovery occurs with a time constant of $270pF \cdot 4.7M\Omega$, or 1.3ms! Feedback around the op amp actually accelerates recovery somewhat but recovery still takes eons compared to the 400ps rise time! Another overdrive mechanism is saturation of the source follower. When saturation occurs, the op amp integrates the error it sees between the input and source follower output, charging its 6.8nF feedback capacitor. Recovery occurs over milliseconds. The seriousness of these overdrive recovery problems is mitigated by the fact that with careful design it can take approximately ±2V to saturate the MOSFET and ±5V to activate the protection diodes. Thus, to overdrive the system, it takes a signal about ten times the full-scale input range of the pre-amp.

I apologize for turning a simple, elegant, single transistor source follower into the "bootstrapped, two-path impedance converter." But as I stated at the beginning, it is the combination of requirements that drives us to such extremes. It is very hard to meet all the requirements at once with a simple circuit. In the next section, I will extend the two-path technique to the attenuator to great advantage. Perhaps there the two-path method will fully justify its complexity.

The Attenuator

I have expended a large number of words and pictures on the impedance converter, so I will more briefly describe the attenuator. I will confine myself to an introduction to the design and performance issues and then illustrate some interesting alternatives for constructing attenuators. The purpose of the attenuator is to reduce the dynamic range requirements placed on the impedance converter and pre-amp. The attenuator must handle stresses as high as ±400V, as well as electrostatic discharge. The attenuator maintains a 1MΩ input resistance on all ranges and attains microwave bandwidths with excellent flatness. No small-signal microwave semiconductors can survive the high input voltages, so high-frequency oscilloscope attenuators are built with all passive components and electromechanical relays for switches.

Figure 7–15 is a simplified schematic of a 1MΩ attenuator. It uses two stages of the well-known "compensated voltage divider" circuit. One stage divides by five and the other by 25, so that division ratios of 1, 5, 25, and 125 are possible. There are two key requirements for the attenuator. First, as shown in Figure 7–3, we must maintain $R_1C_1 = R_2C_2$ in the ÷5 stage to achieve a flat frequency response. A similar requirement holds for the ÷25 stage. Second, the input resistance and capacitance at each stage must match those of the impedance converter and remain very nearly constant, independent of the switch positions. This requirement assures that we maintain attenuation accuracy and flatness for all four combinations of attenuator relay settings.

Dividing by a high ratio such as 125 is similar to trying to build a high-isolation switch; the signal attempts to bypass the divider, causing feed-through problems. If we set a standard for feedthrough of less than one least-significant bit in an 8-bit digital oscilloscope, the attenuator must isolate the input from the output by $20\log_{10}(125 \cdot 2^8) = 90$dB! I once spent two months tracking down such an isolation problem and traced it to wave guide propagation and cavity resonance at 2GHz inside the metallic attenuator cover.

Relays are used for the switches because they have low contact impedance, high isolation, and high withstanding voltages. However, in a realm where 1mm of wire looks like a transmission line, the relays have dreadful parasitics. To make matters worse, the relays are large enough to spread the attenuator out over an area of about 2×3cm. Assuming a propagation velocity of half the speed of light, three centimeters takes 200ps, which is dangerously close to the 700ps rise time of a 500MHz oscilloscope. In spite of the fact that I have said we can have no trans-mission lines in a high-impedance attenuator, we have to deal with them anyway! To deal with transmission line and parasitic reactance effects, a real attenuator includes many termination and damping resistors not shown in Figure 7–15.

Rather than going into extreme detail about the conventional attenuator of Figure 7–15, it would be more interesting to ask if we could somehow eliminate the large and unreliable electromechanical relays. Consider the slightly different implementation of the two-path impedance converter depicted in Figure 7–16. The gate of the depletion MOSFET is self-bi-ased by the 22MΩ resistor so that it operates at zero gate source voltage. If the input and output voltages differ, feedback via the op amp and bipo-lar current source reduces the error to zero. To understand this circuit, it helps to note that the impedance looking into the source of a self-biased FET is very high. Thus the collector of the bipolar current source sees a

Figure 7–15.
A simplified two-stage high-impedance attenuator.

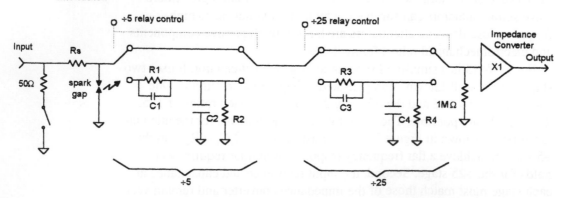

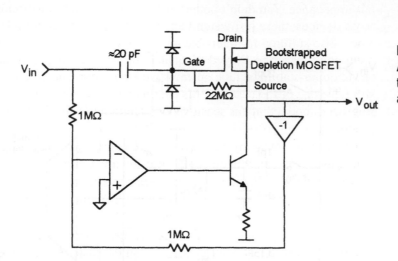

Figure 7–16.
A variation on the two-path impedance converter.

high-impedance load. Slight changes in the op amp output can therefore produce significant changes in the circuit output.

The impedance converter of Figure 7–16 can easily be turned into a fixed attenuator, as shown in Figure 7–17. As before, there is a high-frequency and a low-frequency path, but now each divides by ten. There is an analog multiplier in the feedback path to make fine adjustments to the low-frequency gain. The multiplier matches the low- and high-frequency paths to achieve a high degree of flatness. A calibration procedure determines the appropriate gain for the multiplier.

Now we can build a complete two-path attenuator with switched attenuation, as shown in Figure 7–18 (Roach 1992). Instead of cascading attenuator stages, we have arranged them in parallel. In place of the two double-pole double-throw (DPDT) relays of Figure 7–15, we now need only two single-pole single-throw (SPST) relays. Note that there is no need for a switch in the ÷100 path because any signal within range for

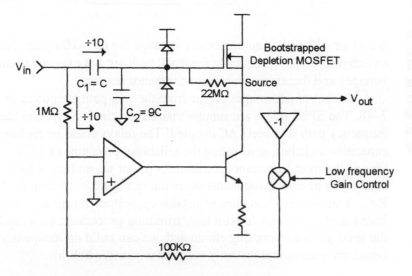

Figure 7–17.
An attenuating impedance converter, or "two-path attenuator."

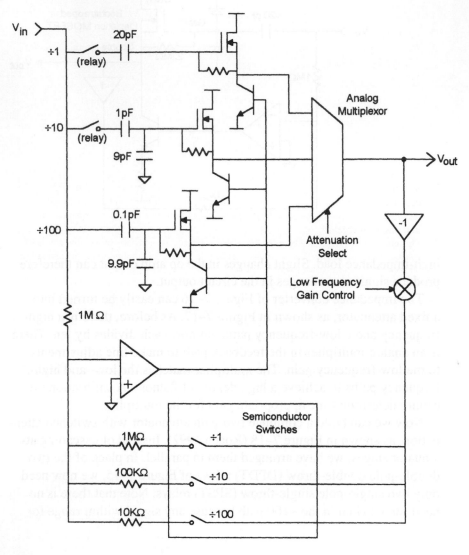

Figure 7–18.
A two-path attenuator and impedance converter using only two SPST electromechanical relays. The protection diodes and some resistors are omitted for clarity.

the ÷1 or ÷10 path is automatically in range for the ÷100 path. The switches in the low-frequency feedback path are not exposed to high voltages and therefore can be semiconductor devices.

A number of advantages accrue from the two-path attenuator of Figure 7–18. The SPST relays are simpler than the original relays, and the high-frequency path is entirely AC coupled! The relays could be replaced with capacitive switches, eliminating the reliability problems of DC contacts. One of the most important contributions is that we no longer have to precisely trim passive components as we did in Figure 7–15 to make $R_1C_1 = R_2C_2$. This feature eliminates adjustable capacitors in printed circuit (PC) board attenuators and difficult laser trimming procedures on hybrids. With the need for laser trimming eliminated, we can build on inexpensive PC board attenuators that formerly required expensive hybrids.

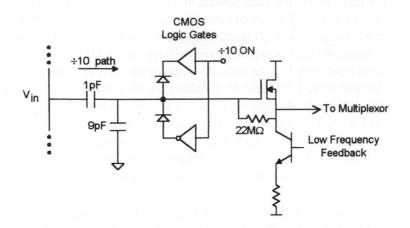

Figure 7–19.
Using the
protection diodes
as switches in the
÷10 path.

We can take the new attenuator configuration of Figure 7–18 further. First observe that we can eliminate the ÷10 relay in Figure 7–18, as shown in Figure 7–19. The diodes are reverse biased to turn the ÷10 path on and forward biased to turn it off. Forward biasing the diodes shorts the 1pF capacitor to ground, thereby shunting the signal and cutting off the ÷10 path. The input capacitance changes by only 0.1pF when we switch the ÷10 path.

Now we are down to one electromechanical relay in the ÷1 path. We can eliminate it by moving the switch from the gate side of the source follower FET to the drain and source, as shown in Figure 7–20. In doing so we have made two switches from one, but that will turn out to be a good trade. With the ÷1 switches closed, the drain and source of the FET are connected to the circuit and the ÷1 path functions in the usual manner. The protection diodes are biased to ±5V to protect the FET.

To cut off the ÷1 path, the drain and source switches are opened, leaving those terminals floating. With the switches open, a voltage change at

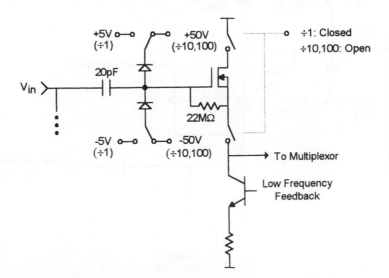

Figure 7–20.
Moving the ÷1
switch from the
high-impedance
input side to the
low-impedance
output side of
the FET.

the input drives the gate, source, and drain of the FET through an equal change via the 20pF input capacitor and the gate-drain and gate-source capacitances. Since all three terminals of the FET remain at the same voltage, the FET is safe from overvoltage stress. Of course, the switches must have very low capacitance in the open state, or capacitive voltage division would allow the terminals of the FET to see differing voltages. In ÷100 mode, the floating FET will see 40V excursions (eight divisions on the oscilloscope screen at 5V per division) as a matter of course. For this reason the ÷1 protection diodes must be switched to a higher bias voltage (±50V) when in the ÷10 and ÷100 modes. The switches that control the voltage on the protection diodes are not involved in the high-frequency performance of the front-end and therefore can be implemented with slow, high-voltage semiconductors.

Can we replace the switches in the drain and source with semiconductor devices? The answer is yes, as Figure 7–21 shows. The relays in the drain and source have been replaced by PIN diodes. PIN diodes are made with a p-type silicon layer (P), an intrinsic or undoped layer (I), and an n-type layer (N). The intrinsic layer is relatively thick, giving the diode high breakdown voltage and extremely low reverse-biased capacitance. A representative packaged PIN diode has 100V reverse breakdown and only 0.08pF junction capacitance. To turn the ÷1 path of Figure 7–21 on, the switches are all set to their "÷1" positions. The PIN diodes are then forward biased, the bipolar transistor is connected to the op amp, and the FET is conducting. To turn the path off, the switches are set to their "÷10,100" positions, reverse-biasing the PIN diodes. Since these switches

Figure 7–21.
Using PIN diodes to eliminate the relays in the ÷1 path.

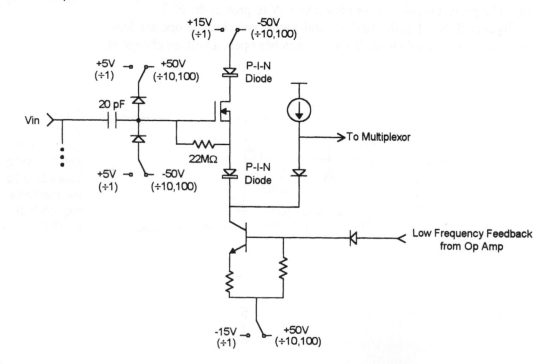

are not involved in the high-frequency signal path, they too can be built with slow, high-voltage semiconductors.

The complete circuit is now too involved to show in one piece on the page of a book, so please use your imagination. We have eliminated all electromechanical switches and have a solid-state oscilloscope front-end. Although I had a great deal of fun inventing this circuit, I do not think it points the direction to future oscilloscope front-ends. Already research is under way on microscopic relays built with semiconductor micro-machining techniques (Hackett 1991). These relays are built on the surface of silicon or gallium arsenide wafers, using photolithography techniques, and measure only 0.5mm in their largest dimension. The contacts open only a few microns, but they maintain high breakdown voltages (100s of volts) because the breakdown voltages of neutral gases are highly nonlinear and not even monotonic for extremely small spacing. The contacts are so small that the inter-contact capacitance in the open state is only a few femtofarads (a femtofarad is 0.001 picofarads). Thus the isolation of the relays is extraordinary! Perhaps best of all, they are electrostatically actuated and consume near zero power. I believe micro-machined relays are a revolution in the wings for oscilloscope front-ends. I eagerly anticipate that they will dramatically improve the performance of analog switches in many applications. Apparently, even a device as old as the electromechanical relay is still fertile ground for a few ambitious inventors!

References

Addis, J. "Versatile Broadband Analog IC." *VLSI Systems Design* (September 1980): 18–31.

Andrews, J., A. Bell, N. Nahman, et. al. "Reference Waveform Flat Pulse Generator." *IEEE Trans. Inst. Meas.* IM-32 (1) (March 1983): 27–32.

Barna, A. "On the Transient Response of Emitter Followers." *IEEE J. Solid State Circuits* (June 1973): 233–235.

Blinchikoff, H. and A. Zverev. *Filtering in the Time and Frequency Domains.* (New York: John Wiley & Sons, 1976).

Evel, E. "DC Stabilized Wideband Amplifier." (Apr. 6, 1971): U.S. Patent #3,573,644.

Hackett, R., L. Larsen, and M. Mendes. "The Integration of Micro-Machine Fabrication with Electronic Device Fabrication on III-IV Semiconductor Materials." IEEE Trans. Comp. Hybrids. and Mfr. Tech. *IEEE Trans. Comp. Hybrids. and Mfr. Tech* (May 1991): 51–54.

Kamath, B., G. Meyer, and P. Gray. "Relationship Between Frequency Response and Settling Time in Operational Amplifiers." *IEEE J. Solid State Circuits* SC-9 (6) (December 1974): 347–352.

Kimura, R. "DC Bootstrapped Unity Gain Buffer." (Apr. 30, 1991): U.S. Patent #5,012,134.

Kozikowski, J. "Analysis and Design of Emitter Followers at High Frequencies." *IEEE Trans. Circuit Theory* (March 1964): 129–136.

Roach, S. "Precision Programmable Attenuator." (Jun. 9, 1992): U.S. Patent #5,121,075.

Rush, K., W. Escovitz, and A. Berger. "High-Performance Probe System for a 1-GHz Digitizing Oscilloscope." *Hewlett-Packard J.* 37(4) (April 1986): 11–19.

Rush, K., S. Draving, and J. Kerley. "Characterizing High Speed Oscilloscopes." *IEEE Spectrum* (Sep. 1990).

Tektronix, Inc. *Instruction Manual for the P6201 Probe*. (Beaverton, Oregon: Tektronix, Inc., 1972).

8. One Trip Down the IC Development Road

· ·

This is the story of the last IC that I developed. I use the word develop rather than design because there is so much more involved in the making of a standard part than just the circuit design and layout. My goal is to give the reader an idea of what is involved in this total development. The majority of this description will be on the evolution of the product definition and the circuit design since that is my major responsibility. I will also describe many of the other important steps that are part of the IC development. To give the reader an idea of what is required, I made an approximate list of the steps involved in the development of an IC.

The steps in the development of a new IC:

1. Definition
2. Circuit design
3. Re-definition
4. More circuit design
5. The first finalizing of the specifications
6. Test system definition
7. Mask design
8. Test system design
9. Waiting for wafers to be made
10. Evaluation
11. Test system debug
12. Redesign (circuit & masks)
13. More waiting
14. Finalizing the test system
15. IC characterization
16. Setting the real specifications
17. Pricing
18. Writing the data sheet
19. Promotion
20. Yield enhancements

Circuit design (steps 2, 4, and 12) is what we usually think of when we talk about IC design. As you can see, it is only a small part of the IC development. At some companies, particularly those that do custom ICs, circuit design is all the design engineers do. In the ideal world of some MBAs, the customer does the definition, the designer makes the IC, the test engineer tests, the market sets the price, and life is a breeze. This simple approach rarely develops an IC that is really new; and the companies that work this way rarely make any money selling ICs.

Most successful IC designers I know are very good circuit designers and enjoy circuit design more than anything else at work. But it is not just their circuit design skills that make these designers successful; it is also their realization that all the steps in the development of an IC must be done properly. These designers do not work to a rigid set of specifications. They learn and understand what the IC specs mean to the customer and how the IC specs affect the system performance. Successful IC designers take the time to do whatever it takes to make the best IC they can.

This is quite different from the custom IC designer who sells design. If you are selling design, it is a disadvantage to beat the customer's spec by too much. If you do the job too well, the customer will not need a new custom IC very soon. But if you just meet the requirement, then in only a year or so the customer will be back for more. This kind of design reminds me of the famous Russian weight lifter who set many world records. For many years he was able to break his own world record by lifting only a fraction of a kilogram more than the last time. He received a bonus every time he set a new world record; his job was setting records. He would be out of a job if he did the best he could every time; so he only did as much as was required.

Product Definition

Where do we get the ideas for new products? From our customers, of course. It is not easy, however. Most customers will tell you what they want, because they are not sure what they need. Also, they do not know what the different IC technologies are capable of and what trade-offs must be made to improve various areas of performance. The way questions are asked often determines the answers. Never say, "Would you like feature XYZ?" Instead say, "What would feature XYZ be worth to you?"

When an IC manufacturer asks a customer, it is often like a grandparent asking a grandchild. The child wants all the things that it cannot get from its parents and knows none of the restrictions that bind the others. The only thing worse would be to have a total stranger do the questioning. That may sound unlikely, but there are companies that have hired non-technical people to ask customers what new products they want. At best, this only results in a very humorous presentation that wastes a lot of people's time.

Talking to customers, applications engineers, and salespeople gives the clues and ideas to a designer for what products will be successful. It is important to pick a product based on the market it will serve. Do not make a new IC because the circuit design is fun or easy. Remember that circuit design is only a small part of the development process. The days of designing a new function that has no specific market should be long gone. Although I have seen some products recently that appear to be solutions looking for problems!

This is not to say that you need marketing surveys with lots of paperwork and calculations on a spreadsheet. These things are often management methods to define responsibility and place blame. It is my experience that the errors in these forms are always in the estimate of the selling price and the size of the market. These inputs usually come from marketing and maybe that is why there is such a high turnover of personnel in semiconductor marketing departments. After all, if the marketers who made the estimates change jobs every three years, no one will ever catch up with them. This is because it typically takes two years for development and two more years to see if the product meets its sales goals.

So with almost no official marketing input, but based on conversations with many people over several years, I began the definition of a new product. I felt there was a market for an IC video fader and that the market was going to grow significantly over the next five years. The driving force behind this growth would be PC based multi-media systems. At the same time I recognized that a fader with only one input driven is a very good adjustable gain amplifier and that is a very versatile analog building block. The main source of this market information was conversations with customers trying to use a transconductance amplifier that I had designed several years earlier in fader and gain control applications.

The Video Fader

The first step is figuring out what a video fader is. The basic fader circuit has two signal inputs, a control input and one output. A block diagram of a fader is shown in Figure 8–1. The control signal varies the gain of the two inputs such that at one extreme the output is all one input and at the other extreme it is the other input. The control is linear; i.e., for the control signal at 50%, the output is the sum of one half of input 1 and one half of input 2. If both inputs are the same, the output is independent of the control signal. Of course implementing the controlled potentiometer is the challenging part of the circuit design.

The circuit must have flat response (0.1dB) from DC to 5MHz and low differential gain and phase (0.1% & 0.1 degree) for composite video applications. For computer RGB applications the -3dB bandwidth must be at least 30MHz and the gain accuracy between parts should be better than 3%. The IC should operate on supply voltages from ±5V to ±15V, since there are still a lot of systems today on ±12V even though the trend is to ±5V. Of course if the circuit could operate on a single +5V supply, that would be ideal for the PC based multi-media market.

The control input can be in many forms. Zero to one or ten volts is common as are bipolar signals around zero. Some systems use current inputs or resistors into the summing node of an op amp. In variable gain amplifier applications often several control inputs are summed together.

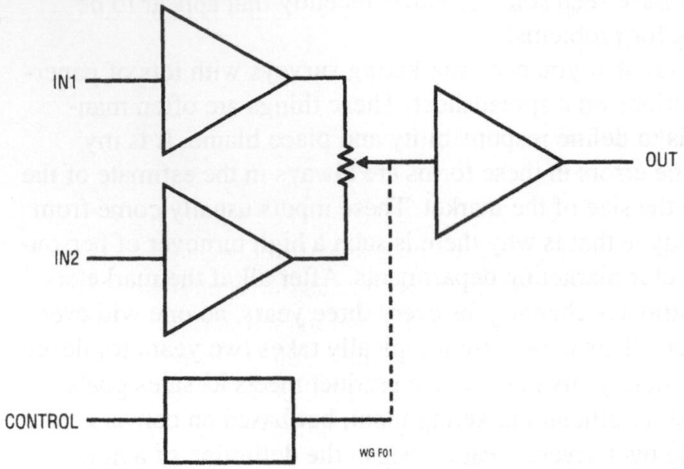

Figure 8–1.
Basic fader circuit.

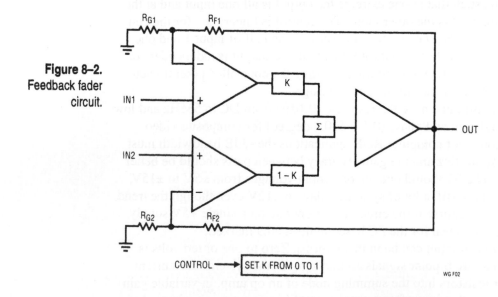

In order to make a standard IC that is compatible with as many systems as possible, it is desirable to make the control input user defined. At the same time it is important that the IC not require a lot of external parts.

To make the circuit more immune to errors in the potentiometer circuit, we can take feedback from the output back to both inputs. Figure 8–2 shows this feedback and replaces the potentiometer with the mathematical equivalent blocks: K, 1-K, and summation. Now the output is better controlled, since the value of K does not determine the total gain, only the ratio of the two input signals at the output. The gain is set by the feedback resistors and, to a smaller degree, the openloop gain of the amplifiers.

Figure 8–2.
Feedback fader
circuit.

Circuits

At this point it is time to look at some actual circuits. Do we use voltage feedback or current feedback? Since the current feedback topology has inherently better linearity and transient response, it seemed a natural for the input stages. One customer showed me a class A, current feedback circuit being implemented with discrete transistors. Figure 8–3 shows the basic circuit. For the moment we will not concern ourselves with how the control signal, V_C, is generated to drive the current steering pairs. Notice that the fader is operating inverting; for AC signals this is not usually a problem, but video signals are uni-polar and another inversion would eventually be needed. I assumed that the inverting topology was chosen to reduce the amount of distortion generated by the bias resistors, R_{B1} and R_{B2}, in the input stages.

Since transistors are smaller than resistors in an IC, I intended to replace the bias resistors with current sources. Therefore my circuit could operate non-inverting as well as inverting, and as a bonus the circuit would have good supply rejection. The complementary bipolar process that I planned to use would make class AB implementations fairly straightforward. I began my circuit simulations with the circuit of Figure 8–4; notice that there are twice as many components compared to the discrete circuit and it is operating non-inverting.

After a bit of tweaking the feedback resistor values and the compensation capacitor, the circuit worked quite well. The transistor sizes and

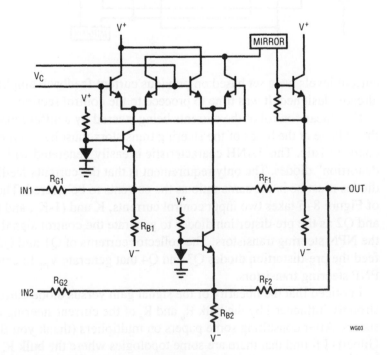

Figure 8–3.
Discrete design, class A current feedback fader.

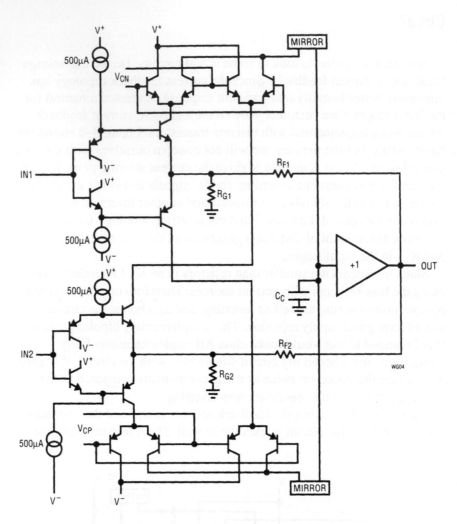

Figure 8–4.
Class AB current
feedback fader.

current levels were set based on previous current feedback amplifiers already designed. It was time to proceed to the control section.

For linear control of the currents being steered by a differential pair, the voltage at the bases of the steering transistors must have a nonlinear characteristic. This TANH characteristic is easily generated with "pre-distortion" diodes. The only requirement is that the currents feeding the diodes must be in the same ratio as the currents to be steered. The circuit of Figure 8–5 takes two input control currents, K and (1-K), and uses Q1 and Q2 as the pre-distortion diodes to generate the control signal V_{CN} for the NPN steering transistors. The collector currents of Q1 and Q2 then feed the pre-distortion diodes Q3 and Q4 that generate V_{CP} to control the PNP steering transistors.

I noticed that the linearity of the signal gain versus diode current is strongly influenced by the bulk R_b and R_e of the current steering transistors. After consulting some papers on multipliers (thank you Barry Gilbert) I found that there are some topologies where the bulk R_b and R_e of the pre-distortion diodes compensate the equivalent in the steering

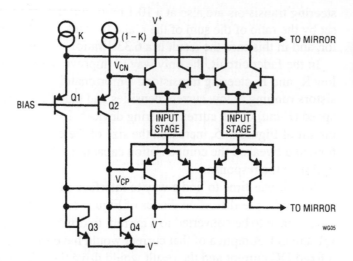

Figure 8–5.
Basic circuit to drive the steering transistors.

transistors. Unfortunately, in my circuit I am using PNPs to drive NPNs and vice versa. In order to match the pre-distortion diodes to the steering transistors, a more complicated circuit was required. I spent a little time and added a lot more transistors to come up with a circuit where the pre-distortion diodes for the NPN steering transistors were NPNs, and the same for the PNPs. Imagine my surprise when it didn't solve the linearity problem. I have not included this circuit because I don't remember it; after all, it didn't work.

So I had to learn a little more about how my circuit really worked. In the fader circuit, the DC current ratio in the steering transistors is not important; the small signal current steering sets the ratio of the two inputs. Figure 8–6 shows a simplified circuit of the pre-distortion diodes and the steering transistors. The diodes and transistors are assumed perfect with 18Ω resistors in series with the emitters to represent the bulk R_b and R_e of the devices. The control currents are at a 10:1 ratio; the DC currents in the

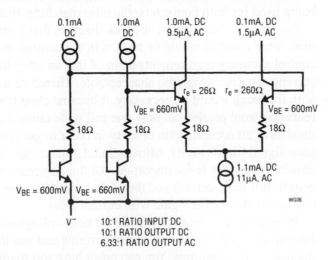

Figure 8–6.
Bulk resistance problems in steering.

steering transistors are also at a 10:1 ratio. But the small signal steering is set by the ratio of the sum of the r_e and the bulk resistance in each transistor, and in this case the result is a 6.33:1 ratio!

In the fader circuit, the only way to improve the gain accuracy is with low R_b and R_e steering transistors. Unfortunately this requires larger transistors running at low current densities and that significantly reduces the speed (F-tau) of the current steering devices. I went back to the simpler circuit of Figure 8–5, increased the size of the current steering transistors, and tweaked the compensation capacitor and feedback resistors to optimize the response.

Now it was time to find a way to interface the external control signal(s) to the pre-distortion diodes of Figure 8–5. The incoming signal would have to be converted to a current to drive the pre-distortion diodes, Q1 and Q3. A replica of that current would have to be subtracted from a fixed DC current and the result would drive the other pre-distortion diodes, Q2 and Q4.

I did not want to include an absolute reference in this product for several reasons. An internal reference would have to be available for the external control circuitry to use, in order not to increase the errors caused by multiple references. Therefore it would have to be capable of significant output drive and tolerant of unusual loading. In short, the internal reference would have to be as good as a standard reference. The inaccuracy of an internal reference would add to the part-to-part variations unless it was trimmed to a very accurate value. Both of these requirements would increase the die size and/or the pin count of the IC. Lastly, there is no standard for the incoming signals, so what value should the reference be?

I decided to require that an external reference, or "full scale" voltage, would be applied to the part. With an external full scale and control voltage, I could use identical circuits to convert the two voltages into two currents. The value of the full scale voltage is not critical because only the ratio between it and the control voltage matters. With the same circuit being used for both converters, the ratio matching should be excellent.

Figure 8–7 shows the basic block diagram that I generated to determine what currents would be needed in the control section. The gain control accuracy requirements dictated that an open loop voltage-to-current converter would be unacceptable. Therefore a simple op amp with feedback would be necessary. It became clear that two control currents (I_C) were needed but only one full scale current (I_{FS}) was. Mirror #1 must have an accurate gain of unity in order to generate the proper difference signal for mirror #3. Mirrors 2 and 3 must match well, but their absolute accuracy is not important. All three mirrors must operate from zero to full scale current and therefore cannot have resistive degeneration that could change their gain with current level.

In order to use identical circuits for both voltage-to-current converters, I decided to generate two full scale currents and use the extra one to bias the rest of the amplifiers. You can never have too many bias currents available.

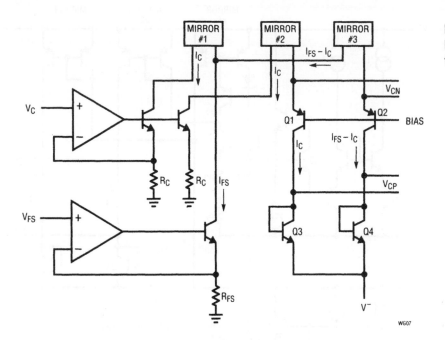

Figure 8–7.
Block diagram of
the control circuit.

The block diagram of Figure 8–7 became the circuit of Figure 8–8 after several iterations. The common mode range of the simple op amp includes the negative supply and the circuit has sufficient gain for the job. Small current sharing resistors, R1, R2, R3, and R4, were added to improve the high current matching of the two output currents and eliminate the need for the two R_C resistors. The small resistors were scaled so they could be used for short circuit protection with Q5 and Q6 as well.

Mirror #1 is a "super diode" connection that reduces base current errors by beta; the diode matches the collector emitter voltages of the matched transistors. Identical mirrors were used for #2 and #3 so that any errors would ratio out. Since these mirrors feed the emitters of the pre-distortion cascodes Q1 and Q2, their output impedance is not critical and they are not cascoded. This allows the bias voltage at the base of Q1 and Q2 to be only two diode drops below the supply, maximizing the common mode range of the input stages.

While evaluating the full circuit, I noticed that when one input was supposed to be off, its input signal would leak through to the output. The level increased with frequency, as though it was due to capacitive feedthrough. The beauty of SPICE came in handy now. I replaced the current steering transistors with ideal devices and still had the problem. Slowly I came to the realization that the feedthrough at the output was coming from the feedback resistor. In a current feedback amplifier, the inverting input is driven from the non-inverting input by a buffer amp and therefore the input signal is always present at the inverting input. Therefore the amount of signal at the output is just the ratio of the feedback resistor to the amplifier output impedance. Of course the output impedance rises with frequency because of the single pole compensation necessary to keep

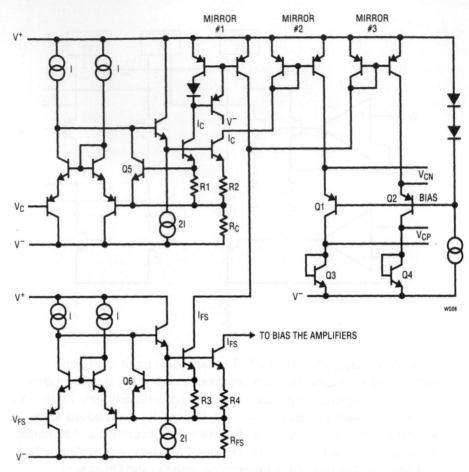

Figure 8–8.
The control circuit.

the amplifier stable. The basic current feedback topology I had chosen was
the feedthrough problem. Now it was obvious why the discrete circuit was
operating inverting. The problem goes away when the non-inverting input
is grounded because then the inverting input has very little signal on it.

Redefinition

At this point I realized I must go back to the beginning and look at voltage feedback. I started with the basic folded cascode topology and
sketched out the circuit of Figure 8–9. It seemed to work and there were
no feedthrough problems. It also appeared to simplify the control requirements, since there were no PNPs to steer. While working with this
circuit I realized that the folded cascode transistors, Q7 and Q8, could be
used as the steering devices, and sketched out Figure 8–10. This looked
great since it had fewer devices in the signal path and therefore better
bandwidth. The only downside I could see was the critical matching of
the current sources; all eight current sources are involved in setting the
gain. While I was pondering how to get eight current sources coming

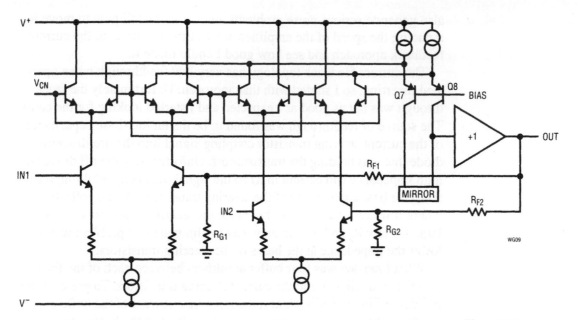

Figure 8–9.
Voltage feedback
fader.

from opposite supplies to match, I decided to run a transient response to determine how much input degeneration was required.

The bottom fell out! When the fader is set for 10% output, the differential input voltage is 90% of the input signal! This means that the *open loop* linearity of the input stage must be very good for signals up to one volt or more. To get signal linearity of 0.1% would require over a volt of degeneration. With that much degeneration in each input stage, the mismatch in offset voltage between the two would be tens of millivolts and that would show up as control feedthrough. Big degeneration resistors

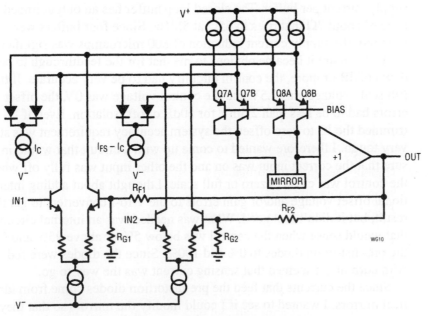

Figure 8–10.
Voltage feedback
with cascode
steering.

also generate serious noise problems and cause the tail pole to move in, reducing the speed of the amplifier. It was time to retreat to the current feedback approach and see how good I could make it.

The current feedback topology has very low feedthrough when operated inverting, so I started with that approach. Unfortunately the feedthrough was not as good as I expected and I started looking for the cause. The source of feedthrough was found to be the emitter-base capacitance of the current steering transistor coupling signal into the pre-distortion diode that was holding the transistor off. Unfortunately the off diode was high impedance (no current in it) so the signal then coupled through the collector base capacitance of the steering transistor into the collector, where it was not supposed to be. Since the steering transistors had to be large for low R_b and R_e, the only way to eliminate this problem was to lower the impedance at the bases of the steering transistors.

What I needed was four buffer amplifiers between each of the four pre-distortion diodes and the current steering transistors. To preserve the pre-distortion diodes' accuracy, the input bias current of the buffers needed to be less than one microamp. The offset of the buffers had to be less than a diode drop in order to preserve the input stage common mode range so that the circuit would work on a single 5V supply. Lastly, the output impedance should be as low as possible to minimize the feedthrough.

The first buffer I tried was a cascode of two emitter followers, as shown on the left in Figure 8–11. By varying the currents in the followers and looking at the overall circuit feedthrough, I determined that the output impedance of the buffers needed to be less than 75Ω for an acceptable feedthrough performance of 60dB at 5MHz. I then tried several closed loop buffers to see if I could lower the supply current. The circuit shown in Figure 8–11 did the job and saved about 200 microamps of supply current per buffer. The closed loop buffer has an output impedance of about 7Ω that rises to 65Ω at 5MHz. Since four buffers were required, the supply current reduction of 800 microamps was significant.

At this point it became obvious to me that for the feedthrough to be down 60dB or more, the control circuitry had to be very accurate. If the full scale voltage was 2.5V and the control voltage was 0V, the offset errors had to be less than 2.5mV for 60dB of off isolation. Even if I trimmed the IC to zero offset, the system accuracy requirement was still very tough. I therefore wanted to come up with a circuit that would insure that the correct input was on and the other input was fully off when the control was close to zero or full scale. I thought about adding intentional offset voltage and/or gain errors to the V-to-I converters to get this result, but it didn't feel good. What was needed was an internal circuit that would sense when the control was below 5% or above 95% and force the pre-distortion diodes to 0% and 100%. Since the diodes were fed with currents, it seemed that sensing current was the way to go.

Since the currents that feed the pre-distortion diodes come from identical mirrors, I wanted to see if I could modify the mirrors so that they

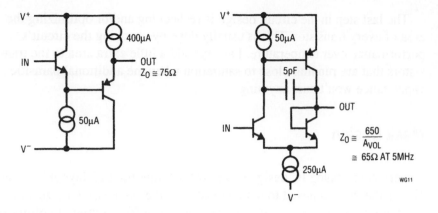

Figure 8–11.
Open- and closed-loop buffers.

would turn off at low currents. This would work at both ends of the control signal because one mirror is always headed towards zero current. The first thought was to put in a small fixed current that subtracted from the input current. This would add an offset near zero (good) and a gain error everywhere else (bad). Now if I could turn off the offset current when the output current was on, it would be perfect. Current mirrors #2 and #3 in Figure 8–8 were each modified to be as shown in Figure 8–12. The offset current is generated by Q9. A small ratio of the output current is used to turn off Q9 by raising its emitter. The ratios are set such that the output goes to zero with the input at about 5% of full scale. The nice thing about this mirror is that the turn-off circuit has no effect on mirror accuracy for inputs of 10% or more. The diode was added to equalize the collector-base voltage of all the matching transistors.

At this point the circuit was working very well in the inverting mode and I went back to non-inverting to see how the feedthrough looked. Since the output impedance of the amplifier determines the feedthrough performance, I eliminated all the output stage degeneration resistors. I set the output quiescent current at 2.5 milliamps so the output devices would be well up on their F-tau curve and the open loop output impedance would be well under 10 Ohms. The feedthrough was still 60dB down at 5MHz. I added a current limit circuit that sensed the output transistors' collector current, and the circuit topology was finalized.

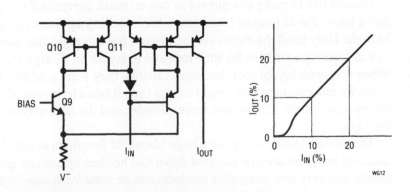

Figure 8–12.
Mirror with a turn-off.

The last step in the circuit design is rechecking and/or optimizing the area of every transistor. This is usually done by checking the circuit's performance over temperature. I always add a little extra area to the transistors that are running close to saturation when the additional parasitic capacitance won't hurt anything.

Mask Design

Experienced analog IC designers know how important IC layout is. Transistors that are supposed to match must have the same emitter size and orientation as well as the same temperature. The fader output amplifier is capable of driving a cable and generating significant thermal gradients in the IC. For this reason I put both input stages on one end of the die next to the current steering devices and put the output stage at the other end. The bias circuits and the control op amps went in the middle. The best way to minimize thermal feedback is distance. The 14-pin SO package set the maximum die size and the pad locations.

The IC process used had only one layer of metalization and therefore I provided the mask designer with an estimate of where "cross-unders" would be needed. For those of you not familiar with the term "cross-under," I will explain. A cross-under is a small resistor, usually made of N+, inserted in a lead so that it can "cross-under" another metal trace. Normally these cross-unders are inserted in the collectors of transistors, since a little extra resistance in the collector has minimal effect.

The fader circuit, with over 140 transistors and very few resistors, was clearly going to have a lot of cross-unders. I was resigned that both supplies would have many cross-unders; in order for the circuit to work properly, the voltage drops introduced by the cross-unders must not disturb the circuit. For example, the current mirrors will common mode out any variation in supply voltage as long as all the emitters are at the same voltage. This is easy to do if the emitters all connect together on one trace and then that trace connects to the supply. As mask design progresses, it is important that each cross-under added to the layout be added to the schematic and that circuit simulation is re-checked. Time spent before the silicon comes out to insure that the circuit works is well spent.

I would like to make a comment or two on mask design and the time that it takes. For as long as I can remember, speeding up mask design has been the Holy Grail. Many, including myself, have thought that some new tool or technique will cut the time required to layout an IC significantly. When computer layout tools became available, they were sold as a productivity enhancement that would cut the time it takes to layout ICs. The reality was that the ICs became more complex and the time stayed about the same.

The analog ASIC concept of a huge library of functions available as standard cells that are just plopped down and hooked up sounds great; except that very few innovative products can be done with standard functions. What typically happens is that each new product requires modifications to the "standard" cells or needs some new standard cells. You're right back at transistor level optimizing the IC. Of course no one ever plans for the extra time that this transistor level optimization takes, so the project gets behind schedule.

The "mono-chip" or "master-chip" idea is often used to speed up development. This technique uses just the metal layer(s) to make the new product; a large standard IC with many transistors and resistors is the common base. The trade-off for time saved in mask design is a larger die size. The argument is often made that if the product is successful, a full re-layout can be done to reduce die size and costs. Of course, this would then require all the effort that should have been done in the first place. I would not argue to save time and money up front because I did not expect my part to be successful!

In summary, mask design is a critical part of analog IC development and must be considered as important as any other step. Doing a poor job of mask design will hurt performance and that will impact the success of a product much more than the extra time in development.

Testing

IC automatic test system development is an art that combines analog hardware and software programming. We cannot sell performance that we cannot test. It is much easier to measure IC performance on the bench than in an automatic handler. In successful companies, the good test development engineers are well respected.

The fader IC requires that the closed loop AC gain be measured very accurately. The gain is trimmed at wafer sort by adjusting the value of resistor R_C. This trim is done with the control input fixed and the linearity of the circuit determines the gain accuracy elsewhere. The errors due to the bulk resistance of the steering transistors have no effect at 50% gain; therefore it seemed like the best place to trim the gain.

While characterizing the parts from the first wafer, I noticed that there were a few parts that had more error than I expected at 90% gain. I also determined that these parts would be fine if I had trimmed them at 90%. It was also true that the parts that were fine at 90% would not suffer from being trimmed at 90%. So, I changed my mind as to where the circuit was to be trimmed and the test engineer modified the sort program. More wafers were sorted and full characterization began.

Setting the data sheet limits is a laborious process that seems like it should be simpler. The designer and product engineer go over the distribution plots from each test to determine the maximum and minimum limits. In a perfect world we would have the full process spread represented in these distributions. Even with a "design of experiments" run that should give us the full spread of process variations, we will come up short of information. It's Murphy's law. This is where the designer's knowledge of which specs are important, and which are not, comes into play. It makes no sense to "over spec" a parameter that the customer is not concerned about because later it could cause a yield problem. On the other hand, it is important to spec all parameters so that any "sports" (oddball parts) are eliminated, since they are usually caused by defects and will often act strangely. The idea is to have all functional parts meet spec if they are normal.

Data Sheets

The data sheet is the most important sales tool the sales people have. Therefore it is important that the data sheet is clear and accurate. A good data sheet is always late. I say this based on empirical data, but there seems to be a logical explanation. The data sheet is useless unless it has all the minimums and maximums that guarantee IC performance; as soon as those numbers are known, the part is ready to sell and we need the data sheet. Of course it takes time to generate the artwork and print the data sheet and so it is late. One solution to this problem is to put out an early, but incomplete, data sheet and then follow it a few months later with a final, complete one.

Analog ICs usually operate over a wide range of conditions and the typical curves in the data sheet are often used to estimate the IC performance under conditions different from those described in the electrical table. The generation of these curves is time consuming and, when done well, requires a fair amount of thought. Human nature being what it is, most people would rather read a table than a graph, even though a table is just an abbreviated version of the data. As a result, the same information is often found in several places within the data sheet. I am often amazed at how inconsistent some data sheets are; just for fun, compare the data on the front page with the electrical tables and the graphs.

Beware of typical specs that are much better than the minimums and maximums. I once worked with a design engineer who argued that the typical value should be the average of the distribution; he insisted that the typical offset voltage of his part was zero even though the limits were ±4mV. Most companies have informal definitions of "typical", and it often varies from department to department. George Erdi added a note to several dual op amp data sheets defining the typical value as the value that would yield 60% based on the distributions of the individual amplifiers. I like and use this definition but obviously not everyone does, since I often see typicals that are 20 times better than the limits! Occasionally the limits are based on automatic testing restrictions and the typicals are real; for example, CMOS logic input leakage current is less than a few nanoamps, but the resolution of the test system sets the limit at 1 microamp.

Summary

Since you are still reading, I hope this long-winded trip was worth it. The development of an IC is fun and challenging. I spent most of this article describing the circuit design because I like circuit design. I hope, however, that I have made it clear how important the other parts of the development process are. There are still more phases of development that I have not mentioned; pricing, press releases, advertising, and applications support are all part of a successful new product development. At the time of this writing, the video fader had not yet reached these phases. Since I am not always accurate at describing the future, I will not even try. Those of you who want to know more about the fader should see the LT1251 data sheet.

At this time I would like to thank all of the people who made the video fader a reality and especially Julie Brown for mask design, Jim Sousae for characterization, Dung (Zoom) Nguyen for test development, and Judd Murkland in product engineering. It takes a team to make things happen and this is an excellent one.

9. Analog Breadboarding

Introduction

While there is no doubt that computer analysis is one of the most valuable tools that the analog designer has acquired in the last decade or so, there is equally no doubt that analog circuit models are not perfect and must be verified with hardware. If the initial test circuit or "breadboard" is not correctly constructed it may suffer from malfunctions which are not the fault of the design but of the physical structure of the breadboard itself. This chapter considers the art of successful breadboarding of high-performance analog circuits.

The successful breadboarding of an analog circuit which has been analyzed to death in its design phase has the reputation of being a black art which can only be acquired by the highly talented at the price of infinite study and the sacrifice of a virgin or two. Analog circuitry actually obeys the very simple laws we learned in the nursery: Ohm's Law, Kirchoff's Law, Lenz's Law and Faraday's Laws. The problem, however, lies in Murphy's Law.

Murphy's Law is the subject of many engineering jokes, but in its simplest form, "If Anything Can Go Wrong—It Will!", it states the simple truth that physical laws do not cease to operate just because we have overlooked or ignored them. If we adopt a systematic approach to breadboard

Figure 9–1.

MURPHY'S LAW

Whatever can go wrong, will go wrong.

Buttered toast, dropped on a sandy floor,
falls butter side down.

The basic principle behind Murphy's Law is that
all physical laws always apply -
when ignored or overlooked they do not stop working.

---

construction it is possible to consider likely causes of circuit malfunction without wasting very much time.

In this chapter we shall consider some simple issues which are likely to affect the success of analog breadboards, namely resistance (including skin effect), capacitance, inductance (both self inductance and mutual inductance), noise, and the effects of careless current routing. We shall then discuss a breadboarding technique which allows us to minimize the problems we have discussed.

Resistance

As an applications engineer I shall be relieved when room-temperature superconductors are finally invented, as too many engineers suppose that they are already available, and that copper is one of them. The assumption that any two points connected by copper are at the same potential completely overlooks the fact that copper is resistive and its resistance is often large enough to affect analog and RF circuitry (although it is rarely important in digital circuits).

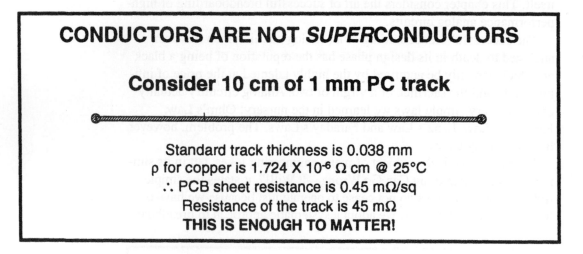

CONDUCTORS ARE NOT *SUPER*CONDUCTORS

Consider 10 cm of 1 mm PC track

Standard track thickness is 0.038 mm
ρ for copper is 1.724 X 10⁻⁶ Ω cm @ 25°C
∴ PCB sheet resistance is 0.45 mΩ/sq
Resistance of the track is 45 mΩ
THIS IS ENOUGH TO MATTER!

Figure 9–2.

The diagram in Figure 9–2 shows the effect of copper resistance at DC and LF. At HF, matters are complicated by "skin effect." Inductive effects cause HF currents to flow only in the surface of conductors. The skin depth (defined as the depth at which the current density has dropped to 1/e of its value at the surface) at a frequency f is

$$\frac{1}{\sqrt{\mu\sigma\pi f}}$$

where μ is the permittivity of the conductor, and σ is its conductivity in Ohm-meters. $\mu = 4\pi\times10^{-7}$ henry/meter except for magnetic materials, where $\mu=4\mu_r\pi\times10^{-7}$ henry/meter (μ_r is the relative permittivity). For the purposes of resistance calculation in cases where the skin depth is less than one-fifth the conductor thickness, we can assume that all the HF current flows in a layer the thickness of the skin depth, and is uniformly distributed.

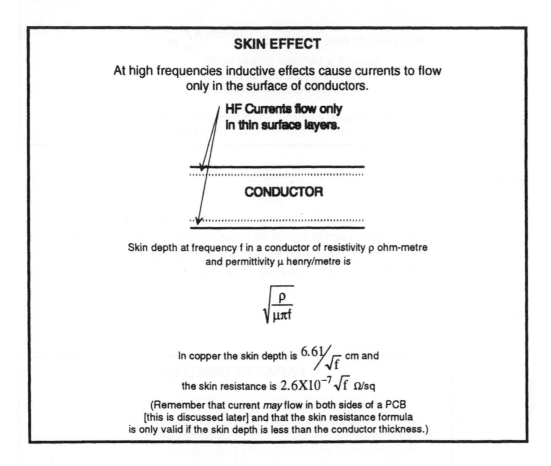

SKIN EFFECT

At high frequencies inductive effects cause currents to flow only in the surface of conductors.

HF Currents flow only in thin surface layers.

CONDUCTOR

Skin depth at frequency f in a conductor of resistivity ρ ohm-metre and permittivity μ henry/metre is

$$\sqrt{\frac{\rho}{\mu\pi f}}$$

In copper the skin depth is $\dfrac{6.61}{\sqrt{f}}$ cm and

the skin resistance is $2.6 \times 10^{-7}\sqrt{f}$ Ω/sq

(Remember that current *may* flow in both sides of a PCB [this is discussed later] and that the skin resistance formula is only valid if the skin depth is less than the conductor thickness.)

Figure 9–3.

Skin effect has the effect of increasing the resistance of conductors at quite modest frequencies and must be considered when deciding if the resistance of wires or PC tracks will affect a circuit's performance. (It also affects the behavior of resistors at HF.)

Capacitance

Good HF analog design must incorporate stray capacitance. Wherever two conductors are separated by a dielectric there is capacitance. The formulae for parallel wires, concentric spheres and cylinders, and other more exotic structures may be found in any textbook but the commonest structure, found on all PCBs, is the parallel plate capacitor.

CAPACITANCE

Wherever two conductors are separated by a dielectric
(including air or a vacuum) there is capacitance.

For a parallel plate capacitor $C = \dfrac{.0885E_r A}{d}$ pF

where A is the plate area in sq.cm
d is the plate separation in cm
& E_r is the dielectric constant

Epoxy PCB material is often 1.5 mm thick and $E_r = 4.7$
Capacity is therefore approximately 2.8 pf/sq.cm

Figure 9–4.

When stray capacitance appears as parasitic capacity to ground it can
be minimized by careful layout and routing, and incorporated into the
design. Where stray capacity couples a signal where it is not wanted the
effect may be minimized by design but often must be cured by the use of
a Faraday shield.

Figure 9–5.

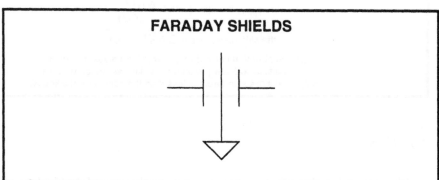

FARADAY SHIELDS

Capacitively coupled noise can be very effectively shielded
by a grounded conductive shield, known as a Faraday Shield.
But it must be grounded or it increases the problem.
For this reason coil and quartz crystal cans should always be grounded.

If inductance is to be minimized the lead and PC track length of capacitors must be kept as small as possible. This does not mean just generally
"short," but that the inductance in the actual circuit function must be minimal. Figure 9–6 shows both a common mistake (the leads of the capacitor C1 are short, but the decoupling path for IC1 is very long) and the

CAPACITOR LEADS MUST BE SHORT

Figure 9–6.

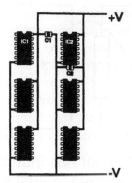

Although the leads of C1 are short the HF decoupling path of IC1 is far too long. The decoupling path of IC2 is ideal.

correct way to decouple an IC (IC2 is decoupled by C2 with a very short decoupling path).

Inductors

Any length of conductor has inductance and it can matter. In free space a 1cm length of conductor has inductance of 7–10nH (depending on diameter), which represents an impedance of 4–6Ω at 100MHz. This may be large enough to be troublesome, but badly routed conductors can cause worse problems as they form, in effect, single turn coils with quite substantial inductance.

INDUCTANCE

Figure 9–7.

Any conductor has some inductance
A straight wire of length L and radius R (both mm & L>>R)

has inductance $0.2L\left[\ln\left(\dfrac{2L}{R}\right) - .75\right]$ nH

A strip of conductor of length L, width W and thickness H (mm)
has inductance

$$0.2L\left[\ln\left(\frac{2L}{W+H}\right) + 0.2235\left(\frac{W+H}{L}\right) + 0.5\right]\text{ nH}$$

1 cm of thin wire or PC track is somewhere between 7 and 10 nH

Figure 9–8.

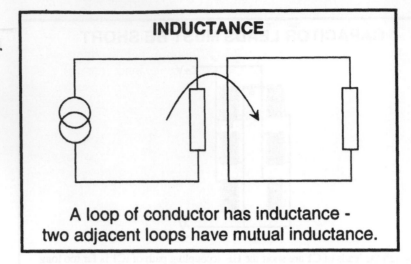

INDUCTANCE

A loop of conductor has inductance -
two adjacent loops have mutual inductance.

If two such coils are close to each other we must consider their mutual inductance as well as their self-inductance. A change of current in one will induce an EMF in the other. Defining the problem, of course, at once suggests cures: reducing the area of the coils by more careful layout, and increasing their separation. Both will reduce mutual inductance, and reducing area reduces self inductance too.

It is possible to reduce inductive coupling by means of shields. At LF shields of mu-metal are necessary (and expensive, heavy and vulnerable to shock, which causes loss of permittivity) but at HF a continuous Faraday shield (mesh will not work so well here) blocks magnetic fields too, provided that the skin depth at the frequency of interest is much less

Figure 9–9.

INDUCTANCE

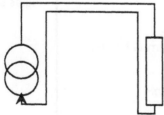
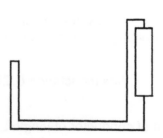

Inductance is reduced by reducing loop area -
mutual inductance is reduced by reducing loop area
and increasing separation.

Since the magnetic fields around coils are dipole fields they attenuate with the *cube* of the
distance - so increasing separation is a very effective way of reducing mutual inductance.

Figure 9–10.

MAGNETIC SHIELDS

At LF magnetic shielding requires Mu-Metal which is heavy, expensive and vulnerable to shock.

At HF a conductor provides effective magnetic shielding provided the skin depth is less than the conductor thickness.
PC foil is an effective magnetic shield above 10-20 MHz.

---

than the thickness of the shield. In breadboards a piece of copper-clad board, soldered at right angles to the ground plane, can make an excellent HF magnetic shield, as well as being a Faraday shield.

Magnetic fields are dipole fields, and therefore the field strength diminishes with the *cube* of the distance. This means that quite modest separation increases attenuation a lot. In many cases physical distance is all that is necessary to reduce magnetic coupling to acceptable levels.

Grounds

Kirchoff's Law tells us that return currents in ground are as important as signal currents in signal leads. We find here another example of the "superconductor assumption"—too many engineers believe that all points marked with a ground symbol on the circuit diagram are at the same potential. In practice ground conductors have resistance and inductance—and potential differences. It is for this reason that such breadboarding techniques as matrix board, prototype boards (the ones where you poke component leads into holes where they are gripped by phosphor-bronze contacts) and wire-wrap have such poor performance as analog prototyping systems.

The best analog breadboard arrangement uses a "ground plane"—a layer of continuous conductor (usually copper-clad board). A ground

Figure 9–11.

KIRCHOFF'S LAW

The net current at any point in a circuit is zero.
OR
What flows in flows out again.
OR
Current flows in circles.
THEREFORE
All signals are differential.
AND
Ground impedance matters.

---

plane has minimal resistance and inductance, but its impedance may still be too great at high currents or high frequencies. Sometimes a break in a ground plane can configure currents so that they do not interfere with each other; sometimes physical separation of different subsystems is sufficient.

Figure 9–12.

GROUND PLANE BREADBOARD

The breadboard ground consists of a single layer of continuous metal, usually (unetched) copper-clad PCB material. In theory all points on the plane are at the same potential, but in practice it may be necessary to configure ground currents by means of breaks in the plane, or careful placement of sub-systems. Nevertheless ground plane is undoubtedly the most effective ground technique for analog breadboards.

Figure 9–13.

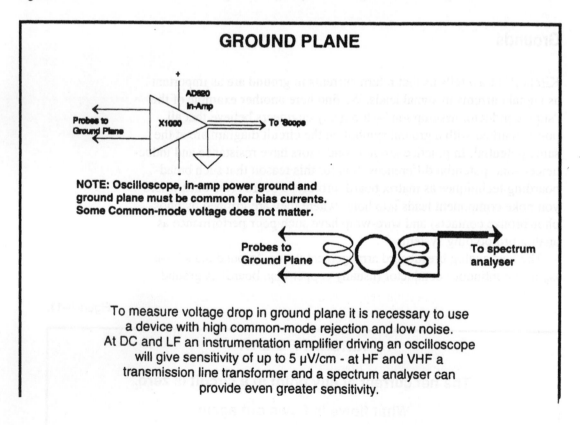

GROUND PLANE

Probes to Ground Plane

AD620 In-Amp
X1000
To 'Scope

NOTE: Oscilloscope, in-amp power ground and ground plane must be common for bias currents. Some Common-mode voltage does not matter.

Probes to Ground Plane

To spectrum analyser

To measure voltage drop in ground plane it is necessary to use a device with high common-mode rejection and low noise. At DC and LF an instrumentation amplifier driving an oscilloscope will give sensitivity of up to 5 µV/cm - at HF and VHF a transmission line transformer and a spectrum analyser can provide even greater sensitivity.

It is often easy to deduce where currents flow in a ground plane, but in complex systems it may be difficult. Breadboards are rarely that complex, but if necessary it is possible to measure differential voltages of as little as 5μV on a ground plane. At DC and LF this is done by using an instrumentation amplifier with a gain of 1,000 to drive an oscilloscope working at 5 mV/cm. The sensitivity at the input terminals of the inamp is 5μV/cm; there will be some noise present on the oscilloscope trace, but it is quite possible to measure ground voltages of the order of 1μV with such simple equipment. It is important to allow a path for the bias current of the inamp, but its common-mode rejection is so good that this bias path is not critical.

The upper frequency of most inamps is 25–50kHz (the AD830 is an exception—it works up to 50 MHz at low gains, but not at ×1,000). Above LF a better technique is to use a broadband transmission line transformer to remove common-mode signals. Such a transformer has little or no voltage gain, so the signal is best displayed on a spectrum analyzer, with μV sensitivity, rather than on an oscilloscope, which only has sensitivity of 5mV or so.

Decoupling

The final issue we must consider before discussing the actual techniques of breadboarding is decoupling. The power supplies of HF circuits must be short-circuited together and to ground at all frequencies above DC. (DC short-circuits are undesirable for reasons which I shall not bother to discuss.) At low frequencies the impedance of supply lines is (or should be) low and so decoupling can be accomplished by relatively few electrolytic capacitors, which will not generally need to be very close to the parts of the circuit they are decoupling, and so may be shared among several parts of a system. (The exception to this is where a component draws a large LF current, when a local, dedicated, electrolytic capacitor should be used.)

At HF we cannot ignore the impedance of supply leads (as we have already seen in Figure 9–6) and ICs must be individually decoupled with low inductance capacitors having short leads and PC tracks. Even 2–3mm of extra lead/track length may make the difference between the success and failure of a circuit layout.

DECOUPLING

Supplies must be short-circuited to each other and to ground at *all* frequencies.
(But not at DC.)

Figure 9–14.

Figure 9–15.

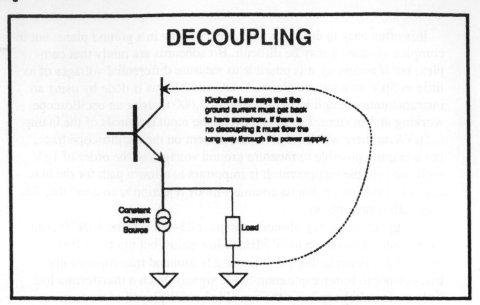

DECOUPLING

Kirchoff's Law says that the ground current must get back to here somehow. If there is no decoupling it must flow the long way through the power supply.

Constant Current Source

Load

Where the HF currents of a circuit are mostly internal (as is the case with many ADCs) it is sufficient that we short-circuit its supplies at HF so that it sees its supplies as stiff voltage sources at all frequencies. When it is driving a load, the decoupling must be arranged to ensure that the total loop in which the load current flows is as small as possible. Figure 9–15 shows an emitter follower without supply decoupling—the HF current in the load must flow through the power supply to return to the output stage (remember that Kirchoff's Law says, in effect, that currents must flow in circles). Figure 9–16 shows the same circuit with proper supply decoupling.

This principle is easy enough to apply if the load is adjacent to the circuit driving it. Where the load must be remote it is much more difficult, but there are solutions. These include transformer isolation and the use of a transmission line. If the signal contains no DC or LF compo-

Figure 9–16.

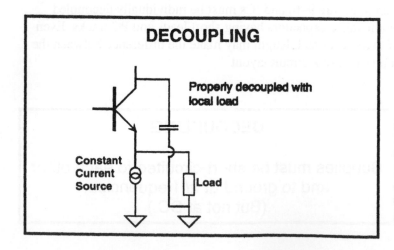

DECOUPLING

Properly decoupled with local load

Constant Current Source

Load

nents, it may be isolated with a transformer close to the driver. Such an arrangement is shown in Figure 9–17. (The nature of the connection from the transformer to the load may present its own problems—but supply decoupling is not one of them.)

A correctly terminated transmission line constrains HF signal currents so that, to the supply decoupling capacitors, the load appears to be adjacent to the driver. Even if the line is not precisely terminated, it will constrain the majority of the return current and is frequently sufficient to prevent ground current problems.

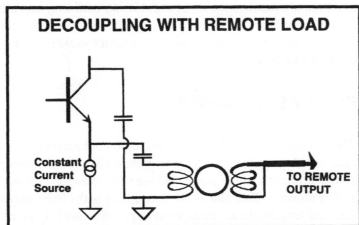

Figure 9–17.

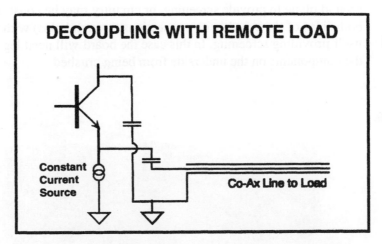

Figure 9–18.

Breadboarding Principles

Having considered issues of resistance, capacitance, and inductance, it is clear that breadboards must be designed to minimize the adverse effects of these phenomena. The basic principle of a breadboard is that it is a temporary structure, designed to test the performance of a circuit or system, and must therefore be easy to modify.

There are many commercial breadboarding systems, but almost all of them are designed to facilitate the breadboarding of digital systems, where noise immunities are hundreds of millivolts or more. (We shall discuss the exception to this generality later.) Matrix board (Veroboard, etc.), wire-wrap, and plug-in breadboard systems (Bimboard, etc.) are, without exception, unsuitable for high performance or high frequency analog breadboarding. They have too high resistance, inductance and capacitance. Even the use of IC sockets is inadvisable. (All analog engineers should practice the art of unsoldering until they can remove an IC from a breadboard [or a plated-through PCB] without any damage to the board or the device—solder wicks and solder suckers are helpful in accomplishing this.)

Practical Breadboarding

The most practical technique for analog breadboarding uses a copper-clad board as a ground plane. The ground pins of the components are soldered directly to the plane, and the other components are wired together above it. This allows HF decoupling paths to be very short indeed. All lead lengths should be as short as possible, and signal routing should separate high-level and low-level signals. Ideally the layout should be similar to the layout to be used on the final PCB.

Pieces of copper-clad may be soldered at right angles to the main ground plane to provide screening, or circuitry may be constructed on both sides of the board (with connections through holes) with the board itself providing screening. In this case the board will need legs to protect the components on the underside from being crushed.

Figure 9–19.

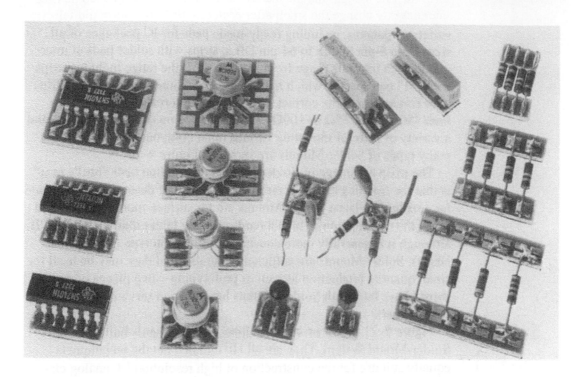

Figure 9–20.

When the components of a breadboard of this type are wired point-to-point in the air (a type of construction strongly advocated by Robert A. Pease of National Semiconductor[1] and sometimes known as "bird's nest" construction) there is always the risk of the circuitry being crushed and resulting short-circuits; also, if the circuitry rises high above the ground plane, the screening effect of the ground plane is diminished and interaction between different parts of the circuit is more likely. Nevertheless the technique is very practical and widely used because the circuit may so easily be modified.

However, there is a commercial breadboarding system which has most of the advantages of "bird's nest over a ground plane" (robust ground, screening, ease of circuit alteration, low capacitance, and low inductance) and several additional advantages: it is rigid, components are close to the ground plane, and where necessary node capacitances and line impedances can be calculated easily. This system was invented by Claire R. Wainwright and is made by WMM GmbH in the town of Andechs in Bavaria and is available throughout Europe and most of the world as "Mini-Mount" but in the USA (where the trademark "Mini-Mount" is the property of another company) as the "Wainwright Solder-Mount System."[2] (There is also a monastery at Andechs where they brew what is arguably the best beer in Germany.)

Solder-Mounts consist of small pieces of PCB with etched patterns on one side and contact adhesive on the other. They are stuck to the ground plane and components are soldered to them. They are available in a wide variety of patterns, including ready-made pads for IC packages of all sizes from 8-pin SOICs to 64-pin DILs, strips with solder pads at intervals (which intervals range from .040" to .25"; the range includes strips with 0.1" pad spacing which may be used to mount DIL devices), strips with conductors of the correct width to form microstrip transmission lines (50Ω, 60Ω, 75Ω or 100Ω) when mounted on the ground plane, and a variety of pads for mounting various other components. A few of the many types of Solder-Mounts are shown in Figure 9–20.

The main advantage of Solder-Mount construction over "bird's nest" is that the resulting circuit is far more rigid, and, if desired, may be made far smaller (the latest Solder-Mounts are for surface-mount devices and allow the construction of breadboards scarcely larger than the final PCB, although it is generally more convenient if the prototype is somewhat larger). Solder-Mounts are sufficiently durable that they may be used for small quantity production as well as prototyping—two pieces of equipment I have built with Solder-Mounts have been in service now for over twenty years.

Figure 9–21 shows several examples of breadboards built with the Solder-Mount System. They are all HF circuits, but the technique is equally suitable for the construction of high resolution LF analog circuitry. A particularly convenient feature of Solder-Mounts at VHF is the ease with which it is possible to make a transmission line.

If a conductor runs over a ground plane it forms a microstrip transmission line. The Solder-Mount System has strips which form microstrip lines when mounted on a ground plane (they are available with impedances of 50Ω, 60Ω, 75Ω and 100Ω). These strips may be used as transmission lines, for impedance matching, or simply as power buses. (Glass fiber/epoxy PCB is somewhat lossy at VHF and UHF, but the losses will probably be tolerable if microstrip runs are short.)

It is important to realize that current flow in a microstrip transmission line is constrained by inductive effects. The signal current flows only on the side of the conductor next to the ground plane (its skin depth is calculated in the normal way) and the return current flows only directly beneath the signal conductor, not in the entire ground plane (skin effect naturally limits this current, too, to one side of the ground plane). This is helpful in separating ground currents, but increases the resistance of the circuit.

It is clear that breaks in the ground plane under a microstrip line will force the return current to flow around the break, increasing impedance. Even worse, if the break is made to allow two HF circuits to cross, the two signals will interact. Such breaks should be avoided if at all possible. The best way to enable two HF conductors on a ground plane to cross without interaction is to keep the ground plane continuous and use a microstrip on the other side of the ground plane to carry one of the signals past the other (drill a hole through the ground plane to go to the other side of the board). If the skin depth is much less than the ground plane thickness the interaction of ground currents will be negligible.

Figure 9–21

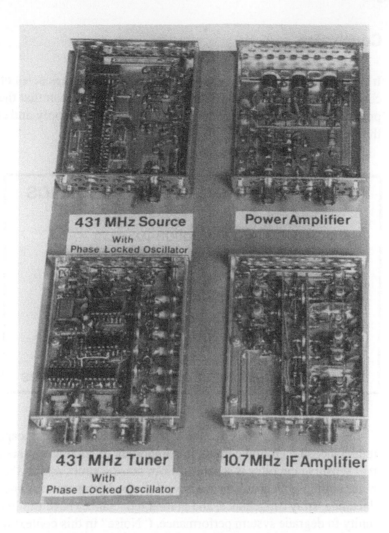

Figure 9–22.

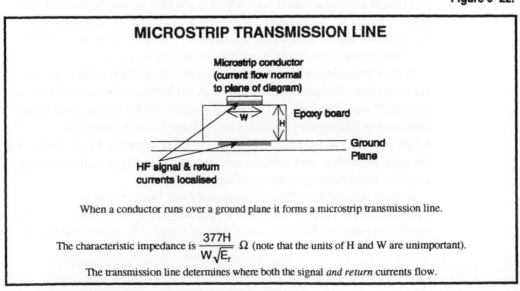

MICROSTRIP TRANSMISSION LINE

Microstrip conductor
(current flow normal
to plane of diagram)

Epoxy board

Ground
Plane

HF signal & return
currents localised

When a conductor runs over a ground plane it forms a microstrip transmission line.

The characteristic impedance is $\dfrac{377H}{W\sqrt{E_r}}$ Ω (note that the units of H and W are unimportant).

The transmission line determines where both the signal *and return* currents flow.

117

Conclusion

It is not possible in a short chapter to discuss all the intricacies of successful analog breadboard construction, but we have seen that the basic principle is to remember all the laws of nature which apply and consider their effects on the design.

Figure 9–23.

SUCCESSFUL ANALOG BREADBOARDS
Pay attention to:
Resistance
Capacitance
Inductance
Decoupling
Ground
&
Separating sensitive circuits from noisy ones

In addition to the considerations of resistance, skin effect, capacitance, inductance and ground current, it is important to configure systems so that sensitive circuitry is separated from noise sources and so that the noise coupling mechanisms we have described (common resistance/inductance, stray capacitance, and mutual inductance) have minimal opportunity to degrade system performance. ("Noise" in this context means a signal we want [or which somebody wants] in a place where we don't want it; not natural noise like thermal, shot or popcorn noise.) The general rule is to have a signal path which is roughly linear, so that outputs are physically separated from inputs and logic and high level external signals only appear where they are needed. Thoughtful layout is important, but in many cases screening may be necessary as well.

A final consideration is the power supply. Switching power supplies are ubiquitous because of their low cost, high efficiency and reliability, and small size. But they can be a major source of HF noise, both broadband and at frequencies harmonically related to their switching frequency. This noise can couple into sensitive circuitry by all the means we have discussed, and extreme care is necessary to prevent switching supplies from ruining system performance.

Prototypes and breadboards frequently use linear supplies or even batteries, but if a breadboard is to be representative of its final version it should be powered from the same type of supply. At some time during

Figure 9–24.

SWITCHING POWER SUPPLIES

Generate noise at every frequency under the Sun (and some interstellar ones as well).

Every mode of noise transmission is present.

If you must use them you should filter, screen, keep them far away from sensitive circuits, and still worry!

---

development, however, it is interesting (and frightening, and helpful) to replace the switching supply with a battery and observe the difference in system performance.

Figure 9–25.

OBEY THE LAW

Unexpected behaviour of analog circuitry is almost always due to the designer overlooking one of the basic laws of electronics. Remember and obey Ohm, Faraday, Lenz, Maxwell, Kirchoff **and MURPHY.**

"Murphy always was an optimist" - Mrs. Murphy.

References

1. Robert A. Pease, *Troubleshooting Analog Circuits* (Butterworth-Heinemann, 1991).
2. Wainwright Instruments Inc., 7770 Regents Rd., #113 Suite 371, San Diego, CA 92122 (619) 558 1057 Fax: (619) 558 1019.

    WMM GmbH, Wainwright Mini-Mount-System, Hartstraße, 28C, D-82346 Andechs-Frieding, Germany, (+49)8152-3162 Fax: (+49)8152-4025.

10. Who Wakes the Bugler?

Introduction: T-Coils in Oscilloscope Vertical Systems

Few engineers realize the level of design skill and the care that is needed to produce an oscilloscope, the tool that the industry uses and trusts. To be really effective, the analog portion of a vertical channel of the oscilloscope should have a bandwidth greater than the bandwidth of the circuit being probed, and the transient response should be near perfect. A vertical amplifier designer is totally engrossed in the quest for this unnatural fast-and-perfect step-response. The question becomes, "How do 'scope designers make vertical amplifier circuits both faster and cleaner than the circuits being probed?" After all, the designers of both circuits basically have the same technology available.

One of many skillful tricks has been the application of precise, special forms of the T-coil section. I'll discuss these T-coil applications in Tektronix oscilloscopes from a personal and a historical perspective, and also from the viewpoint of an oscilloscope vertical amplifier designer. Two separate stand-alone pages contain "cookbook" design formulas, response functions, and related observations.

The T-coil section is one of the most fun, amazing, demanding, capable, and versatile circuits I have encountered in 'scopes. Special forms

Figure 10–1.
The T-coil Section.

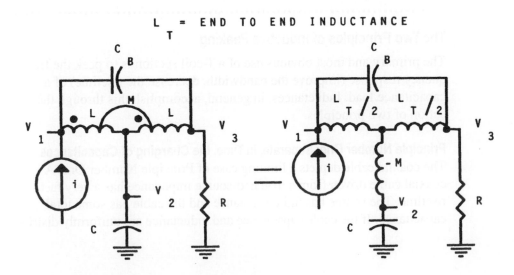

L_T = END TO END INDUCTANCE

of this basic circuit block are used with precision and finesse to do the following:

Peak capacitive loads
Peak amplifier interstages
Form "loop-thru" circuits
Equalize nonlinear phase
Transform capacitive terminations to resistive terminations
Form distributed deflectors in cathode ray tubes
Form artificial delay line sections
Form distributed amplifier sections

I have successfully used T-coils in all of these applications except the last two. Recently, however, some successful designers from the '40s and '50s shared their experiences with those two applications.

....

Over My Head

While on a camping trip in Oregon in 1961, I stopped at Tektronix and received an interview and a job offer the same day. Tektronix wanted me. They were at a stage where they needed to exploit transistors to build fast, high-performance 'scopes. I had designed a 300MHz transistor amplifier while working at Sylvania. In 1961, that type of experience was a rare commodity. Actually, I had designed a wideband 300MHz IF amplifier that only achieved 200MHz. What we (Sylvania) used was a design that my technician came up with that made 300MHz. So I arrived at this premier oscilloscope company feeling somewhat of a fraud. I was more than just a bit intimidated by the Tektronix reputation and the distributed amplifiers and artificial delay lines and all that "stuff" that really worked. The voltage dynamic range, the transient response cleanliness, and DC response requirements for a vertical output amplifier made my low-power, 50 Ohm, 300MHz IF amplifier seem like child's play. Naturally, I was thrown immediately into the job of designing high-bandwidth oscilloscope transistor vertical-output amplifiers. I felt like a private, fresh out of basic training, on the front lines in a war.

....

The Two Principles of Inductive Peaking

The primary and most obvious use of a T-coil section is to peak the frequency response (improve the bandwidth, decrease the risetime) of a capacitance load. Inductances, in general, accomplish this through the action of two principles.

Principle Number One: Separate, in Time, the Charging of Capacitances

The coaxial cable depicts a limiting case of Principle Number One. A coaxial cable driven from a matched-source impedance has a very fast risetime. The source has finite resistance and the cable has some total capacitance. If the cable capacitance and inductance are uniformly distrib-

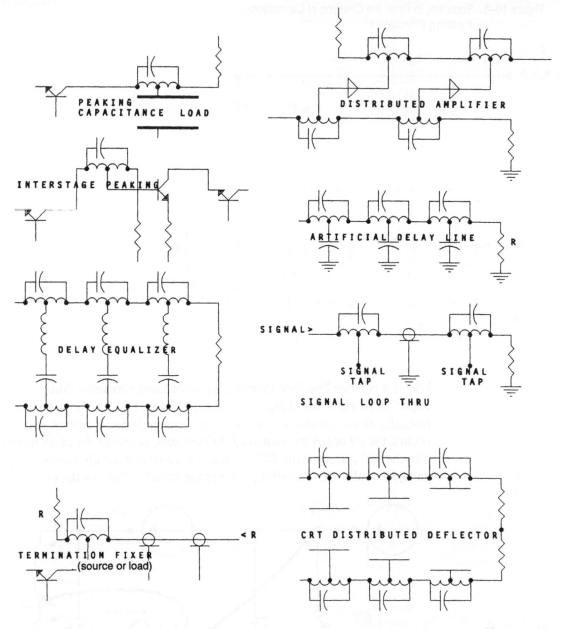

Figure 10–2.
The Versatile T-coil.

uted and the cable is situated in the proper impedance environment, the bandwidth is $>> 1/2\pi RC_{cable}$ and the risetime $<< 2.2\ RC_{cable}$. The distributed inductance in the line has worked with the distributed capacitance to spread out, in time, the charging of this capacitance. A pi-section LC filter could also demonstrate Principle Number One, as could a distributed amplifier.

Figure 10–3. Separate, In Time, the Charging of Capacitances.
Peaking Principle 1

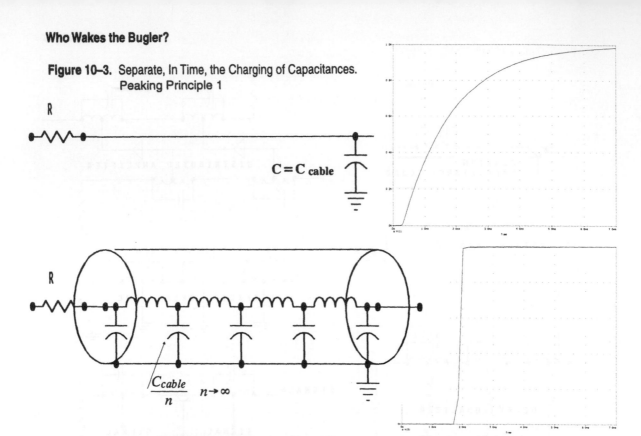

Principle Number Two: Don't Waste Current Feeding a Resistor When a Capacitor Needs to Be Charged In Figure 10–4 a helpful elf mans the normally closed switch in series with the resistor. When a current step occurs, the elf opens the switch for RC seconds, allowing the capacitor to take the full current. After RC seconds, the capacitor has charged to a voltage equal to IR. The elf then closes the switch, allowing the current

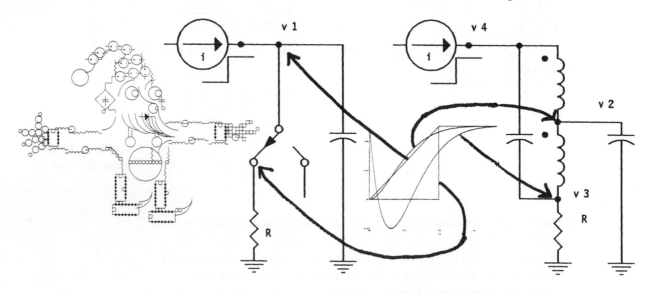

Figure 10–4.
Don't Waste Time Feeding a Resistor When a Capacitor Needs to be Charged.
Peaking Principle 2

to feed the resistor, also producing a voltage equal to IR. No current is wasted in the resistor while the capacitor is charging.

A current step applied to the constant-resistance bridged T-coil yields the same capacitor voltage risetime, 0.8 RC, as the elf circuit. In both cases, during the rise of voltage on the capacitor, the voltage waveform on the termination resistor is negative, zero, or at least low. Without the helpful elf, or without the T-coil, the risetime would have been 2.2 RC. With these risetime enhancers, the risetime is lowered to 0.8 RC. This is a risetime improvement factor of 2.75. If there are two or more capacitor lumps, Principle Number One can combine with Principle Number Two to obtain even higher risetime improvement factors.

When both principles are working optimally, reflections, overshoot, and ringing are avoided or controlled. This is a matter of control of energy flow in and out of the T-coil section reactances. A T-coil needs to be tuned or tolerated. In the constant-resistance T-coil section, given a load capacitance, there is only one set of values for the inductance, mutual inductance, and bridging capacitance which will satisfy one set of specifications of the driving point resistance (may imply reflection coefficient) and desired damping factor (relates to step response overshoot).

T-Coils Peaking Capacitance Loads

A cathode ray tube (CRT) electrostatic deflection plate pair is considered a pure capacitance load. In the '50s and '60s, T-coils were often used in deflection plate drive circuits. Usually a pentode-type tube was used as the driver, rather than a transistor, because of the large voltage swing required. The pentode output looked like a capacitive high-impedance source. A common technique was to employ series peaking of the driver capacitance, cascaded with T-coiled CRT deflection plate capacitance.

The 10–MHz Tektronix 3A6

The 3A6 vertical deflection amplifier works really hard. The 3A6 plug-in was designed to operate in the 560 series mainframes, where the plug-ins drove the CRT deflection plates directly. The deflection sensitivity was poor (20 volts per division) and the capacitance was high. To cover the display screen linearly and allow sufficient overscan, the output beam power tube on each side had to traverse at least 80 volts. The T-coils on the 3A6 made the bandwidth and dynamic range possible without burning up the large output vacuum tubes.

A Real T-Coil Response

A vertical-output deflection-amplifier designer has a unique situation—the amplifier output is on the screen—no other monitor is needed. This is the case with the 3A6 circuit shown here. The input test signal is clean and fast. The frequency and step response of the entire vertical system is dominated by the "tuning" of the T-coil L384 and its opposite-side

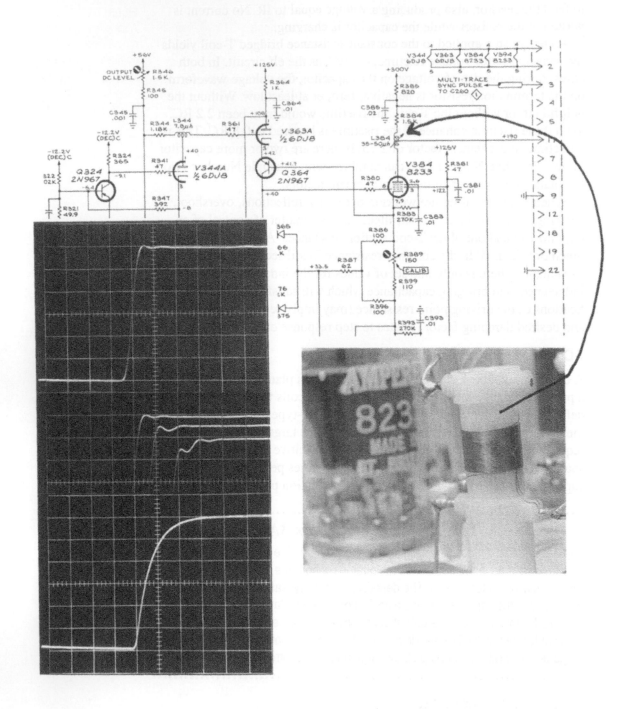

Figure 10–5.
Step Response Waveforms 3A6 T-coil Peaking.

FACT SHEET FOR
CONSTANT-RESISTANCE T-COILS

L_T - END TO END INDUCTANCE

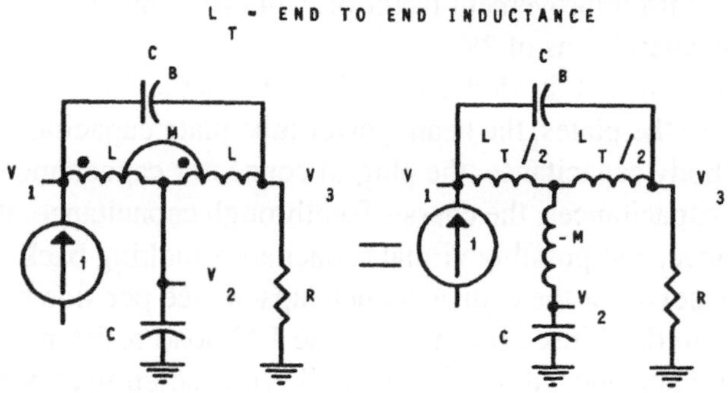

damping factor

$$\delta = \frac{1}{2}\sqrt{\frac{(1+k)}{(1-k)}}$$

$L_T = 2L + 2M$ $k = coupling\ coefficient$ $k = \dfrac{M}{L}$ and $\dfrac{M}{L_T} = \dfrac{k}{2(1+k)}$

If $L_T = R^2C$ and $C_B = \dfrac{(1-k)C}{4(1+k)}$

Then $\dfrac{v_1}{i} = R$ **the Constant Resistance Property**

and $\dfrac{v_2}{i} = \dfrac{R}{1 + \dfrac{RCs}{2} + \dfrac{(1-k)R^2C^2s^2}{4(1+k)}}$ **a Quadratic (2 pole) Response at v_2**

and $\dfrac{v_3}{i} = R\ \dfrac{1 - \dfrac{RCs}{2} + \dfrac{(1-k)R^2C^2s^2}{4(1+k)}}{1 + \dfrac{RCs}{2} + \dfrac{(1-k)R^2C^2s^2}{4(1+k)}}$ **an ALL PASS response at v_3**

v_2 step response overshoot

k = .6 (CRITICAL DAMPING)	0.0%
k = .5 (FLAT DELAY)	0.4 %
k = .333 (FLAT AMPLITUDE)	4.3%
k = 0.0 (high frequency DELAY BOOST)	16.0%

SPECIAL NOTE ON m-DERIVED T-COILS.

The m-derived t-coils arise from m-derived filter theory. They _do not have the constant-resistance_ property. The total inductance $= R^2C$. They have no bridging capacitance. They do not have a simple quadratic (2 pole) response. The value of "m" implies a coupling coefficient $k = \dfrac{m^2-1}{m^2+1}$

Figure 10–6.
Fact Sheet on Constant Resistance T-coils.

127

counterpart. The bottom picture shows the response when the coils (L384 and its mate) were disabled. (All three terminals of each coil were shorted together.) This reveals that, without the coils, the response looks very much like a single-time-constant response. The middle picture illustrates the progression of tuning after the shorts are removed. The powdered iron slugs in the coil forms are adjusted to optimize the response. The top picture shows the best response. The 10–to-90% risetime of the beginning waveform is 75 nanoseconds, and in the final waveform it drops to 28 nanoseconds. This is a ratio of risetimes of 2.6—near the theoretical bandwidth improvement factor of 2.74. The final waveform has peak-to-peak aberrations of 2%.

The total capacitance at the deflector node includes the deflection plates, the wires to the plates, the beam power tube plate capacitance, the wiring and coil body capacitance, the plug-in connector capacitance, the mounting point capacitances, the chassis feedthrough capacitance, the resistor capacitance, and possibly virtual capacitance looking back into the tube. We can solve for the equivalent net capacitance per side by working back from the 75nsec risetime and the 1.5k load resistance. This yields about 23pF per side. Although each coil is one solenoidal winding, it actually performs as two coils. The coil end connected to the tube plate works as a series peaking coil, and the remainder as the actual T-coil.

L344, which is also a T-coil, appears upstream in the 3A6 schematic fragment. Notice that the plate feeds the center tap of this coil. This is an application of reciprocity (Look in your old circuit textbook!). If the driving device output capacitance is significantly greater than the load capacitance, it may be appropriate to use this connection.

Distributed Amplifiers in Oscilloscopes

The idea of a distributed amplifier goes back to a British "Patent Specification" by W.S. Percival in 1936. In August 1948, Ginzton, Hewlett, Jasberg, and Noe published a classic paper on distributed amplifiers in the "Proceedings of IRE." At about the same time, Bill Hewlett (yes, of HP) and Logan Belleville (of Tektronix) met at Yaws Restaurant in Portland. Bill Hewlett described the new distributed amplifier concepts (yes, he "penciled out" the idea on a napkin!). In 1948, from August through October, Howard Vollum and Richard Rhiger built a distributed amplifier under a government contract. This amplifier was intended for use in a high-resolution ground radar. It had about a 6nsec risetime and a hefty output swing. In order to measure the new amplifier's performance, Vollum and Rhiger had outboarded it on the side of an early 511 'scope, directly feeding the deflectors.

It soon became clear that what the government and industry really needed was a very fast oscilloscope. I am not sure of the details or sequence of events, but Tektronix—Howard Vollum's two-year-old company—was making history. Vollum, Belleville, and Rhiger developed the 50MHz 517 oscilloscope, an oscilloscope with a distributed amplifier in the vertical deflection path. Vollum and Belleville had successfully refined the distributed amplifier enough to satisfy this oscilloscope vertical amplifier application. The product was successful and order

Figure 10–7.
1948 Experiment—
Outboarded
Distributed
Amplifier.

rates exceeded Tek's ability to manufacture. Logan left Tektronix in the early '50s and Vollum and Rhiger were left managing this new big company. John Kobbe, Cliff Moulton, and Bill Polits, as well as other key electrical circuit designers, took up where Vollum, Belleville, and Rhiger had left off. Other distributed amplifiers were designed for other 'scopes during the '50s, including the 540 series at 30MHz and the 580 series at 100MHz.

Manufacturing Distributed Amplifier Oscilloscopes

The whole idea of using a distributed amplifier as an oscilloscope vertical amplifier is rather incredible to me. Obtaining a very fast, clean step response is a hard job. When T-coils are employed, the job is even harder. When they are employed wholesale, as in a distributed amplifier, they are "fussy squared or tripled." The tuning of an oscilloscope distributed amplifier and/or an artificial delay line is tricky. Tuning is done in the time domain, with clues about where and in which direction to adjust, coming from observations of the "glitches" in the step response. If the use of a distributed amplifier in the vertical channel of an oscilloscope was proposed in today's business climate, it would be declared "unmanufacturable." It would never see the light of day. However, the Tektronix boom expansion in the '50s occurred largely through the development, manufacture, and sale of distributed amplifier 'scopes.

The 100MHz 580 series was the last use of distributed amplifiers in Tektronix 'scope vertical systems. Dual triodes, low cathode connection inductance, cross-coupled capacitance neutralization, and distributed deflectors in the CRT helped to achieve this higher bandwidth.

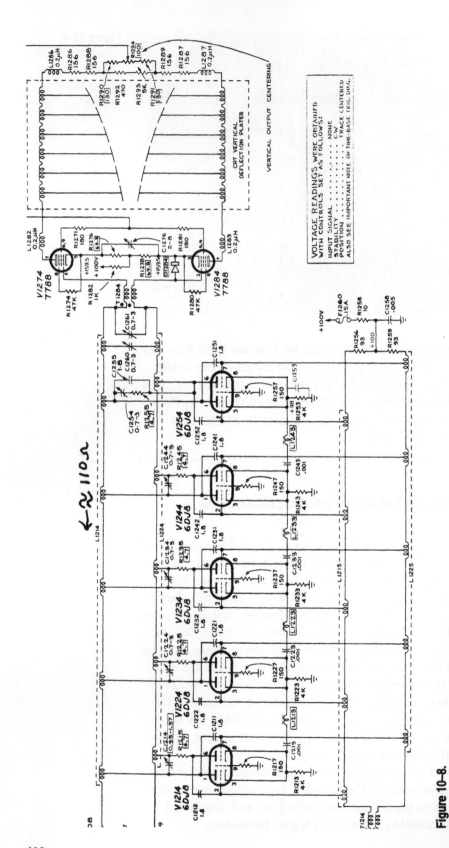

Figure 10–8.
Tektronix 585 Distributed Amplifier Vertical Output.

Distributed Deflector for a Cathode Ray Tube

In 1961, Cliff Moulton's 1GHz 519 'scope led the bandwidth race. This instrument had no vertical amplifier. The input was connected to a 125-ohm transmission line which directly fed a single-ended distributed deflection system. Schematics in Figures 10–8 and 10–9 show somewhat pictorially what a distributed deflector looks like. The 519 deflector is not shown. Within the CRT envelope was a meander line distributed deflection plate. Tuning capacitors were located at the sharp bends of the meander line. The line was first tuned as a mechanical assembly and later incorporated into the CRT envelope.

Terminated distributed deflector structures create a resistive driving-point impedance in place of one lumped capacitance. They also synchronize the signal travel along the deflection plate to the velocity of the electron beam speeding through the deflection plate length. If a distributed deflector is not used, deflection sensitivity is lost at high frequency due to transit time. Relative sensitivity is

$$\frac{\sin \frac{f}{f_{tx}}}{\frac{f}{f_{tx}}}$$ where f is frequency and f_{tx} is an inverse transit time function.

This is usually significant at 100MHz and above, and therefore distributed deflectors show up in 'scopes with bandwidths of 100MHz or higher. Various ingenious structures have been used to implement distributed deflectors. All could be modeled as assemblies of T-coils. The effective electron beam deflection response is a function of all of the T-coil tap voltages properly delayed and weighted.

Theoretical and Pragmatic Coil Proportions

The basis for the earliest T-coil designs was m-derived[1] filter theory. The delay lines and the distributed amplifier seemed to work best when the coils were proportioned—as per the classic Jasberg-Hewlett paper[2]—at m = 1.27 (coupling coefficient = 0.234). This corresponds to a coil length slightly longer than the diameter. In the design phase, there was an intelligent juggling of coil proportions based on the preshoot-overshoot behavior of the amplifier or delay line. The trial addition of bridging capacitance invariably led to increased step response aberrations.

---

1. m-derived filters were outcomes of image-parameter filter theory of the past. The parameter "m" determined the shape of the amplitude and phase response. "m"=1.27 approximated flat delay response. Filters could not be exactly designed, using this theory, because the required termination was not realizable.

2. This classic paper described both the m-derived T-coil section and, very briefly, the constant-resistance T-coil section. The use of these sections in distributed amplifiers was the main issue and nothing was mentioned of other uses.

In contrast with the artificial delay lines and the distributed amplifiers, the individual peaking applications usually needed a coil with more coupling (k = 0.4 to 0.5), which was realized by a coil shorter than its diameter. When the coil value is near or below 100 nanohenries, the goal is then to get as much coupling as possible so that the lead inductance of the center tap connection can be overcome. Flat pancake or sandwich coils of thin PC board material, thin films, or thick films are used to achieve high coupling.

The Importance of Stray Capacitance in T-Coils

The stray interwinding capacitance of a T-coil can be crudely modeled by one bridging capacitance C_{bs} across the whole coil. It is defined by the coil self-resonance frequency "f_{res}."

$$C_{bs} = \frac{1}{(2\pi f_{res})^2 L_T}$$

where L_T is the coil total inductance. If C_B is the required bridging capacitance for constant-resistance proportions, then $C_x = C_b - C_{bs}$ needs to be added. This is an effective working approximation. The recent coils built for high-frequency 50 Ohm circuits usually need additional bridging capacitance. On the other hand, the old nominally m-derived circuits never needed any added bridging capacitance. They were high-impedance circuits with very large coils and probably had enough effective bridging from the stray interwinding capacitance. They were probably constant-resistance coils in disguise. Capacitance to ground of the coil body is always a significant factor also.

Interstage Peaking

The Tektronix L and K units of the '50s were good examples of interstage T-coil peaking. The T-coils were used to peak, not the preamp input or the output, but in the middle of the amplifier. The interstage bandwidth was boosted well above the

$$f_{interstage} = \frac{1}{2\pi R_L C_{total}} = \frac{g_m}{gain\ 2\pi C_{total}} < \frac{g_m}{gain\ 2\pi C_{subtotal}} = \frac{f_r}{gain}$$

The individual pre-amp bandwidths are 60MHz. This is amazing because the effective f_t of the tubes was only 200MHz or so. Both inductive peaking and f_t doubling techniques were needed to "hot rod" these plug-ins to this bandwidth.

T-Coils in Transistor Interstages

The 150MHz 454 evolved from the 50MHz 453 oscilloscope by adding distributed deflection plates to the cathode ray tube and, among other things, using a new output amplifier. This amplifier employed T-coil peaking in the interstages. The T-coil design was based on a lossless virtual capacitance, a very big approximation. This virtual capacitance at the base was dominated by the transformation of the emitter feedback admittance into the base. The emitter feedback cascode connection made two transistors function more like a pentode. The initial use of transistors in the early '60s showed us that, most of the time, vacuum tube techniques didn't work with "those blasted transistors." After all, vacuum tubes had a physical capacitance that was measurable on an "off" tube; transistors had this "virtual capacitance thing"! The conventional thinking in the design groups at Tek in the early and mid '60s was that inductive peaking and transistor high-fidelity pulse amplifiers were not compatible. Despite this, the T-coils and transistors did work, the 454 worked, and the 454 was a "cash cow" for Tektronix for several years. Since then, ICs have displaced discrete transistors and the 'scope bandwidths translated upwards, with and without T-coils. The fastest amplifiers, however, are always produced with the aid of some T-coil configuration.

Figure 10–9.
Tektronix 454
Vertical-Output
Amplifier and
Interstage T-coil.

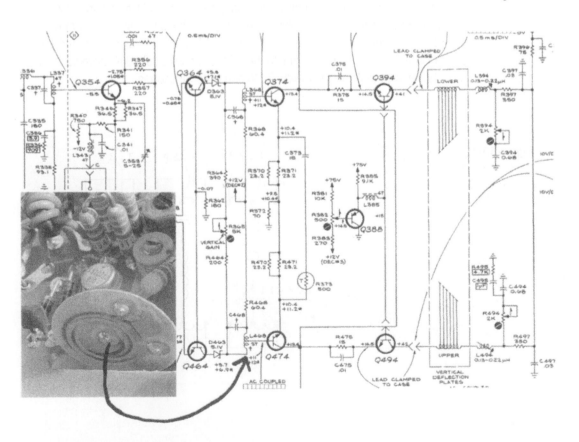

Phase Compensation with T-Coils

The portable 453 needed a compact delay line for the vertical system that didn't require tuning. Kobbe had designed and developed a balanced-counterwound delay line for the 580 series of 'scopes. We made it still smaller. This delay line worked well at 50MHz, and had reasonably low loss at 150MHz. Unfortunately, the step response revealed a preshoot problem. The explanation in the frequency domain is nonlinear phase response. High-frequency delay was insufficient, and one could see it as preshoot in the step response. Three sections of a constant-resistance-balanced T-coil structure added enough high-frequency delay to clean up the preshoot, and even speed the risetime by moving high frequencies into their "proper time slot." T-coil sections can provide delay boost at high frequencies if the T-coil section is proportioned differently from that of the peaking application. A negative value for "k" is usually appropriate and is realized by adding a separate inductor in the common leg.

Integrated Circuits

In the late '60s, when the 454A was being developed, George Wilson, head of the new Tektronix Integrated Circuits Group at that time, wanted to promote the design of an integrated circuit vertical amplifier. I rebuffed him, saying, "We can never use ICs in vertical amplifiers because they have too much substrate capacitance, too much collector resistance, and too low an f_t." I was correct at the time, but dead wrong in the long run. In the '70s, Tektronix pushed IC development in parallel with the high-bandwidth 7000 series oscilloscopes.

Figure 10–10.
Correcting
Insufficient High-
Frequency Delay.

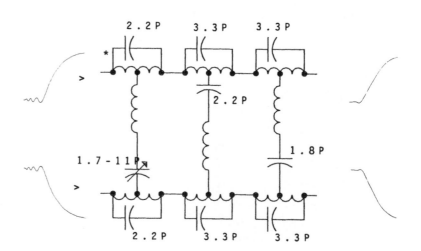

I stopped my slide into obsolescence in 1971 by doing a little downward mobility. I left the small portable oscilloscope group I headed, and joined George Wilson in the IC group as a designer. This foresight on my part was most uncharacteristic.

....

T-Coils with Integrated Circuit Vertical Amplifiers

The initial use of integrated circuits in the vertical amplifiers of Tektronix 'scopes supplied a huge bandwidth boost, but not just because of the high f_t. New processes included thin film resistors that allowed designers to put the small value emitter feedback resistors on the chip, thus eliminating the connection inductance in the emitters of transistors. That emitter inductance had made a brick wall limit in bandwidth for discrete transistor amplifiers. That wall was pretty steep, starting in the 150–200MHz area. In order to have flat, non ripple, frequency response at VHF and UHF, the separately packaged vertical amplifier stages needed to operate in a terminated transmission line environment. T-coils were vital to achieve this environment. Thor Hallen derived formulas for a minimum VSWR T-coil. Packaging and bond wire layout made constant-resistance T-coil design impossible. Hallen's T-coil incorporated and enhanced the base connection inductance. The Tektronix 7904 achieved 500MHz bandwidth by using all of the above, along with 3GHz transistors and an ft-doubler amplifier circuit configuration.

In 1979, the 1GHz 7104 employed many of the 7904 techniques but, in addition, had 8GHz f_t transistors, thin film conductors on substrates, and a package design having transmission line interconnects. It also had a much more sensitive cathode ray tube. Robert Ross had earlier developed formulas for a constant-resistance T-coil to drive a non-pure capacitor (a series capacitor-resistance combination). John Addis and Winthrop Gross made use of the Ross type T-coils (patterned with the thin film conductor) to successfully peak the stages and terminate the inter-chip transmission lines.

I have lumped Thor Hallen's and Bob Ross's T-coils together in a class I call "lossy capacitor T-coils."

Dual Channel Hybrid with T-Coils

In 1988, the digitizing 1GHz Tektronix 11402 was introduced. A fast real-time cathode ray tube deflection amplifier was no longer needed. T-coils were employed, however, in the 11A72 dual-channel plug-in pre-amp hybrid (Figure 10–12), where all of the two-channel analog signal processing took place. The T-coils peaked frequency response and minimized input reflections in the 50 Ohm input system. As in the 7904 'scope, Hallen used a design technique for the T-coils that minimized VSWR. To realize this schematic, a T-coil was needed which had

Two Types of Lossy Capacitor T-coils

ROSS CONSTANT-RESISTANCE T-COIL

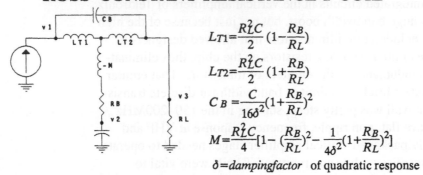

$$L_{T1}=\frac{R_L^2 C}{2}(1-\frac{R_B}{R_L})$$

$$L_{T2}=\frac{R_L^2 C}{2}(1+\frac{R_B}{R_L})$$

$$C_B=\frac{C}{16\delta^2}(1+\frac{R_B}{R_L})^2$$

$$M=\frac{R_L^2 C}{4}[1-(\frac{R_B}{R_L})^2-\frac{1}{4\delta^2}(1+\frac{R_B}{R_L})^2]$$

$\delta=dampingfactor$ of quadratic response

$\dfrac{v_1}{i}=R_L$ The Constant-Resistance property

$\dfrac{v_2}{i_{in}}=\dfrac{R_L}{1+\dfrac{(R_L+R_B)}{2}Cs+R_L^2 CC_B s^2}$ **Two Pole Response**

HALLEN MINIMUM VSWR T-COIL

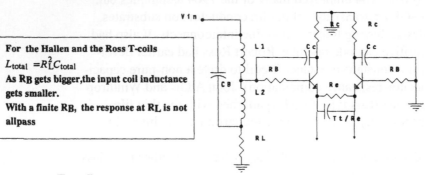

For the Hallen and the Ross T-coils
$L_{total}=R_L^2 C_{total}$
As R_B gets bigger, the input coil inductance gets smaller.
With a finite R_B, the response at R_L is not allpass

$$L_{total}=R_L^2[\frac{T_T}{R_e}+(\frac{R_c}{R_e}+1)\,C_c]$$

$$L_1=\frac{L_{total}}{2}\,[\,1+\frac{1}{R_L}\,(\,\frac{R_e C_c T_t(R_c+2R_B)}{(T_t+R_e C_c+R_c C_c)^2}-\frac{2R_B T_t+R_e R_c C_c}{T_t+R_e C_c+R_c C_c}\,)\,]$$

$$L_2=L_{total}-L_1$$

$$C_B=\frac{1}{R_L^2}\,[\,\frac{R_L C_c T_t(R_c+2R_B)(L_1-L_2)}{(T_t+R_e C_c+R_c C_c)(L_1+L_2)}+\frac{2R_B R_c C_c T_t}{T_t+R_e C_c+R'_c C_c}+\frac{L_1 L_2}{L_1+L_2}\,]$$

Figure 10–11.
Two Types of Lossy
Capacitor T-coils.

enough mutual inductance to cancel the bond wire inductance that would be in series with its center tap. The remaining net branch inductances then had to match Hallen's values. To guide the physical layout of this coil, I used a three-dimensional inductance calculation program. This program was used iteratively. The two "G" patterns on the multilayer thick film hybrid are the top layer of these input T-coils. The major dimension of these coils is 0.05 inches. In between the chips are coils which "tune out" the collector capacitance of the transistor of each output channel. These coils are formed by multiple-layer runs and bond wire "loopbacks."

Afterglow

Conspicuous by its absence is a discussion of wideband amplifier configurations and how they operate. I have referred to f_t-doublers and current doublers without explanation. I had to really restrain myself to avoid that topic for the sake of brevity. The ultimate bandwidth limit of high-fidelity pulse amplifiers depends on the power gain capability (expressed by an f_{MAX}, for example) of the devices, and the power gain requirements of the amplifier. To approach this ultimate goal requires the sophisticated use of inductors to shape the response. For bipolar transistors, the f_t-doubler configurations and single-stage feedback amplifiers, combined with inductive peaking, do a very good job.

I hope this chapter has raised your curiosity about the circuit applications of the T-coil section. I have not written this chapter like a textbook

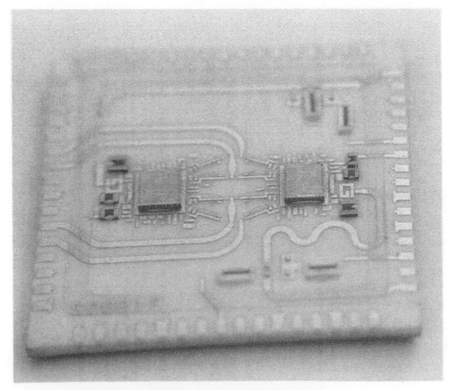

Figure 10–12.
11A72 1.5GHz
Multilayer Hybrid
with Thick Film
T-coils.

and I am hoping that my assertions and derivation results are challenged by the reader. To get really radical, breadboard a real circuit! A less fun but easier way to verify circuit behavior is via SPICE or a similar simulator program. Keep in mind, while you are doing this, that most of the very early design took place without digital computer simulators. Frequency- and impedance-scaled simulations took place though, with physical analog models.

I'm grateful to the many knowledgeable folks who talked with me recently and added considerable information, both technical and historical. These included Gene Andrews, Phil Crosby, Logan Belleville, Dean Kidd, John Kobbe, Jim Lamb, Cliff Moulton, Oscar Olson, Ron Olson, and Richard Rhiger. If this chapter has errors, however, don't blame these guys; any mistakes are my own.

Bob Ross and Thor Hallen have been sources of insight on these topics over many years and have been ruthless in their rigorous analyses, helping me in my work immensely.

Finally, I leave you with my mother's and Socrates' advice, "Moderation in all things." Might I add, "Just do it!" If these Tek guys had waited for proper models of all known effects and proper theory before doing something, we would still be waiting. Everything can be tidied up in hindsight but, in fact, the real circuits in the real products are often more complicated than our simple schematics and were realized by a lot of theory, intuition, and especially smart, hard, and sometimes long work. I am proud of all of this heritage and the small part I played in it.

11. Tripping the Light Fantastic

Introduction

Where do good circuits come from, and what is a good circuit? Do they only arrive as lightning bolts in the minds of a privileged few? Are they synthesized, or derived after careful analysis? Do they simply evolve? What is the role of skill? Of experience? Of luck? I can't answer these weighty questions, but I do know how the best circuit I ever designed came to be.

What is a good circuit, anyway? Again, that's a fairly difficult question, but I can suggest a few guidelines. Its appearance should be fundamentally simple, although it may embody complex and powerful theoretical elements and interactions. That, to me, is the essence of elegance. The circuit should also be widely utilized. An important measure of a circuit's value is if lots of people use it, and are satisfied after they have done so. Finally, the circuit should also generate substantial revenue. The last time I checked, they still charge money at the grocery store. My employer is similarly faithful about paying me, and, in both cases, it's my obligation to hold up my end of the bargain.

So, those are my thoughts on good circuits, but I never addressed the statement at the end of the first paragraph. How did my best circuit come to be? That's a long story. Here it is.

The Postpartum Blues

Towards the end of 1991 I was in a rut. I had finished a large high-speed amplifier project in August. It had required a year of constant, intense, and sometimes ferocious effort right up to its conclusion. Then it was over, and I suddenly had nothing to do. I have found myself abruptly disconnected from an absorbing task before, and the result is always the same. I go into this funky kind of rut, and wonder if I'll ever find anything else interesting to do, and if I'm even capable of doing anything anymore.

Portions of this text have appeared in the January 6, 1994 issue of *EDN* magazine and publications of Linear Technology Corporation. They are used here with permission.

I've been dating me a long time, so this state of mind doesn't promote quite the panic and urgency it used to. The treatment is always the same. Keep busy with mundane chores at work, read, cruise electronic junk stores, fix things and, in general, look available so that some interesting problem might ask me to dance. During this time I can do some of the stuff I completely let go while I was immersed in whatever problem owned me. The treatment always seems to work, and usually takes a period of months. In this case it took exactly three.

What's a Backlight?

Around Christmas my boss, Bob Dobkin, asked me if I ever thought about the liquid crystal display (LCD) backlights used in portable computers. I had to admit I didn't know what a backlight was. He explained that LCD displays require an illumination source to make the display readable, and that this source consumed about half the power in the machine. Additionally, the light source, a form of fluorescent lamp, requires high-voltage, high-frequency AC drive. Bob was wondering how this was done, with what efficiency, and if we couldn't come up with a better way and peddle it. The thing sounded remotely interesting. I enjoy transducer work, and that's what a light bulb is. I thought it might be useful to get my hands on some computers and take a look at the backlights. Then I went off to return some phone calls, attend to other housekeeping type items, and, basically, maintain my funk.

A Call from Some Guy Named Steve

Three days later the phone rang. The caller, a guy named Steve Young from Apple Computer, had seen a cartoon (Figure 11–1) I stuck on the back page of an application note in 1989. Since the cartoon invited calls, he was doing just that. Steve outlined several classes of switching power supply problems he was interested in. The application was portable computers, and a more efficient backlight circuit was a priority. Dobkin's interest in backlights suddenly sounded a lot less academic.

This guy seemed like a fairly senior type, and Apple was obviously a prominent computer company. Also, he was enthusiastic, seemed easy to work with and quite knowledgeable. This potential customer also knew what he wanted, and was willing to put a lot of front end thinking and time in to get it. It was clear he wasn't interested in a quick fix; he wanted true, "end-to-end" system oriented thinking.

What a customer! He knew what he wanted. He was open and anxious to work, had time and money, and was willing to sweat to get better solutions. On top of all that, Apple was a large and successful company with excellent engineering resources. I set up a meeting to introduce him to Dobkin and, hopefully, get something started.

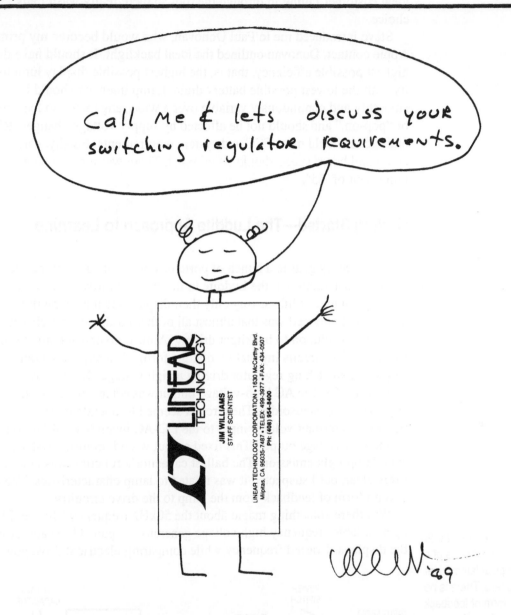

Figure 11–1.
This invitation appeared in a 1989 application note. Some guy named Steve Young from Apple Computer took me up on it. (Reproduced with permission of Linear Technology Corporation)

The meeting went well, things got defined, and I took the backlight problem. I still wasn't enthralled with backlights, but here was an almost ideal customer falling in through the roof so there really wasn't any choice.

Steve introduced me to Paul Donovan, who would become my primary Apple contact. Donovan outlined the ideal backlight. It should have the highest possible efficiency, that is, the highest possible display luminosity with the lowest possible battery drain. Lamp intensity should be smoothly and continuously variable over a wide range with no hysteresis, or "pop-on," and should not be affected by supply voltage changes. RF emissions should meet FCC and system requirements. Finally, parts count and board space should be minimal. There was a board height requirement of .25".

Getting Started—The Luddite Approach to Learning

I got started by getting a bunch of portable computers and taking them apart. I must admit that the Luddite in me enjoyed throwing away most of the computers while saving only their display sections. One thing I immediately noticed was that almost all of them utilized a purchased, board-level solution to backlight driving. Almost no one actually built the function. The circuits invariably took the form of an adjustable output step-down switching regulator driving a high voltage DC-AC inverter (Figure 11–2). The AC high-voltage output was often about 50kHz, and approximately sinusoidal. The circuits seemed to operate on the assumption that a constant voltage input to the DC-AC inverter would produce a fixed, high voltage output. This fixed output would, in turn, produce constant lamp light emission. The ballast capacitor's function was not entirely clear, but I suspected it was related to lamp characteristics. There was no form of feedback from the lamp to the drive circuitry.

Was there something magic about the 50kHz frequency? To see, I built up a variable-frequency high voltage generator (Figure 11–3) and drove the displays. I varied frequency while comparing electrical drive power

Figure 11–2. Architecture of a typical lamp driver board. There is no form of feedback from the lamp.

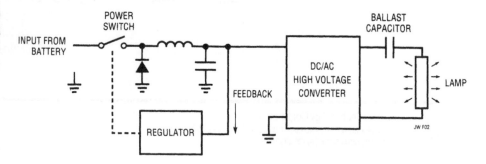

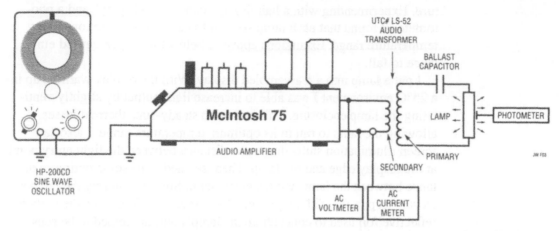

Figure 11–3.
Variable frequency
high-voltage test
setup for evaluating
lamp frequency
sensitivity.

to optical emission. Lamp conversion efficiency seemed independent of frequency over a fairly wide range. I did, however, notice that higher frequencies tended to introduce losses in the wiring running to the lamp. These losses occurred at all frequencies, but became pronounced above about 100kHz or so. Deliberately introducing parasitic capacitances from the wiring or lamp to ground substantially increased the losses. The lesson was clear. The lamp wiring was an inherent and parasitic part of the circuit, and any stray capacitive path was similarly parasitic.

Armed with this information I returned to the computer displays. I modified things so that the wire length between the inverter board and display was minimized. I also removed the metal display housing in the lamp area. The result was a measurable decrease in inverter drive power for a given display intensity. In two machines the improvement approached 20%! My modifications weren't very practical from a mechanical integrity viewpoint, but that wasn't relevant. Why hadn't these computers been originally designed to take advantage of this "free" efficiency gain?

Playing around with Light Bulbs

I removed lamps from the displays. They all appeared to have been installed by the display vendor, as opposed to being selected and purchased by the computer manufacturer. Even more interesting was that I found identical backlight boards in different computers driving different types of lamps. There didn't seem to be any board changes made to accommodate the various lamps. Now, I turned my attention to the lamps.

The lamps seemed to be pretty complex and wild animals. I noticed that many of them took noticeable time to arrive at maximum intensity. Some types seemed to emit more light than others for a given input power. Still others had a wider dynamic range of intensities than the rest, although all had a seemingly narrow range of intensity control. Most striking was that every lamp's emissivity varied with ambient temperature. Experimenting with a hair dryer, a can of "cold spray" and a photometer, I found that each lamp seemed to have an optimum operating temperature range. Excursions above or below this region caused emittance to fall.

I put a lamp into a reassembled display. With the display warmed up in a 25°C environment I was able to increase light output by slightly ventilating the lamp enclosure. This increased steady-state thermal losses, allowing the lamp to run in its optimum temperature range. I also saw screen illumination shifts due to the distance between the light entry point at the display edge and the lamp. There seemed to be some optimum distance between the lamp and the entry point. Simply coupling the lamp as closely as possible did not provide the best results. Similarly, the metallic reflective foil used to concentrate the lamp's output seemed to be sensitive to placement. Additionally, there was clearly a trade-off between benefits from the foil's optical reflection and its absorption of high voltage field energy. Removing the foil decreased input energy for a given lamp emission level. I could watch input power rise as I slipped the foil back along the lamp's length. In some cases, with the foil fully replaced, I could draw sparks from it with my finger!

I also assembled lamps, displays, and inverter boards in various unoriginal combinations. In some cases I was able to increase light output, at lower input power drain, over the original "as shipped" configuration.

Grandpa Would Have Liked It

I tried a lot of similarly simple experiments and slowly developed a growing suspicion that nobody, at least in my sample of computers, was making any serious attempt at optimizing (or they did not know how to optimize) the backlight. It appeared that most people making lamps were simply filling tubes up with gas and shipping them. Display manufacturers were dropping these lamps into displays and shipping them. Computer vendors bought some "backlight power supply" board, wired it up to the display, took whatever electrical and optical efficiency they got, and shipped the computer.

If I allowed this conclusion, several things became clear. Development of an efficient backlight required an interdisciplinary approach to address a complex problem. There was worthwhile work to be done. I could contribute to the electronic portion, and perhaps the thermal design, but the optical engineering was beyond me. It was not, however, beyond Apple's resources. Apple had some very good optical types. Working together, it seemed we had a chance to build a better backlight with its attendant display quality and battery life advantages. Apple would get a more saleable product and my company would develop a valued customer. And, because the whole thing was beginning to get interesting, I could get out of my rut. The business school types would call this "synergistic" or "win-win." Other people who "do lunch" a lot on company money would call it "strategic partnering." My grandfather would have called it "such a deal."

Goals for the backlight began to emerge. For best overall efficiency, the display enclosure, optical design, lamp, and electronics had to be simultaneously considered. My job was the electronics, although I met regularly with Paul Donovan, who was working on the other issues. In particular, I was actively involved in setting lamp specifications and evaluating lamp vendors.

The electronics should obviously be as efficient as possible. The circuit should be physically compact, have a low parts count, and assemble easily. It should have a wide, continuous dimming range with no hysteresis or "pop-on," and should meet all RF and system emission requirements. Finally, it must regulate lamp intensity against wide power supply shifts, such as when the computer's AC adapter is plugged in.

Help from Dusty Circuits

Where, I wondered, had I seen circuitry which contained any or all of these characteristics? Nowhere. But, one place to start looking was oscilloscopes. Although oscilloscope circuits do not accomplish what I needed to do, oscilloscope designers use high frequency sine wave conversion to generate the high voltage CRT supply. This technique minimizes noise and reduces transformer and capacitor size. Additionally, by doing the conversion at the CRT, long high voltage runs from the main power supply are eliminated.

I looked at the schematic of the high voltage converter in a Tektronix 547 (Figure 11–4). The manual's explanation (Figure 11–5) says the capacitor (C808) and transformer primary form a resonant tank circuit. More subtly, the "transformer primary" also includes the complex impedance reflected back from the secondary and its load. But that's a detail for this circuit and for now. A CRT is a relatively linear and benign load. The backlight's loading characteristics would have to be evaluated and matched to the circuit.

This CRT circuit could not be used to drive a fluorescent backlight tube in a laptop computer. For one reason, this circuit is not very efficient. It does not have to be. A 547 pulls over 500 watts, so efficiency in this circuit was not a big priority. Latter versions of this configuration were transistorized (Figure 11–6, Tektronix 453), but used basically the same architecture. In both circuits the resonating technique is employed, and a feedback loop enforces voltage regulation. For another reason, the CRT requires the high voltage to be rectified to DC. The backlight requires AC, eliminating the rectifier and filter. And, the CRT circuit had no feedback. Some form of feedback for the fluorescent lamp seemed desirable.

The jewel in the CRT circuit, however, was the resonating technique used to create the sine wave. The transformer does double duty. It helps create the sine wave while simultaneously generating the high voltage.

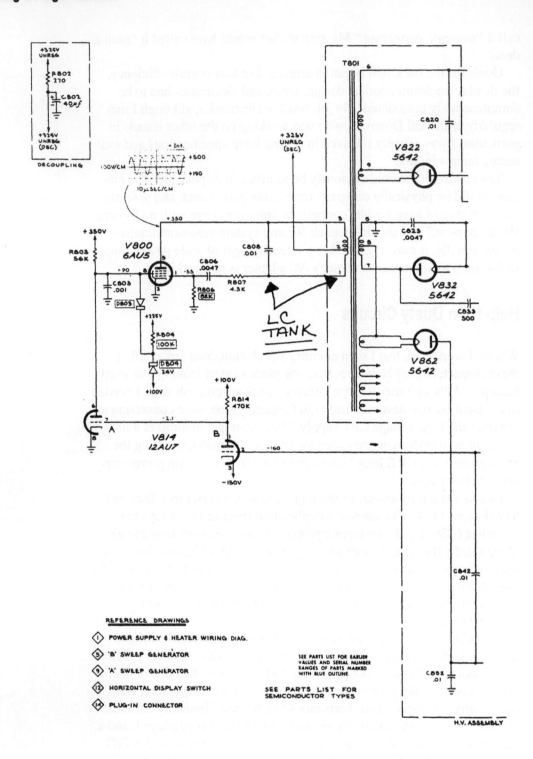

REFERENCE DRAWINGS

1. POWER SUPPLY & HEATER WIRING DIAG.
5. 'B' SWEEP GENERATOR
9. 'A' SWEEP GENERATOR
12. HORIZONTAL DISPLAY SWITCH
14. PLUG-IN CONNECTOR

SEE PARTS LIST FOR EARLIER VALUES AND SERIAL NUMBER RANGES OF PARTS MARKED WITH BLUE OUTLINE.

SEE PARTS LIST FOR SEMICONDUCTOR TYPES

TYPE 547 OSCILLOSCOPE

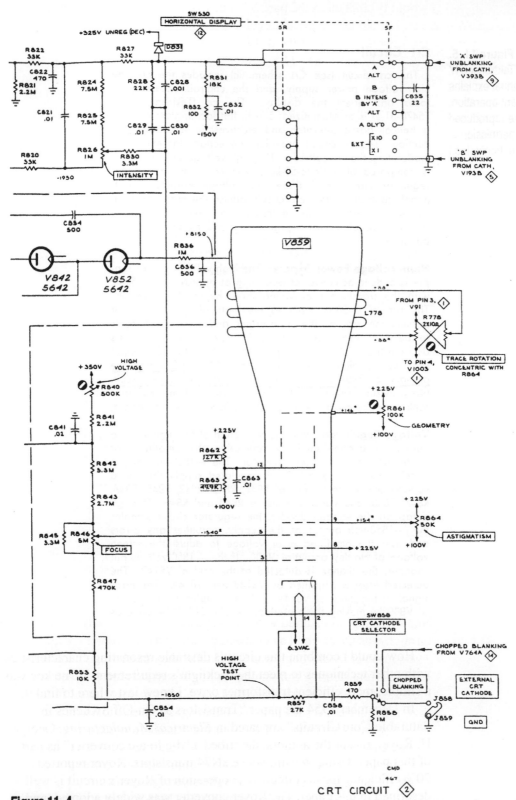

Figure 11–4.
CRT supply used in Tektronix 547. C808 resonates with transformer, creating sine wave drive. (Figure reproduced with permission of Tektronix, Inc.)

Figure 11–5.
Tektronix 547
manual explains
resonant operation.
(Figure reproduced
with permission of
Tektronix, Inc.)

Crt Circuit

The crt circuit (see Crt schematic) includes the crt, the high-voltage power supply, and the controls necessary to focus and orient the display. The crt (Tektronix Type T5470-31-2) is an aluminized, 5-inch, flat-faced, glass crt with a helical post-accelerator and electrostatic focus and deflection. The crt circuit provides connections for externally modulating the crt cathode. The high-voltage power supply is composed of a dc-to-50-kc power converter, a voltage-regulator circuit, and three high-voltage outputs. Front-panel controls in the crt circuit adjust the trace rotation (screwdriver adjustment), intensity, focus, and astigmatism. Internal controls adjust the geometry and high-voltage output level.

High-Voltage Power Supply. The high-voltage power supply is a dc-to-ac converter operating at approximately 50 kc with the transformer providing three high-voltage outputs. The use of a 50-kc input to the high-voltage transformer permits the size of the transformer and filter components to be kept small. A modified Hartley oscillator converts dc from the +325-volt unregulated supply to the 50-kc input required by high-voltage transformer T801. C808 and the primary of T801 form the oscillator resonant tank circuit. No provisions are made for precise tuning of the oscillator tank since the exact frequency of oscillation is not important.

Voltage Regulation. Voltage regulation of the high-voltage outputs is accomplished by regulating the amplitude of oscillations in the Hartley oscillator. The —1850-volt output is referenced to the +350-volt regulated supply through a voltage divider composed of R841, R842, R843, R845, R846, R847, R853, and variable resistors R840 and R846. Through a tap on the voltage divider, the regulator circuit samples the —1850-volt output of the supply, amplifies any errors and uses the amplified error voltage to adjust the screen voltage of Hartley oscillator V800. If the —1850-volt output changes, the change is detected at the grid of V814B. The detected error is amplified by V814B and V814A. The error signal at the plate of V814A is direct coupled to the screen of V800 by making the plate-load resistor of V814A serve as

How could I combine this circuit's desirable resonating characteristics with other techniques to meet the backlight's requirements? One key was a simple, more efficient transformer drive. I knew just where to find it.

In December 1954 the paper "Transistors as On-Off Switches in Saturable-Core Circuits" appeared in *Electrical Manufacturing*. George H. Royer, one of the authors, described a "d-c to a-c converter" as part of this paper. Using Westinghouse 2N74 transistors, Royer reported 90% efficiency for his circuit. The operation of Royer's circuit is well described in this paper. The Royer converter was widely adopted, and used in designs from watts to kilowatts. It is still the basis for a wide variety of power conversion.

Royer's circuit is not an LC resonant type. The transformer is the sole energy storage element and the output is a square wave. Figure 11–7 is a conceptual schematic of a typical converter. The input is applied to a self-oscillating configuration composed of transistors, a transformer, and a biasing network. The transistors conduct out of phase switching (Figure 11–8: Traces A and C are Q1's collector and base, while Traces B and D are Q2's collector and base) each time the transformer saturates. Transformer saturation causes a quickly rising, high current to flow (Trace E).

This current spike, picked up by the base drive winding, switches the transistors. This phase opposed switching causes the transistors to exchange states. Current abruptly drops in the formerly conducting transistor and then slowly rises in the newly conducting transistor until saturation again forces switching. This alternating operation sets transistor duty cycle at 50%.

The photograph in Figure 11–9 is a time and amplitude expansion of Figure 11–8's Traces B and E. It clearly shows the relationship between transformer current (Trace B, Figure 11–9) and transistor collector voltage (Trace A, Figure 11–9).[1]

The Royer has many desirable elements which are applicable to backlight driving. Transformer size is small because core utilization is efficient. Parts count is low, the circuit self-oscillates, it is efficient, and output power may be varied over a wide range. The inherent nature of operation produces a square wave output, which is not permissible for backlight driving.

Adding a capacitor to the primary drive (Figure 11–10) should have the same resonating effect as in the Tektronix CRT circuits. The beauty of this configuration is its utter simplicity and high efficiency. As loading (e.g., lamp intensity) is varied the reflected secondary impedance changes, causing some frequency shift, but efficiency remains high.

The Royer's output power is controllable by varying the primary drive current. Figure 11–11 shows a way to investigate this. This circuit works well, except that the transistor current sink operates in its linear region, wasting power. Figure 11–12 converts the current sink to switch mode operation, maintaining high efficiency. This is obviously advantageous to the user, but also a good deal for my employer. I had spent the last six months playing with light bulbs, reminiscing over old oscilloscope circuits, taking arcane thermal measurements, and similar dalliances. All the while faithfully collecting my employer's money. Finally, I had found a place to actually sell something we made. Linear Technology (my employer) builds a switching regulator called the LT1172. Its features include a high power open collector switch, trimmed reference, low quiescent current, and shutdown capability. Additionally, it is available in an 8 pin surface-mount package, a must for board space considerations. It was also an ideal candidate for the circuit's current sink portion.

---

1. The bottom traces in both photographs are not germane and are not referenced in the discussion.

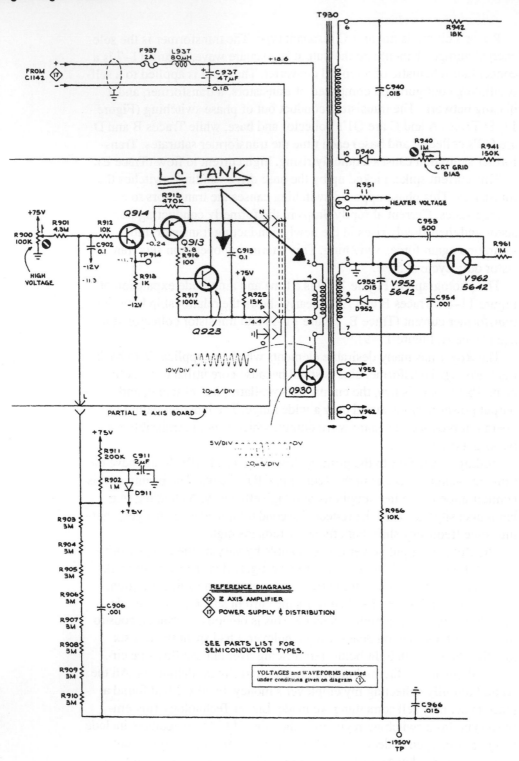

LC TANK

PARTIAL Z AXIS BOARD

REFERENCE DIAGRAMS

⬢15 Z AXIS AMPLIFIER

⬢17 POWER SUPPLY & DISTRIBUTION

SEE PARTS LIST FOR
SEMICONDUCTOR TYPES.

VOLTAGES and WAVEFORMS obtained
under conditions given on diagram ①.

TYPE 453 OSCILLOSCOPE

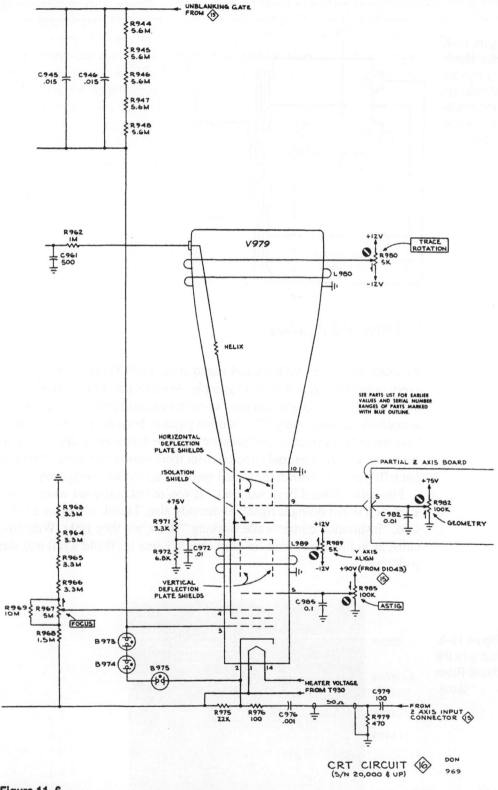

Figure 11–6.
Later model Tektronix 453 is transistorized version of 547's resonant approach. (Figure reproduced with permission of Tektronix, Inc.)

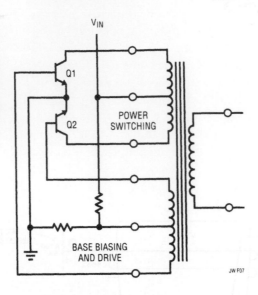

Figure 11–7.
Conceptual classic
Royer converter.
Transformer ap-
proaching satura-
tion causes
switching.

Of Rafts and Paddles

At about this stage I sat back and stared at the wall. There comes a time in every project where you have to gamble. At some point the analytics and theorizing must stop and you have to commit to an approach and start actually doing something. This is often painful, because you never really have enough information and preparation to be confidently decisive. There are never any answers, only choices. But there comes this time when your gut tells you to put down the pencil and pick up the soldering iron.

Physicist Richard Feynman said, "If you're not confused when you start, you're not doing it right." Somebody else, I think it was an artist, said, "Inspiration comes while working." Wow, are they right. With circuits, as in life, never wait for your ship to come in. Build a raft and start paddling.

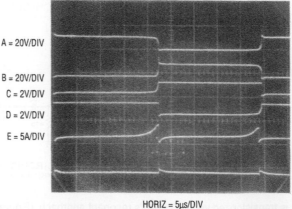

Figure 11–8.
Waveforms for the
classic Royer
circuit.

A = 20V/DIV

B = 20V/DIV

C = 2V/DIV

D = 2V/DIV

E = 5A/DIV

HORIZ = 5µs/DIV

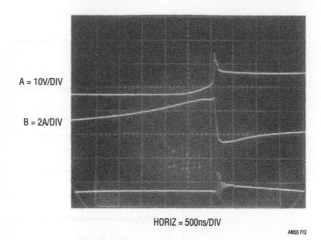

A = 10V/DIV

B = 2A/DIV

HORIZ = 500ns/DIV

AN55 F13

Figure 11–9.
Detail of transistor switching. Turn-off (Trace A) occurs just as transformer heads into saturation (Trace B).

Everything was still pretty fuzzy, but I had learned a few things. A practical, highly efficient LCD backlight design is a classic study of compromise in a transduced electronic system. Every aspect of the design is interrelated, and the physical embodiment is an integral part of the electrical circuit. The choice and location of the lamp, wires, display housing, and other items have a major effect on electrical characteristics. The greatest care in every detail is required to achieve a practical, high efficiency LCD backlight. Getting the lamp to light is just the beginning!

A good place to start was to reconsider the lamps. These "Cold Cathode Fluorescent Lamps" (CCFL) provide the highest available efficiency for converting electrical energy to light. Unfortunately, they are optically and electrically highly nonlinear devices.

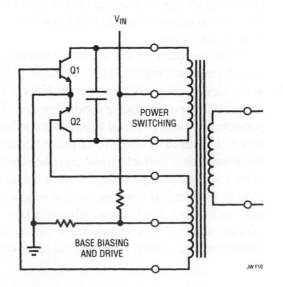

V_IN

Q1

Q2

POWER SWITCHING

BASE BIASING AND DRIVE

JW F10

Figure 11–10.
Adding the resonating capacitor to the Royer.

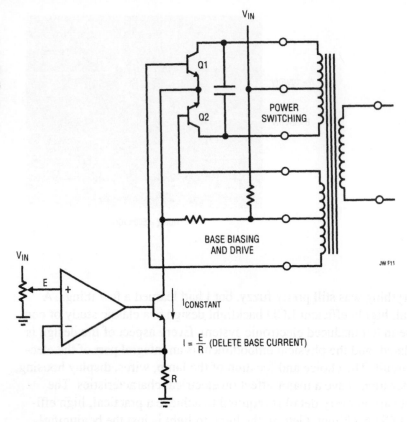

Figure 11–11.
Current sink permits controlling Royer power, but is inefficient.

$I = \dfrac{E}{R}$ (DELETE BASE CURRENT)

Cold Cathode Fluorescent Lamps (CCFLs)

Any discussion of CCFL power supplies must consider lamp characteristics. These lamps are complex transducers, with many variables affecting their ability to convert electrical current to light. Factors influencing conversion efficiency include the lamp's current, temperature, drive waveform characteristics, length, width, gas constituents, and the proximity to nearby conductors.

These and other factors are interdependent, resulting in a complex overall response. Figures 11–13 through 11–16 show some typical characteristics. A review of these curves hints at the difficulty in predicting lamp behavior as operating conditions vary. The lamp's current and temperature are clearly critical to emission, although electrical efficiency may not necessarily correspond to the best optical efficiency point. Because of this, both electrical and photometric evaluation of a circuit is often required. It is possible, for example, to construct a CCFL circuit with 94% electrical efficiency which produces less light output than an approach with 80% electrical efficiency (see Appendix C, "A Lot of Cut-off Ears and No Van Goghs—Some Not-So-Great Ideas"). Similarly, the performance of a very well matched lamp-circuit combination can be

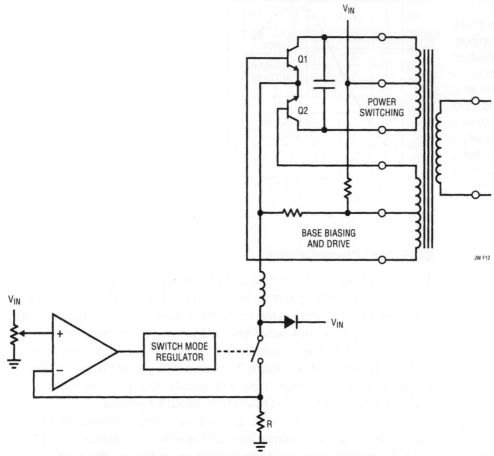

POWER
SWITCHING

Q1

Q2

BASE BIASING
AND DRIVE

JW F12

V_{IN}

SWITCH MODE
REGULATOR

R

Figure 11–12.
Switched mode
current sink re-
stores efficiency.

severely degraded by a lossy display enclosure or excessive high voltage
wire lengths. Display enclosures with too much conducting material near
the lamp have huge losses due to capacitive coupling. A poorly designed
display enclosure can easily degrade efficiency by 20%. High voltage
wire runs typically cause 1% loss per inch of wire.

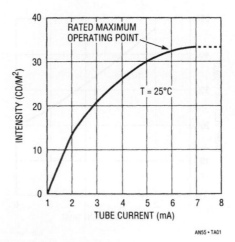

Figure 11–13.
Emissivity for a
typical 6mA lamp;
curve flattens badly
above 6mA.

AN55 • TA01

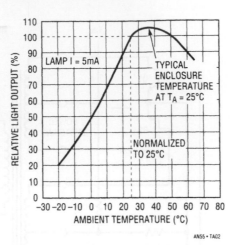

Figure 11–14.
Ambient temperature effects on emissivity of a typical 5mA lamp. Lamp and enclosure must come to thermal steady state before measurements are made.

CCFL Load Characteristics

These lamps are a difficult load to drive, particularly for a switching regulator. They have a "negative resistance" characteristic; the starting voltage is significantly higher than the operating voltage. Typically, the start voltage is about 1000V, although higher and lower voltage lamps are common. Operating voltage is usually 300V to 400V, although other lamps may require different potentials. The lamps will operate from DC, but migration effects within the lamp will quickly damage it. As such, the waveform must be AC. No DC content should be present.

Figure 11–17A shows an AC driven lamp's characteristics on a curve tracer. The negative resistance induced "snapback" is apparent. In Figure 11–17B, another lamp, acting against the curve tracer's drive, produces oscillation. These tendencies, combined with the frequency compensation problems associated with switching regulators, can cause severe loop instabilities, particularly on start-up. Once the lamp is in its operating region it assumes a linear load characteristic, easing stability criteria. Lamp operating frequencies are typically 20kHz to 100kHz and a sine-

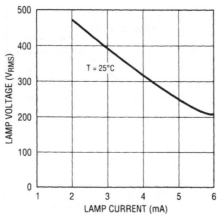

Figure 11–15.
Current vs. voltage for a lamp in the operating region.

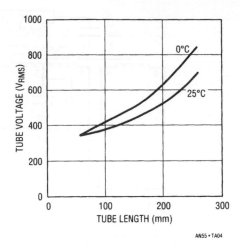

Figure 11–16.
Running voltage vs.
lamp length at two
temperatures.
Start-up voltages
are usually 50% to
200% higher over
temperature.

like waveform is preferred. The sine drive's low harmonic content minimizes RF emissions, which could cause interference and efficiency
degradation. A further benefit of the continuous sine drive is its low crest
factor and controlled risetimes, which are easily handled by the CCFL.
CCFL's RMS current-to-light output efficiency is degraded by high crest
factor drive waveforms.[2]

CCFL Power Supply Circuits

Figure 11–18's circuit meets CCFL drive requirements. Efficiency is
88% with an input voltage range of 4.5V to 20V. This efficiency figure
will be degraded by about 3% if the LT1172 V_{IN} pin is powered from the
same supply as the main circuit V_{IN} terminal. Lamp intensity is continuously and smoothly variable from zero to full intensity. When power is

Figure 11–17.
Negative resistance
characteristic for
two CCFL lamps.
"Snap-back" is
readily apparent,
causing oscillation
in 11–17B. These
characteristics
complicate power
supply design.

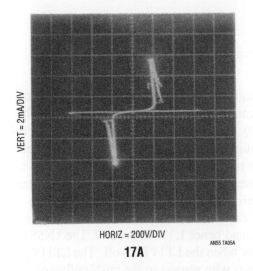

17A

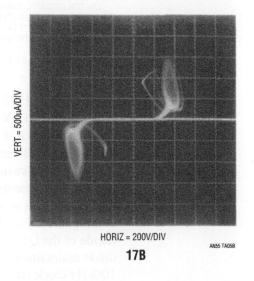

17B

2. See Appendix C, "A Lot of Cut-off Ears and No Van Goghs—Some Not-So-Great Ideas."

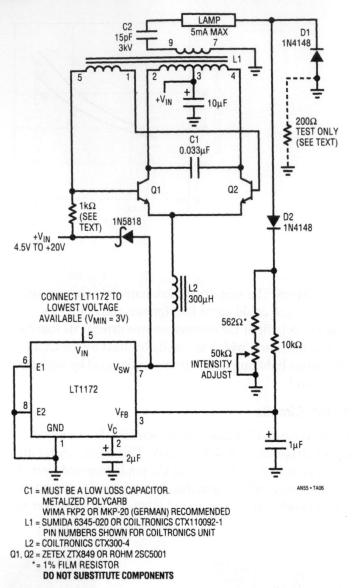

Figure 11–18.
An 88% efficiency cold cathode fluorescent lamp (CCFL) power supply.

C1 = MUST BE A LOW LOSS CAPACITOR.
   METALIZED POLYCARB
   WIMA FKP2 OR MKP-20 (GERMAN) RECOMMENDED
L1 = SUMIDA 6345-020 OR COILTRONICS CTX110092-1
   PIN NUMBERS SHOWN FOR COILTRONICS UNIT
L2 = COILTRONICS CTX300-4
Q1, Q2 = ZETEX ZTX849 OR ROHM 2SC5001
* = 1% FILM RESISTOR
DO NOT SUBSTITUTE COMPONENTS

COILTRONICS (305) 781-8900, SUMIDA (708) 956-0666

applied the LT1172 switching regulator's feedback pin is below the device's internal 1.2V reference, causing full duty cycle modulation at the V_{SW} pin (Trace A, Figure 11–19). L2 conducts current (Trace B) which flows from L1's center tap, through the transistors, into L2. L2's current is deposited in switched fashion to ground by the regulator's action.

L1 and the transistors comprise a current driven Royer class converter which oscillates at a frequency primarily set by L1's characteristics (including its load) and the .033μF capacitor. LT1172 driven L2 sets the magnitude of the Q1-Q2 tail current, and hence L1's drive level. The 1N5818 diode maintains L2's current flow when the LT1172 is off. The LT1172's 100kHz clock rate is asynchronous with respect to the push-pull converter's (60kHz) rate, accounting for Trace B's waveform thickening.

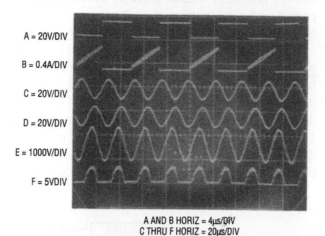

A = 20V/DIV

B = 0.4A/DIV

C = 20V/DIV

D = 20V/DIV

E = 1000V/DIV

F = 5VDIV

A AND B HORIZ = 4µs/DIV
C THRU F HORIZ = 20µs/DIV
TRIGGERS FULLY INDEPENDENT

AN55 TA07

Figure 11–19.
Waveforms for the cold cathode fluorescent lamp power supply. Note independent triggering on Traces A and B, and C through F.

The .033µF capacitor combines with L1's characteristics to produce sine wave voltage drive at the Q1 and Q2 collectors (Traces C and D, respectively). L1 furnishes voltage step-up, and about 1400V p-p appears at its secondary (Trace E). Current flows through the 15pF capacitor into the lamp. On negative waveform cycles the lamp's current is steered to ground via D1. Positive waveform cycles are directed, via D2, to the ground referred 562Ω–50k potentiometer chain. The positive half-sine appearing across the resistors (Trace F) represents ½ the lamp current. This signal is filtered by the 10k–1µF pair and presented to the LT1172's feedback pin. This connection closes a control loop which regulates lamp current. The 2µF capacitor at the LT1172's V_C pin provides stable loop compensation. The loop forces the LT1172 to switch-mode modulate L2's average current to whatever value is required to maintain a constant current in the lamp. The constant current's value, and hence lamp intensity, may be varied with the potentiometer. The constant current drive allows full 0%–100% intensity control with no lamp dead zones or "pop-on" at low intensities. Additionally, lamp life is enhanced because current cannot increase as the lamp ages. This constant current feedback approach contrasts with the open loop, voltage type drive used by other approaches. It greatly improves control over the lamp under all conditions.

This circuit's 0.1% line regulation is notably better than some other approaches. This tight regulation prevents lamp intensity variation when abrupt line changes occur. This typically happens when battery powered apparatus is connected to an AC powered charger. The circuit's excellent line regulation derives from the fact that L1's drive waveform never changes shape as input voltage varies. This characteristic permits the simple 10kΩ–1µF RC to produce a consistent response. The RC averaging characteristic has serious error compared to a true RMS conversion, but the error is constant and "disappears" in the 562Ω shunt's value. The base drive resistor's value (nominally 1kΩ) should be selected to provide full V_{CE} saturation without inducing base overdrive or beta starvation. A procedure for doing this is described in the following section, "General Measurement and Optimization Considerations."

Figure 11–20's circuit is similar, but uses a transformer with lower copper and core losses to increase efficiency to 91%. The trade-off is slightly larger transformer size. Value shifts in C1, L2, and the base drive resistor reflect different transformer characteristics. This circuit also features shutdown via Q3 and a DC or pulse width controlled dimming input. Figure 11–21, directly derived from Figure 11–20, produces 10mA output to drive color LCDs at 92% efficiency. The slight efficiency improvement comes from a reduction in LT1172 "housekeeping" current as a percentage

Figure 11–20.
A 91% efficient CCFL supply for 5mA loads features shutdown and dimming inputs.

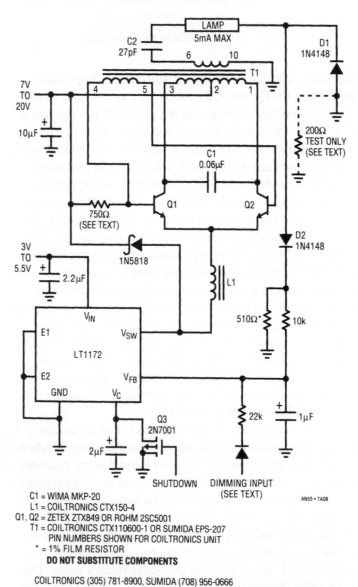

C1 = WIMA MKP-20
L1 = COILTRONICS CTX150-4
Q1, Q2 = ZETEX ZTX849 OR ROHM 2SC5001
T1 = COILTRONICS CTX110600-1 OR SUMIDA EPS-207
PIN NUMBERS SHOWN FOR COILTRONICS UNIT
* = 1% FILM RESISTOR
DO NOT SUBSTITUTE COMPONENTS

COILTRONICS (305) 781-8900, SUMIDA (708) 956-0666

AN55 • TA08

of total current drain. Value changes in components are the result of higher power operation. The most significant change involves driving two tubes. Accommodating two lamps involves separate ballast capacitors but circuit operation is similar. Two lamp designs reflect slightly different loading back through the transformer's primary. C2 usually ends up in the 10pF to 47pF range. Note that C2A and B appear with their lamp loads in parallel across the transformer's secondary. As such, C2's value is often smaller than in a single tube circuit using the same type lamp. Ideally the transformer's secondary current splits evenly between the C2-lamp branches, with the total load current being regulated. In practice, differences between C2A and B and differences in lamps and lamp wiring layout preclude a perfect current split. Practically, these differences are small, and the

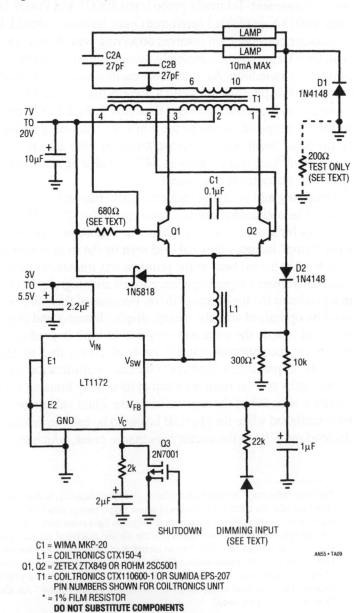

Figure 11–21.
A 92% efficient CCFL supply for 10mA loads features shutdown and dimming inputs. Two lamps are typical of color displays.

C1 = WIMA MKP-20
L1 = COILTRONICS CTX150-4
Q1, Q2 = ZETEX ZTX849 OR ROHM 2SC5001
T1 = COILTRONICS CTX110600-1 OR SUMIDA EPS-207
    PIN NUMBERS SHOWN FOR COILTRONICS UNIT
* = 1% FILM RESISTOR
DO NOT SUBSTITUTE COMPONENTS

AN55 • TA09

COILTRONICS (305) 781-8900, SUMIDA (708) 956-0666

lamps appear to emit equal amounts of light. Layout and lamp matching can influence C2's value. Some techniques for dealing with these issues appear in the section "Layout Issues."

General Measurement and Optimization Considerations

Several points should be kept in mind when observing operation of these circuits. L1's high voltage secondary can only be monitored with a wide-band, high voltage probe fully specified for this type of measurement. *The vast majority of oscilloscope probes will break down and fail if used for this measurement.* Tektronix probe types P6007 and P6009 (acceptable) or types P6013A and P6015 (preferred) must be used to read L1's output.

Another consideration involves observing waveforms. The LT1172's switching frequency is completely asynchronous from the Q1-Q2 Royer converter's switching. As such, most oscilloscopes cannot simultaneously trigger and display all the circuit's waveforms. Figure 11–19 was obtained using a dual beam oscilloscope (Tektronix 556). LT1172 related Traces A and B are triggered on one beam, while the remaining traces are triggered on the other beam. Single beam instruments with alternate sweep and trigger switching (e.g., Tektronix 547) can also be used, but are less versatile and restricted to four traces.

Obtaining and verifying high efficiency[3] requires some amount of diligence. The optimum efficiency values given for C1 and C2 are typical, and will vary for specific types of lamps. An important realization is that the term "lamp" includes the total load seen by the transformer's secondary. This load, reflected back to the primary, sets transformer input impedance. The transformer's input impedance forms an integral part of the LC tank that produces the high voltage drive. Because of this, circuit efficiency must be optimized with the wiring, display housing and physical layout arranged exactly the same way they will be built in production. Deviations from this procedure will result in lower efficiency than might otherwise be possible. In practice, a "first cut" efficiency optimization with "best guess" lead lengths and the intended lamp in its display housing usually produces results within 5% of the achievable figure. Final values for C1 and C2 may be established when the physical layout to be used in production has been decided on. C1 sets the circuit's resonance point, which varies to some

---

3. The term "efficiency" as used here applies to electrical efficiency. In fact, the ultimate concern centers around the efficient conversion of power supply energy into light. Unfortunately, lamp types show considerable deviation in their current-to-light conversion efficiency. Similarly, the emitted light for a given current varies over the life and history of any particular lamp. As such, this publication treats "efficiency" on an electrical basis; the ratio of power removed from the primary supply to the power delivered to the lamp. When a lamp has been selected, the ratio of primary supply power to lamp-emitted light energy may be measured with the aid of a photometer. This is covered in Appendix B, "Photometric Measurements." See also Appendix D, "Perspectives on Efficiency."

extent with the lamp's characteristics. C2 ballasts the lamp, effectively buffering its negative resistance characteristic. Small values of C2 provide the most load isolation, but require relatively large transformer output voltage for loop closure. Large C2 values minimize transformer output voltage, but degrade load buffering. Also, C1's "best" value is somewhat dependent on the lamp type used. Both C1 and C2 must be selected for given lamp types. Some interaction occurs, but generalized guidelines are possible. Typical values for C1 are 0.01μF to .15μF. C2 usually ends up in the 10pF to 47pF range. C1 must be a low-loss capacitor and substitution of the recommended devices is not recommended. A poor quality dielectric for C1 can easily degrade efficiency by 10%. C1 and C2 are selected by trying different values for each and iterating towards best efficiency. During this procedure, ensure that loop closure is maintained by monitoring the LT1172's feedback pin, which should be at 1.23V. Several trials usually produce the optimum C1 and C2 values. Note that the highest efficiencies are not necessarily associated with the most esthetically pleasing waveshapes, particularly at Q1, Q2, and the output.

Other issues influencing efficiency include lamp wire length and energy leakage from the lamp. The high voltage side of the lamp should have the smallest practical lead length. Excessive length results in radiative losses, which can easily reach 3% for a 3 inch wire. Similarly, no metal should contact or be in close proximity to the lamp. This prevents energy leakage, which can exceed 10%.[4]

It is worth noting that a custom designed lamp affords the best possible results. A jointly tailored lamp-circuit combination permits precise optimization of circuit operation, yielding highest efficiency.

Special attention should be given to the layout of the circuit board, since high voltage is generated at the output. The output coupling capacitor must be carefully located to minimize leakage paths on the circuit board. A slot in the board will further minimize leakage. Such leakage can permit current flow outside the feedback loop, wasting power. In the worst case, long term contamination build-up can increase leakage inside the loop, resulting in starved lamp drive or destructive arcing. It is good practice for minimization of leakage to break the silk screen line which outlines transformer T1. This prevents leakage from the high voltage secondary to the primary. Another technique for minimizing leakage is to evaluate and specify the silk screen ink for its ability to withstand high voltages.

---

4. A very simple experiment quite nicely demonstrates the effects of energy leakage. Grasping the lamp at its low-voltage end (low field intensity) with thumb and forefinger produces almost no change in circuit input current. Sliding the thumb-forefinger combination towards the high-voltage (higher field intensity) lamp end produces progressively greater input currents. Don't touch the high-voltage lead or you may receive an electrical shock. Repeat: Do not touch the high-voltage lead or you may receive an electrical shock.

Efficiency Measurement

Once these procedures have been followed efficiency can be measured. Efficiency may be measured by determining lamp current and voltage. Measuring current involves measuring RMS voltage across a temporarily inserted 200Ω .1% resistor in the ground lead of the negative current steering diode. The lamp current is

$$Ilamp = \frac{ERMS}{200} \times 2$$

The ×2 factor is necessitated because the diode steering dumps the current to ground on negative cycles. The 200Ω value allows the RMS meter to read with a scale factor numerically identical to the total current. Once this measurement is complete, the 200Ω resistor may be deleted and the negative current steering diode again returned directly to ground. Lamp RMS voltage is measured at the lamp with a properly compensated high voltage probe. Multiplying these two results gives power in watts, which may be compared to the DC input supply E × I product. In practice, the lamp's current and voltage contain small out of phase components but their error contribution is negligible.

Both the current and voltage measurements require a wideband true RMS voltmeter. The meter must employ a thermal type RMS converter— the more common logarithmic computing type based instruments are inappropriate because their bandwidth is too low.

The previously recommended high voltage probes are designed to see a 1MΩ–10pF–22pF oscilloscope input. The RMS voltmeters have a 10 meg Ω input. This difference necessitates an impedance matching network between the probe and the voltmeter. Details on this and other efficiency measurement issues appear in Appendix A, "Achieving Meaningful Efficiency Measurements."

Layout

The physical layout of the lamp, its leads, the display housing, and other high voltage components, is an integral part of the circuit. Poor layout can easily degrade efficiency by 25%, and higher layout induced losses have been observed. Producing an optimal layout requires attention to how losses occur. Figure 11–22 begins our study by examining potential parasitic paths between the transformer's output and the lamp. Parasitic capacitance to AC ground from any point between the transformer output and the lamp creates a path for undesired current flow. Similarly, stray coupling from any point along the lamp's length to AC ground induces parasitic current flow. All parasitic current flow is wasted, causing the circuit to produce more energy to maintain the desired current flow in D1 and D2. The high-voltage path from the transformer to the display housing should be as short as possible to minimize losses. A good rule of thumb is

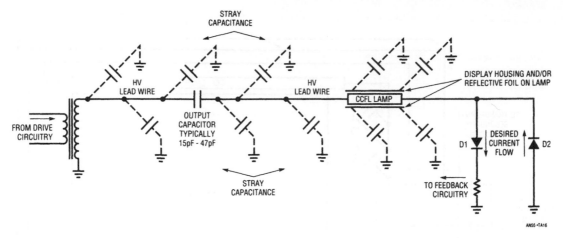

Figure 11–22.
Loss paths due to stray capacitance in a practical LCD installation. Minimizing these paths is essential for good efficiency.

to assume 1% efficiency loss per inch of high voltage lead. Any PC board ground or power planes should be relieved by at least ¼" in the high voltage area. This not only prevents losses, but eliminates arcing paths.

Parasitic losses associated with lamp placement within the display housing require attention. High voltage wire length within the housing must be minimized, particularly for displays using metal construction. Ensure that the high voltage is applied to the shortest wire(s) in the display. This may require disassembling the display to verify wire length and layout. Another loss source is the reflective foil commonly used around lamps to direct light into the actual LCD. Some foil materials absorb considerably more field energy than others, creating loss. Finally, displays supplied in metal enclosures tend to be lossy. The metal absorbs significant energy and an AC path to ground is unavoidable. Direct grounding of a metal enclosed display further increases losses. Some display manufacturers have addressed this issue by relieving the metal in the lamp area with other materials.

The highest efficiency "in system" backlights have been produced by careful attention to these issues. In some cases the entire display enclosure was re-engineered for lowest losses.

Layout Considerations for Two-Lamp Designs

Systems using two lamps have some unique layout problems. Almost all two lamp displays are color units. The lower light transmission characteristics of color displays necessitate more light. Therefore, display manufacturers use two tubes to produce more light. The wiring layout of these two tube color displays affects efficiency and illumination balance in the lamps. Figure 11–23 shows an "x-ray" view of a typical display. This symmetrical arrangement presents equal parasitic losses. If C1 and C2 and the lamps are matched, the circuit's current output splits evenly and equal illumination occurs.

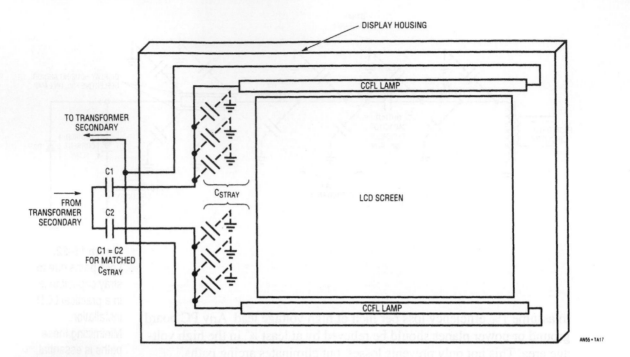

Figure 11–23.
Loss paths for a
"best case" dual
lamp display.
Symmetry pro-
motes balanced
illumination.

Figure 11–24's display arrangement is less friendly. The asymmetrical wiring forces unequal losses, and the lamps receive imbalanced current. Even with identical lamps, illumination may not be balanced. This condition is correctable by skewing C1's and C2's values. C1, because it drives greater parasitic capacitance, should be larger than C2. This tends to equalize the currents, promoting equal lamp drive. It is important to realize that this compensation does nothing to recapture the lost energy—efficiency is still compromised. There is no substitute for minimizing loss paths.

In general, imbalanced illumination causes fewer problems than might be supposed. The effect is very difficult for the eye to detect at high intensity levels. Unequal illumination is much more noticeable at lower levels. In the worst case, the dimmer lamp may only partially illuminate. This phenomenon is discussed in detail in the section "Thermometering."

Feedback Loop Stability Issues

The circuits shown to this point rely on closed loop feedback to maintain the operating point. All linear closed loop systems require some form of frequency compensation to achieve dynamic stability. Circuits operating with relatively low power lamps may be frequency compensated simply by overdamping the loop. Figures 11–18 and 11–20 use this approach. The higher power operation associated with color displays requires more attention to loop response. The transformer produces much higher output

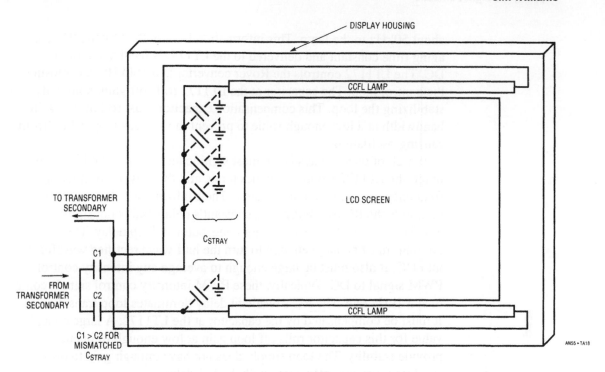

DISPLAY HOUSING

CCFL LAMP

TO TRANSFORMER
SECONDARY

C_{STRAY}

LCD SCREEN

C1

FROM
TRANSFORMER
SECONDARY

C2

C1 > C2 FOR
MISMATCHED
C_{STRAY}

CCFL LAMP

AN55 • TA18

Figure 11–24.
Symmetric losses
in a dual lamp
display. Skewing C1
and C2 values
compensates
imbalanced loss
paths, but not
wasted energy.

voltages, particularly at start-up. Poor loop damping can allow transformer voltage ratings to be exceeded, causing arcing and failure. As such, higher power designs may require optimization of transient response characteristics.

Figure 11–25 shows the significant contributors to loop transmission in these circuits. The resonant Royer converter delivers information at

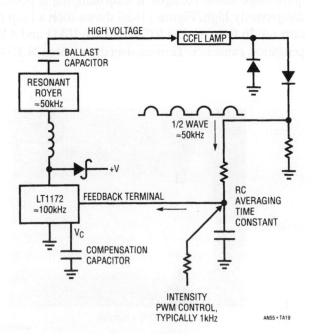

HIGH VOLTAGE

CCFL LAMP

BALLAST
CAPACITOR

RESONANT
ROYER
≈50kHz

1/2 WAVE
≈50kHz

+V

LT1172
≈100kHz

FEEDBACK TERMINAL

RC
AVERAGING
TIME
CONSTANT

V_C

COMPENSATION
CAPACITOR

INTENSITY
PWM CONTROL,
TYPICALLY 1kHz

AN55 • TA19

Figure 11–25.
Delay terms in the
feedback path. The
RC time constant
dominates loop
transmission delay
and must be compensated for stable
operation.

about 50kHz to the lamp. This information is smoothed by the RC averaging time constant and delivered to the LT1172's feedback terminal as DC. The LT1172 controls the Royer converter at a 100kHz rate, closing the control loop. The capacitor at the LT1172 rolls off gain, nominally stabilizing the loop. This compensation capacitor must roll off the gain bandwidth at a low enough value to prevent the various loop delays from causing oscillation.

Which of these delays is the most significant? From a stability viewpoint, the LT1172's output repetition rate and the Royer's oscillation frequency are sampled data systems. Their information delivery rate is far above the RC averaging time constant's delay and is not significant. The RC time constant is the major contributor to loop delay. This time constant must be large enough to turn the half wave rectified waveform into DC. It also must be large enough to average any intensity control PWM signal to DC. Typically, these PWM intensity control signals come in at a 1kHz rate. The RC's resultant delay dominates loop transmission. It must be compensated by the capacitor at the LT1172. A large enough value for this capacitor rolls off loop gain at low enough frequency to provide stability. The loop simply does not have enough gain to oscillate at a frequency commensurate with the RC delay.

This form of compensation is simple and effective. It ensures stability over a wide range of operating conditions. It does, however, have poorly damped response at system turn-on. At turn-on, the RC lag delays feedback, allowing output excursions well above the normal operating point. When the RC acquires the feedback value, the loop stabilizes properly. This turn-on overshoot is not a concern if it is well within transformer breakdown ratings. Color displays, running at higher power, usually require large initial voltages. If loop damping is poor, the overshoot may be dangerously high. Figure 11–26 shows such a loop responding to turn-on. In this case the RC values are 10kΩ and 4.7µf, with a 2µf compensation capacitor. Turn-on overshoot exceeds 3500 volts for over 10

Figure 11–26.
Destructive high voltage overshoot and ring-off due to poor loop compensation. Transformer failure and field recall are nearly certain. Job loss may also occur.

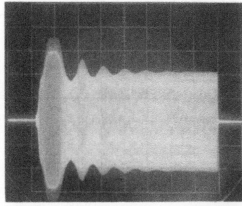

A = 1000V/DIV

HORIZ = 20ms/DIV

AN55 TA20

Figure 11–27.
Poor loop compensation caused this transformer failure. Arc occurred in high voltage secondary (lower right). Resultant shorted turns caused overheating.

milliseconds! Ring-off takes over 100 milliseconds before settling occurs. Additionally, an inadequate (too small) ballast capacitor and excessively lossy layout force a 2000 volt output once loop settling occurs. This photo was taken with a transformer rated well below this figure. The resultant arcing caused transformer destruction, resulting in field failures. A typical destroyed transformer appears in Figure 11–27.

Figure 11–28 shows the same circuit, with the RC values reduced to 10kΩ and 1μf. The ballast capacitor and layout have also been optimized. Figure 11–28 shows peak voltage reduced to 2.2 kilovolts with duration down to about 2 milliseconds. Ring-off is also much quicker, with lower amplitude excursion. Increased ballast capacitor value and wiring layout optimization reduce running voltage to 1300 volts. Figure 11–29's results are even better. Changing the compensation capacitor to a 3kΩ–2μf network introduces a leading response into the loop, allowing faster acquisition. Now, turn-on excursion is slightly lower, but greatly reduced in duration. The running voltage remains the same.

The photos show that changes in compensation, ballast value, and layout result in dramatic reductions in overshoot amplitude and duration. Figure 11–26's performance almost guarantees field failures, while Figures 11–28 and 11–29 do not overstress the transformer. Even with

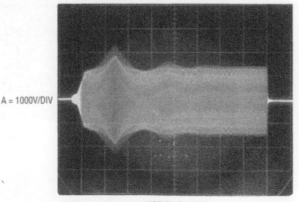

A = 1000V/DIV

HORIZ = 5ms/DIV

Figure 11–28.
Reducing RC time constant improves transient response, although peaking, ring-off, and run voltage are still excessive.

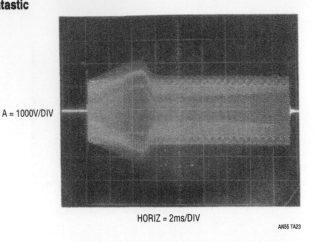

Figure 11–29.
Additional optimization of RC time constant and compensation capacitor reduces turn-on transient. Run voltage is large, indicating possible lossy layout and display.

A = 1000V/DIV

HORIZ = 2ms/DIV

AN55 TA23

the improvements, more margin is possible if display losses can be controlled. Figures 11–26–11–29 were taken with an exceptionally lossy display. The metal enclosure was very close to the foil wrapped lamps, causing large losses with subsequent high turn-on and running voltages. If the display is selected for lower losses, performance can be greatly improved.

Figure 11–30 shows a low loss display responding to turn-on with a 2µf compensation capacitor and 10kΩ-1µf RC values. Trace A is the transformer's output while Traces B and C are the LT1172's Vcompensation and feedback pins, respectively. The output overshoots and rings badly, peaking to about 3000 volts. This activity is reflected by overshoots at the Vcompensation pin (the LT1172's error amplifier output) and the feedback pin. In Figure 11–31, the RC is reduced to 10kΩ–.1µf. This substantially reduces loop delay. Overshoot goes down to only 800 volts—a reduction of almost a factor of four. Duration is also much shorter. The Vcompensation and feedback pins reflect this tighter control. Damping is much better, with slight overshoot induced at turn-on. Further reduction of the RC to 10kΩ–.01µf (Figure 11–32) results in even faster loop capture, but a new problem appears. In Trace A, lamp turn on is so fast that the overshoot does not register in the photo. The

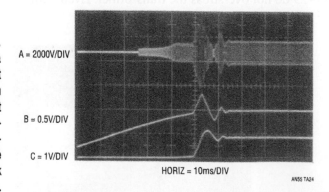

Figure 11–30.
Waveforms for a lower loss layout and display. High voltage overshoot (Trace A) is reflected at compensation node (Trace B) and feedback pin (Trace C).

A = 2000V/DIV

B = 0.5V/DIV

C = 1V/DIV

HORIZ = 10ms/DIV

AN55 TA24

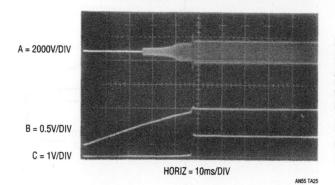

A = 2000V/DIV

B = 0.5V/DIV

C = 1V/DIV

HORIZ = 10ms/DIV

AN55 TA25

Figure 11–31.
Reducing RC time
constant produces
quick, clean loop
behavior. Low loss
layout and display
result in 650 VRMS
running voltage.

Vcompensation (Trace B) and feedback nodes (Trace C) reflect this with
exceptionally fast response. Unfortunately, the RC's light filtering causes
ripple to appear when the feedback node settles. As such, Figure 11–31's
RC values are probably more realistic for this situation.

The lesson from this exercise is clear. The higher voltages involved in
color displays mandate attention to transformer outputs. Under running
conditions, layout and display losses can cause higher loop compliance
voltages, degrading efficiency and stressing the transformer. At turn-on,
improper compensation causes huge overshoots, resulting in possible
transformer destruction. Isn't a day of loop and layout optimization
worth a field recall?

Extending Illumination Range

Lamps operating at relatively low currents may display the "thermometer
effect," that is, light intensity may be nonuniformly distributed along
lamp length. Figure 11–33 shows that although lamp current density is
uniform, the associated field is imbalanced. The field's low intensity,
combined with its imbalance, means that there is not enough energy to
maintain uniform phosphor glow beyond some point. Lamps displaying
the thermometer effect emit most of their light near the positive electrode,
with rapid emission fall-off as distance from the electrode increases.

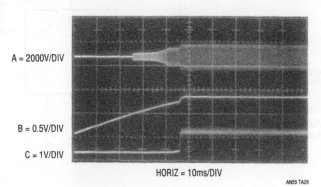

A = 2000V/DIV

B = 0.5V/DIV

C = 1V/DIV

HORIZ = 10ms/DIV

AN55 TA26

Figure 11–32.
Very low RC value
provides even
faster response, but
ripple at feedback
pin (Trace C) is
too high. Figure
11–31 is the best
compromise.

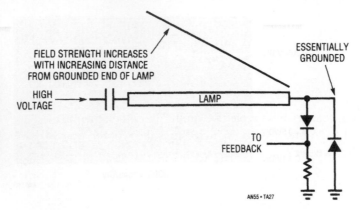

Figure 11–33. Field strength vs. distance for a ground referred lamp. Field imbalance promotes uneven illumination at low drive levels.

Figure 11–34. The "low thermometer" configuration. "Topside sensed" primary derived feedback balances lamp drive, extending dimming range.

Placing a conductor along the lamp's length largely alleviates "thermometering." The trade-off is decreased efficiency due to energy leakage (see Note 4 and associated text). It is worth noting that various lamp types have different degrees of susceptibility to the thermometer effect.

Some displays require an extended illumination range. "Thermometering" usually limits the lowest practical illumination level. One acceptable way to minimize "thermometering" is to eliminate the large

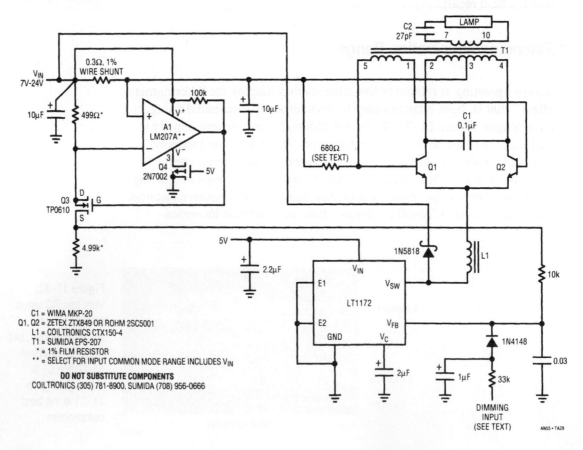

field imbalance. Figure 11–34's circuit does this. This circuit's most significant aspect is that the lamp is fully floating—there is no galvanic connection to ground as in the previous designs. This allows T1 to deliver symmetric, differential drive to the lamp. Such balanced drive eliminates field imbalance, reducing thermometering at low lamp currents. This approach precludes any feedback connection to the now floating output. Maintaining closed loop control necessitates deriving a feedback signal from some other point. In theory, lamp current proportions to T1's or L1's drive level, and some form of sensing this can be used to provide feedback. In practice, parasitics make a practical implementation difficult.[5]

Figure 11–34 derives the feedback signal by measuring Royer converter current and feeding this information back to the LT1172. The Royer's drive requirement closely proportions to lamp current under all conditions. A1 senses this current across the .3Ω shunt and biases Q3, closing a local feedback loop. Q3's drain voltage presents an amplified, single ended version of the shunt voltage to the feedback point, closing the main loop. The lamp current is not as tightly controlled as before, but .5% regulation over wide supply ranges is possible. The dimming in this circuit is controlled by a 1kHz PWM signal. Note the heavy filtering (33kΩ–2μf) outside the feedback loop. This allows a fast time constant, minimizing turn-on overshoot.[6]

In all other respects, operation is similar to the previous circuits. This circuit typically permits the lamp to operate over a 40:1 intensity range without "thermometering." The normal feedback connection is usually limited to a 10:1 range.

The losses introduced by the current shunt and A1 degrade overall efficiency by about 2%. As such, circuit efficiency is limited to about 90%. Most of the loss can be recovered at moderate cost in complexity. Figure 11–35's modifications reduce shunt and A1 losses. A1, a precision micropower type, cuts power drain and permits a smaller shunt value without performance degradation. Unfortunately, A1 does not function when its inputs reside at the V+ rail. Because the circuit's operation requires this, some accommodation must be made.[7]

At circuit start-up, A1's input is pulled to its supply pin potential (actually, slightly above it). Under these conditions, A1's input stage is shut off. Normally, A1's output state would be indeterminate but, for the amplifier specified, it will always be high. This turns off Q3, permitting the LT1172 to drive the Royer stage. The Royer's operation causes Q1's collector swing to exceed the supply rail. This turns on the 1N4148, the BAT-85 goes off, and A1's supply pin rises above the supply rail. This "bootstrapping" action results in A1's inputs being biased within the am-

---

5. See Appendix C, "A Lot of Cut-Off-Ears and No Van Goghs—Some Not-So-Great Ideas," for details.

6. See section "Feedback Loop Stability Issues."

7. In other words, we need a hack.

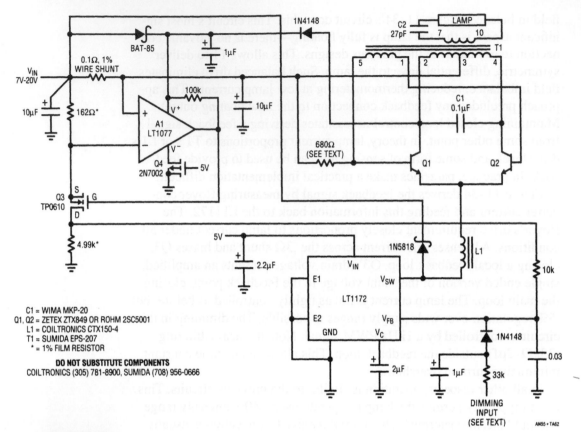

Figure 11–35.
The "low thermometer" circuit using a micropower, precision topside sensing amplifier. Supply bootstrapping eliminates input common mode requirement, permitting a 1.6% efficiency gain.

plifier's common mode range, and normal circuit operation commences. The result of all this is a 1.6% efficiency gain, permitting an overall circuit efficiency of just below 92%.

Epilogue

Our understanding with Apple Computer gave them six months sole use of everything I learned while working with them. After that, we were free to disclose the circuit and most attendant details to anyone else, which we did. It found immediate use in other computers and applications, ranging from medical equipment to automobiles, gas pumps, retail terminals and anywhere else LCD displays are used. The development work consumed about 20 months, ending in August, 1993. Upon its completion I immediately fell into a rut, certain I would never do anything worthwhile again.

References

1. Blake, James W. *The Sidewalks of New York*. (1894).

2. Bright, Pittman, and Royer. "Transistors As On-Off Switches in Saturable Core Circuits." *Electrical Manufacturing* (December 1954): Available from Technomic Publishing, Lancaster, PA.

3. Sharp Corporation. *Flat Panel Displays*. (1991).

4. Kitchen, C., and L. Counts. *RMS-to-DC Conversion Guide*. Analog Devices, Inc. (1986).

5. Williams, Jim. "A Monolithic IC for 100MHz RMS-DC Conversion." Linear Technology Corporation, *Application Note 22* (September 1987).

6. Hewlett-Packard. "1968 Instrumentation. Electronic Analytical-Medical." *AC Voltage Measurement* (1968): 197–198.

7. Hewlett-Packard. *Model 3400RMS Voltmeter Operating and Service Manual*. (1965).

8. Hewlett-Packard. *Model 3403C True RMS Voltmeter Operating and Service Manual*. (1973).

9. Ott, W.E. "A New Technique of Thermal RMS Measurement." *IEEE Journal of Solid State Circuits* (December 1974).

10. Williams, J.M., and T.L. Longman. "A 25MHz Thermally Based RMS-DC Converter." *IEEE ISSCC Digest of Technical Papers* (1986).

11. O'Neill, P.M. "A Monolithic Thermal Converter." *H.P. Journal* (May 1980).

12. Williams, J. "Thermal Technique in Measurement and Control Circuitry," "50MHz Thermal RMS-DC Converter." Linear Technology Corporation, *Application Note 5* (December 1984).

13. Williams, J., and B. Huffman. "Some Thoughts on DC-DC Converters": Appendix A, "The +5 to 10 ±15V Converter—A Special Case." Linear Technology Corporation, *Application Note 29* (October 1988).

14. Baxendall, P.J. "Transistor Sine-Wave LC Oscillators." *British Journal of IEEE* (February 1960): Paper No. 2978E.

15. Williams, J. "Temperature Controlling to Microdegrees." Massachusetts Institute of Technology, Education Research Center (1971): out of print.

16. Fulton, S.P. "The Thermal Enzyme Probe." Thesis, Massachusetts Institute of Technology (1975).

17. Williams, J. "Designer's Guide to Temperature Measurement." *EDN* part II (May 20, 1977).

18. Williams. J. "Illumination Circuitry for Liquid Crystal Displays." Linear Technology Corporation, *Application Note 49* (August 1992).

19. Olsen, J.V. "A High Stability Temperature Controlled Oven." Thesis, Massachusetts Institute of Technology (1974).

20. MIT Reports on Research. *The Ultimate Oven*. (March 1972).

21. McDermott, James. "Test System at MIT Controls Temperature of Microdegrees." *Electronic Design* (January 6, 1972).

22. Williams, Jim. "Techniques for 92% Efficient LCD Illumination." Linear Technology Corporation, *Application Note 55* (August 1993).

Appendix A

Achieving Meaningful Efficiency Measurements

Obtaining reliable efficiency data for the CCFL circuits presents a high order difficulty measurement problem. Establishing and maintaining accurate AC measurements is a textbook example of attention to measurement technique. The combination of high frequency, harmonic laden waveforms and high voltage makes meaningful results difficult to obtain. The choice, understanding, and use of test instrumentation is crucial. Clear thinking is needed to avoid unpleasant surprises![1]

Probes

The probes employed must faithfully respond over a variety of conditions. Measuring across the resistor in series with the CCFL is the most favorable circumstance. This low voltage, low impedance measurement allows use of a standard 1X probe. The probe's relatively high input capacitance does not introduce significant error. A 10X probe may also be used, but frequency compensation issues (discussion to follow) must be attended to.

The high voltage measurement across the lamp is considerably more demanding on the probe. The waveform fundamental is at 20kHz to 100kHz, with harmonics into the MHz region. This activity occurs at peak voltages in the kilovolt range. The probe must have a high fidelity response under these conditions. Additionally, the probe should have low input capacitance to avoid loading effects which would corrupt the measurement. The design and construction of such a probe requires significant attention. Figure 11–A1 lists some recommended probes along with their characteristics. As stated in the text, almost all standard oscilloscope probes will fail[2] if used for this measurement. Attempting to circumvent the probe requirement by resistively dividing the lamp voltage also creates problems. Large value resistors often have significant voltage coefficients and their shunt capacitance is high and uncertain. As such, simple voltage dividing is not recommended. Similarly, common high voltage probes intended for DC measurement will have large errors because of AC effects. The P6013A and P6015 are the favored probes; their 100MΩ input and small capacitance introduces low loading error. The penalty for their 1000X attenuation is reduced output, but the recommended voltmeters (discussion to follow) can accommodate this.

All of the recommended probes are designed to work into an oscilloscope input. Such inputs are almost always 1MΩ paralleled by (typically)

---

1. It is worth considering that various constructors of Figure 11–18 have reported efficiencies ranging from 8% to 115%.
2. That's twice I've warned you nicely.

10pF–22pF. The recommended voltmeters, which will be discussed, have significantly different input characteristics. Figure 11–A2's table shows higher input resistances and a range of capacitances. Because of this the probe must be compensated for the voltmeter's input characteristics. Normally, the optimum compensation point is easily determined and adjusted by observing probe output on an oscilloscope. A known-amplitude square wave is fed in (usually from the oscilloscope calibrator) and the probe adjusted for correct response. Using the probe with the voltmeter presents an unknown impedance mismatch and raises the problem of determining when compensation is correct.

The impedance mismatch occurs at low and high frequency. The low frequency term is corrected by placing an appropriate value resistor in shunt with the probe's output. For a 10MΩ voltmeter input, a 1.1MΩ resistor is suitable. This resistor should be built into the smallest possible BNC equipped enclosure to maintain a coaxial environment. No cable connections should be employed; the enclosure should be placed directly between the probe output and the voltmeter input to minimize stray capacitance. This arrangement compensates the low frequency impedance mismatch. Figure 11–A4 shows the impedance-matching box attached to the high voltage probe.

Correcting the high frequency mismatch term is more involved. The wide range of voltmeter input capacitances combined with the added shunt resistor's effects presents problems. How is the experimenter to know where to set the high frequency probe compensation adjustment? One solution is to feed a known value RMS signal to the probe-voltmeter combination and adjust compensation for a proper reading. Figure 11–A3 shows a way to generate a known RMS voltage. This scheme is simply a standard backlight circuit reconfigured for a constant voltage output. The op amp permits low RC loading of the 5.6K feedback termination without introducing bias current error. The 5.6kΩ value may be series or parallel trimmed for a 300V output. Stray parasitic capacitance in the feedback network affects output voltage. Because of this, all feedback associated nodes and components should be rigidly fixed and the entire circuit built into a small metal box. This prevents any significant change in the parasitic terms. The result is a known $300V_{RMS}$ output.

Now, the probe's compensation is adjusted for a 300V voltmeter indication, using the shortest possible connection (e.g., BNC-to-probe adapter) to the calibrator box. This procedure, combined with the added resistor, completes the probe-to-voltmeter impedance match. If the probe compensation is altered (e.g., for proper response on an oscilloscope) the voltmeter's reading will be erroneous.[3] It is good practice to verify the

---

3. The translation of this statement is to hide the probe when you are not using it. If anyone wants to borrow it, look straight at them, shrug your shoulders, and say you don't know where it is. This is decidedly dishonest, but eminently practical. Those finding this morally questionable may wish to reexamine their attitude after producing a day's worth of worthless data with a probe that was unknowingly readjusted.

Figure 11–A1. Characteristics of some wideband high voltage probes. Output impedances are designed for oscilloscope inputs.

TEKTRONIX PROBE TYPE	ATTENUATION FACTOR	ACCURACY	INPUT RESISTANCE	INPUT CAPACITANCE	RISE TIME	BAND-WIDTH	MAXIMUM VOLTAGE	DERATED ABOVE	DERATED TO AT FREQUENCY	COMPENSATION RANGE	ASSUMED TERMINATION RESISTANCE
P6007	100X	3%	10MΩ	2.2pF	14ns	25MHz	1.5kV	200kHz	700V$_{RMS}$ at 10MHz	15-55pF	1M
P6009	100X	3%	10MΩ	2.5pF	2.9ns	120MHz	1.5kV	200kHz	450V$_{RMS}$ at 40MHz	15-47pF	1M
P6013A	1000X	Adjustable	100MΩ	3pF	7ns	50MHz	12kV	100kHz	800V$_{RMS}$ at 20MHz	12-60pF	1M
P6015	1000X	Adjustable	100MΩ	3pF	1.4ns	250MHz	20kV	100kHz	2000V$_{RMS}$ at 20MHz	12-47pF	1M

Figure 11–A2. Pertinent characteristics of some thermally based RMS voltmeters. Input impedances necessitate matching network and compensation for high voltage probes.

MANUFACTURER AND MODEL	FULL SCALE RANGES	ACCURACY AT 1MHz	ACCURACY AT 100kHz	INPUT RESISTANCE AND CAPACITANCE	MAXIMUM BANDWIDTH	CREST FACTOR
Hewlett-Packard 3400 Meter Display	1mV to 300V, 12 Ranges	1%	1%	0.001V to 0.3V Range = 10M and < 50pF, 1V to 300V Range = 10M and < 20pF	10MHz	10:1 At Full Scale, 100:1 At 0.1 Scale
Hewlett-Packard 3403C Digital Display	10mV to 1000V, 6 Ranges	0.5%	0.2%	10mV and 100mV Range = 20M and 20pF ±10%, 1V to 1000V Range = 10M and 24pF ±10%	100MHz	10:1 At Full Scale, 100:1 At 0.1 Scale
Fluke 8920A Digital Display	2mV to 700V, 7 Ranges	0.7%	0.5%	10M and < 30pF	20MHz	7:1 At Full Scale, 70:1 At 0.1 Scale

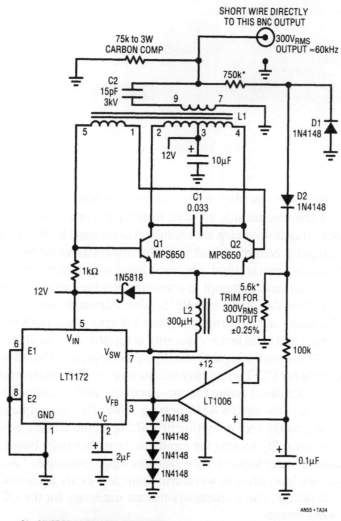

Figure 11–A3.
High voltage RMS calibrator is voltage output version of CCFL circuit.

SHORT WIRE DIRECTLY
TO THIS BNC OUTPUT
300V_{RMS}
OUTPUT ≈60kHz

75k to 3W
CARBON COMP

750k*

C2
15pF
3kV

L1

9   7

5   1   2   3   4

12V   10µF

D1
1N4148

C1
0.033

Q1
MPS650   Q2
MPS650

D2
1N4148

1kΩ   1N5818

12V

5

V_{IN}

E1   LT1172   V_{SW}

E2   V_{FB}

GND   V_C

6

8

1   2

2µF

L2
300µH

5.6k*
TRIM FOR
300V_{RMS}
OUTPUT
±0.25%

100k

+12

LT1006

1N4148
1N4148
1N4148
1N4148

0.1µF

AN55 • TA34

C1 = MUST BE A LOW LOSS CAPACITOR.
    METALIZED POLYCARB
    WIMA FKP2 OR MKP-20 (GERMAN) RECOMMENDED
L1 = SUMIDA 6345-020 OR COILTRONICS CTX110092-1
    PIN NUMBERS SHOWN FOR COILTRONICS UNIT
L2 = COILTRONICS CTX300-4
Q1, Q2 = AS SHOWN OR BCP 56 (PHILLIPS SO PACKAGE)
    * = 1% FILM RESISTOR (10kΩ TO 75kΩ RESISTORS IN SERIES)
DO NOT SUBSTITUTE COMPONENTS

COILTRONICS (305) 781-8900, SUMIDA (708) 956-0666

calibrator box output before and after every set of efficiency measurements. This is done by directly connecting, via BNC adapters, the calibrator box to the RMS voltmeter on the 1000V range.

RMS Voltmeters

The efficiency measurements require an RMS responding voltmeter. This instrument must respond accurately at high frequency to irregular and harmonically loaded waveforms. These considerations eliminate almost all AC voltmeters, including DVMs with AC ranges.

Figure 11–A4.
The impedance
matching box
(extreme left)
mated to the high
voltage probe. Note
direct connection.
No cable is used.

There are a number of ways to measure RMS AC voltage. Three of the most common include average, logarithmic, and thermally responding. Averaging instruments are calibrated to respond to the average value of the input waveform, which is almost always assumed to be a sine wave. Deviation from an ideal sine wave input produces errors. Logarithmically based voltmeters attempt to overcome this limitation by continuously computing the input's true RMS value. Although these instruments are "real time" analog computers, their 1% error bandwidth is well below 300kHz and crest factor capability is limited. Almost all general purpose DVMs use such a logarithmically based approach and, as such, are not suitable for CCFL efficiency measurements. Thermally based RMS voltmeters are direct acting thermo-electronic analog computers. They respond to the input's RMS heating value. This technique is explicit, relying on the very definition of RMS (e.g., the heating power of the waveform). By turning the input into heat, thermally based instruments achieve vastly higher bandwidth than other techniques.[4] Additionally, they are insensitive to waveform shape and easily accommodate large crest factors. These characteristics are necessary for the CCFL efficiency measurements.

Figure 11–A5 shows a conceptual thermal RMS-DC converter. The input waveform warms a heater, resulting in increased output from its associated temperature sensor. A DC amplifier forces a second, identical, heater-sensor pair to the same thermal conditions as the input driven pair. This differentially sensed, feedback enforced loop makes ambient temperature shifts a common mode term, eliminating their effect. Also, although the voltage and thermal interaction is non-linear, the input-output RMS voltage relationship is linear with unity gain.

The ability of this arrangement to reject ambient temperature shifts depends on the heater-sensor pairs being isothermal. This is achievable by thermally insulating them with a time constant well below that of ambient shifts. If the time constants to the heater-sensor pairs are matched, ambient temperature terms will affect the pairs equally in phase and amplitude.

---

4. Those finding these descriptions intolerably brief are commended to references 4, 5, and 6.

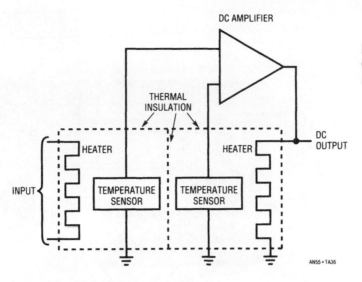

Figure 11–A5.
Conceptual thermal RMS-DC converter.

The DC amplifier rejects this common mode term. Note that, although the pairs are isothermal, they are insulated from each other. Any thermal interaction between the pairs reduces the system's thermally based gain terms. This would cause unfavorable signal-to-noise performance, limiting dynamic operating range.

Figure 11–A5's output is linear because the matched thermal pair's nonlinear voltage-temperature relationships cancel each other.

The advantages of this approach have made its use popular in thermally based RMS-DC measurements.

The instruments listed in Figure 11–A2, while considerably more expensive than other options, are typical of what is required for meaningful results. The HP3400A and the Fluke 8920A are currently available from their manufacturers. The HP3403C, an exotic and highly desirable instrument, is no longer produced but readily available on the secondary market.

Figure 11–A6 shows equipment in a typical efficiency test setup. The RMS voltmeters (photo center and left) read output voltage and current via high voltage (left) and standard 1X probes (lower left). Input voltage is read on a DVM (upper right). A low loss clip-on ammeter (lower right) determines input current. The CCFL circuit and LCD display are in the foreground. Efficiency, the ratio of input to output power, is computed with a hand held calculator (lower right).

Calorimetric Correlation of Electrical Efficiency Measurements

Careful measurement technique permits a high degree of confidence in the accuracy of the efficiency measurements. It is, however, a good idea to check the method's integrity by measuring in a completely different domain. Figure 11–A7 does this by calorimetric techniques. This arrangement, identical to the thermal RMS voltmeter's operation (Figure 11–A5),

Figure 11–A6.
Typical efficiency
measurement
instrumentation.
RMS voltmeters
(center left) mea-
sure output voltage
and current via
appropriate probes.
Clip-on ammeter
(right) gives low
loss input current
readings. DVM
(upper right) mea-
sures input voltage.
Hand calculator
(lower right) is
used to compute
efficiency.

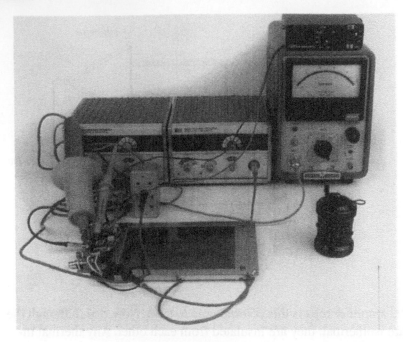

determines power delivered by the CCFL circuit by measuring its load
temperature rise. As in the thermal RMS voltmeter, a differential approach
eliminates ambient temperature as an error term. The differential amplifier's output, assuming a high degree of matching in the two thermal enclosures, proportions to load power. The ratio of the two cells' E × I
products yields efficiency information. In a 100% efficient system, the
amplifier's output energy would equal the power supplies' output.
Practically it is always less, as the CCFL circuit has losses. This term
represents the desired efficiency information.

Figure 11–A7.
Efficiency
determination via
calorimetric mea-
surement. Ratio
of power supply
to output energy
gives efficiency
information.

Figure 11–A8 is similar except that the CCFL circuit board is placed
within the calorimeter. This arrangement nominally yields the same information, but is a much more demanding measurement because far less
heat is generated. The signal-to-noise (heat rise above ambient) ratio is
unfavorable, requiring almost fanatical attention to thermal and instru-

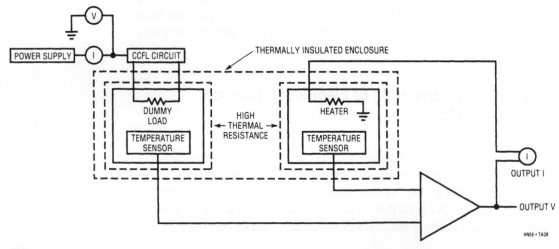

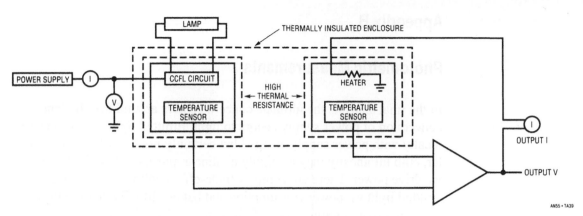

mentation considerations.[5] It is significant that the total uncertainty between electrical and both calorimetric efficiency determinations was 3.3%. The two thermal approaches differed by about 2%. Figure 11–A9 shows the calorimeter and its electronic instrumentation. Descriptions of this instrumentation and thermal measurements can be found in the References section following the main text.

Figure 11–A8. The calorimeter measures efficiency by determining circuit heating losses.

---

5. Calorimetric measurements are not recommended for readers who are short on time or sanity.

Figure 11–A9. The calorimeter (center) and its instrumentation (top). Calorimeter's high degree of thermal symmetry combined with sensitive servo instrumentation produces accurate efficiency measurements. Lower portion of photo is calorimeter's top cover.

Appendix B

Photometric Measurements

In the final analysis the ultimate concern centers around the efficient conversion of power supply energy to light. Emitted light varies monotonically with power supply energy,[1] but certainly not linearly. In particular, bulb luminosity may be highly nonlinear, particularly at high power, vs. drive power. There are complex trade-offs involving the amount of emitted light vs. power consumption and battery life. Evaluating these trade-offs requires some form of photometer. The relative luminosity of lamps may be evaluated by placing the lamp in a light tight tube and sampling its output with photodiodes. The photodiodes are placed along the lamp's length and their outputs electrically summed. This sampling technique is an uncalibrated measurement, providing relative data only. It is, however, quite useful in determining relative bulb emittance under various drive conditions. Figure 11–B1 shows this "glometer," with its uncalibrated output appropriately scaled in "brights." The switches allow various sampling diodes along the lamp's length to be disabled. The photodiode signal conditioning electronics are mounted behind the switch panel.

Calibrated light measurements call for a true photometer. The Tektronix J-17/J1803 photometer is such an instrument. It has been found

Figure 11–B1.
The "glometer" measures relative lamp emissivity. CCFL circuit mounts to the right. Lamp is inside cylindrical housing. Photodiodes (center) convert light to electrical output (lower left) via amplifiers (not visible in photo).

---

1. But not always! It is possible to build highly electrically efficient circuits that emit less light than "less efficient" designs. See Appendix C, "A Lot of Cut-Off Ears and No Van Goghs—Some Not-So-Great Ideas."

particularly useful in evaluating display (as opposed to simply the lamp) luminosity under various drive conditions. The calibrated output permits reliable correlation with customer results.[2] The light tight measuring head allows evaluation of emittance evenness at various display locations. This capability is invaluable when optimizing lamp location and/or ballast capacitor values in dual lamp displays.

Figure 11–B2 shows the photometer in use evaluating a display.

---

2. It is unlikely that customers would be enthusiastic about correlating the "brights" units produced by the aforementioned glometer.

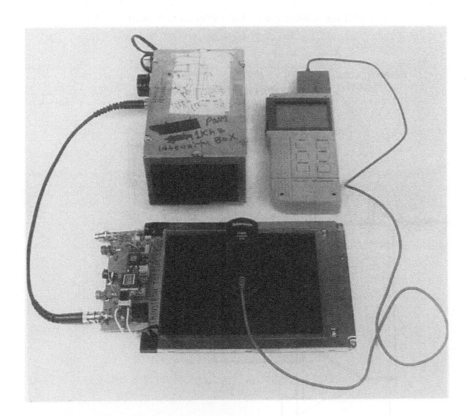

Figure 11–B2.
Apparatus for calibrated photometric display evaluation. Photometer (upper right) indicates display luminosity via sensing head (center). CCFL circuit (left) intensity is controlled by a calibrated pulse width generator (upper left).

Appendix C

A Lot of Cut-Off Ears and No Van Goghs—Some Not-So-Great Ideas

The hunt for a practical CCFL power supply covered (and is still covering) a lot of territory. The wide range of conflicting requirements combined with ill-defined lamp characteristics produces plenty of unpleasant surprises. This section presents a selection of ideas that turned into disappointing breadboards. Backlight circuits are one of the deadliest places for theoretically interesting circuits the author has ever encountered.

Not-So-Great Backlight Circuits

Figure 11–C1 seeks to boost efficiency by eliminating the LT1172's saturation loss. Comparator C1 controls a free running loop around the Royer by on-off modulation of the transistor base drive. The circuit delivers bursts of high voltage sine drive to the lamp to maintain the feedback

Figure 11–C1. A first attempt at improving the basic circuit. Irregular Royer drive promotes losses and poor regulation.

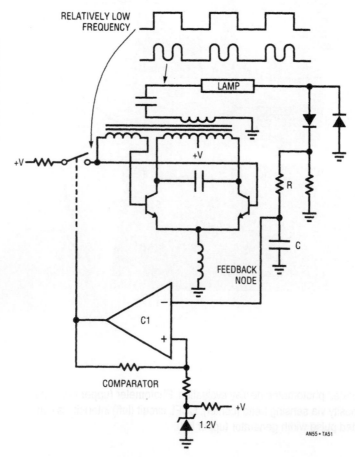

node. The scheme worked, but had poor line rejection, due to the varying waveform vs. supply seen by the RC averaging pair. Also, the "burst" modulation forces the loop to constantly re-start the bulb at the burst rate, wasting energy. Finally, bulb power is delivered by a high crest factor waveform, causing inefficient current-to-light conversion in the bulb.

Figure 11–C2 attempts to deal with some of these issues. It converts the previous circuit to an amplifier-controlled current mode regulator. Also, the Royer base drive is controlled by a clocked, high frequency pulse width modulator. This arrangement provides a more regular waveform to the averaging RC, improving line rejection. Unfortunately the improvement was not adequate. 1% line rejection is required to avoid annoying flicker when the line moves abruptly, such as when a charger is activated. Another difficulty is that, although reduced by the higher frequency PWM, crest factor is still non-optimal. Finally, the lamp is still forced to restart at each PWM cycle, wasting power.

Figure 11–C3 adds a "keep alive" function to prevent the Royer from turning off. This aspect worked well. When the PWM goes low, the Royer is kept running, maintaining low level lamp conduction. This eliminates the continuous lamp restarting, saving power. The "supply correc-

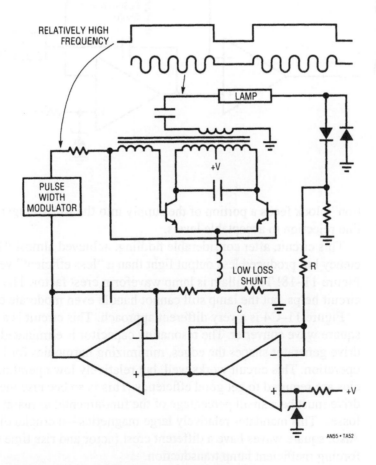

Figure 11–C2.
A more sophisticated failure still has losses and poor line regulation.

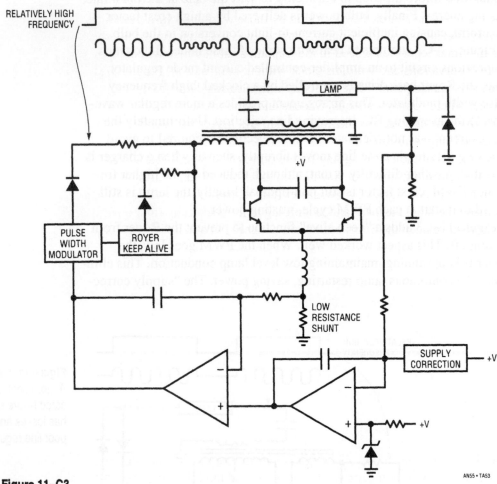

RELATIVELY HIGH
FREQUENCY

LAMP

+V

PULSE
WIDTH
MODULATOR

ROYER
KEEP ALIVE

LOW
RESISTANCE
SHUNT

SUPPLY
CORRECTION   +V

+V

AN55 • TA53

Figure 11–C3.
"Keep alive" circuit
eliminates turn-on
losses and has
94% efficiency.
Light emission is
lower than "less
efficient" circuits.

tion" block feeds a portion of the supply into the RC averager, improving
line rejection to acceptable levels.

This circuit, after considerable fiddling, achieved almost 94% efficiency but produced less output light than a "less efficient" version of
Figure 11–18! The villain is lamp waveform crest factor. The keep alive
circuit helps, but the lamp still cannot handle even moderate crest factors.

Figure 11–C4 is a very different approach. This circuit is a driven
square wave converter. The resonating capacitor is eliminated. The base
drive generator shapes the edges, minimizing harmonics for low noise
operation. This circuit works well, but relatively low operating frequencies are required to get good efficiency. This is so because the sloped
drive must be a small percentage of the fundamental to maintain low
losses. This mandates relatively large magnetics—a crucial disadvantage.
Also, square waves have a different crest factor and rise time than sines,
forcing inefficient lamp transduction.

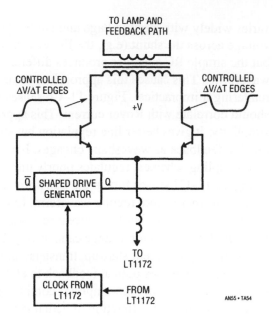

Figure 11–C4.
A non-resonant approach. Slew retarded edges minimize harmonics, but transformer size goes up. Output waveform is also non-optimal, causing lamp losses.

Not-So-Great Primary Side Sensing Ideas

Figures 11–34 and 11–35 use primary side current sensing to control bulb intensity. This permits the bulb to fully float, extending its dynamic operating range. A number of primary side sensing approaches were tried before the "topside sense" won the contest.

Figure 11–C5's ground referred current sensing is the most obvious way to detect Royer current. It offers the advantage of simple signal conditioning—there is no common mode voltage. The assumption that essentially all Royer current derives from the LT1172 emitter pin path is true. Also true, however, is that the waveshape of this path's current

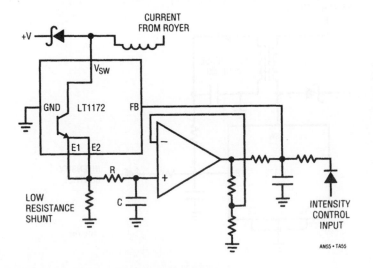

Figure 11–C5.
"Bottom side" current sensing has poor line regulation due to RC averaging characteristics.

189

varies widely with input voltage and lamp operating current. The RMS voltage across the shunt (e.g., the Royer current) is unaffected by this, but the simple RC averager produces different outputs for the various waveforms. This causes this approach to have very poor line rejection, rendering it impractical. Figure 11–C6 senses inductor flux, which should correlate with Royer current. This approach promises attractive simplicity. It gives better line regulation but still has some trouble giving reliable feedback as waveshape changes. It also, in keeping with most flux sampling schemes, regulates poorly under low current conditions.

Figure 11–C7 senses flux in the transformer. This takes advantage of the transformer's more regular waveform. Line regulation is reasonably good because of this, but low current regulation is still poor. Figure 11–C8 samples Royer collector voltage capacitively, but the feedback signal does not accurately represent start-up, transient, and low current conditions.

Figure 11–C9 uses optical feedback to eliminate all feedback integrity problems. The photodiode-amplifier combination provides a DC feedback signal which is a function of actual lamp emission. It forces the lamp to constant emissivity, regardless of environmental or aging factors.

This approach works quite nicely, but introduces some evil problems. The lamp comes up to constant emission immediately at turn-on. There is no warm-up time required because the loop forces emission, instead of current. Unfortunately, it does this by driving huge overcurrents through the lamp, stressing it and shortening life. Typically, 2 to 5 times rated current flows for many seconds before lamp temperature rises, allowing the loop to back down drive. A subtle result of this effect occurs with lamp aging. When lamp emissivity begins to fall off, the loop increases current to correct the condition. This increase in current accelerates lamp aging, causing further emissivity degradation. The resultant downward spiral continues, resulting in dramatically shortened lamp life.

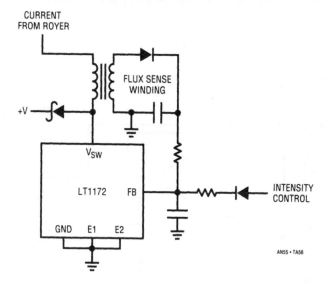

Figure 11–C6. Flux sensing has irregular outputs, particularly at low currents.

Other problems involve increased component count, photodiode mounting, and the requirement for photodiodes with predictable response or some form of trim.

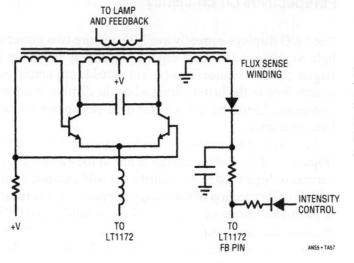

Figure 11–C7. Transformer flux sensing gives more regular feedback, but not at low currents.

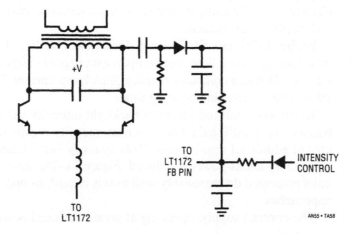

Figure 11–C8. AC couples drive waveform feedback is not reliable at low currents.

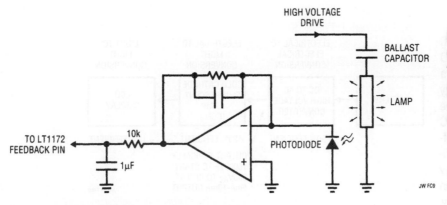

Figure 11–C9. Optically sensed feedback eliminates feedback irregularities, but introduces other problems.

191

Appendix D

Perspectives on Efficiency

The LCD displays currently available require two power sources, a backlight supply and a contrast supply. The display backlight is the single largest power consumer in a typical portable apparatus, accounting for almost 50% of the battery drain when the display intensity control is at maximum. Therefore, every effort must be expended to maximize backlight efficiency.

The backlight presents a cascaded energy attenuator to the battery (Figure 11–D1). Battery energy is lost in the electrical-to-electrical conversion to high voltage AC to drive the cold cathode fluorescent lamp (CCFL). This section of the energy attenuator is the most efficient; conversion efficiencies exceeding 90% are possible. The CCFL, although the most efficient electrical-to-light converter available today, has losses exceeding 80%. Additionally, the light transmission efficiency of present displays is about 10% for monochrome, with color types even lower. Clearly, overall backlight efficiency improvements must come from bulb and display improvements.

Higher CCFL circuit efficiency does, however, directly translate into increased operating time. For comparison purposes Figure 11–20's circuit was installed in a computer running 5mA lamp current. The result was a 19 minute increase in operating time.

Relatively small reductions in backlight intensity can greatly extend battery life. A 20% reduction in screen intensity results in nearly 30 minutes of additional running time. This assumes that efficiency remains reasonably flat as power is reduced. Figure 11–D2 shows that the circuits presented do reasonably well in this regard, as opposed to other approaches.

The contrast supply, operating at greatly reduced power, is not a major source of loss.

Figure 11–D1.
The backlit LCD display presents a cascaded energy attenuator to the battery. DC to AC conversion is significantly more efficient than energy conversions in lamp and display.

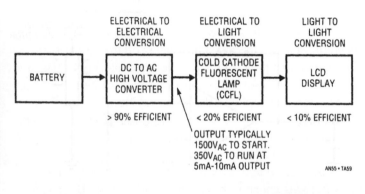

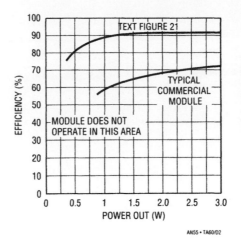

Figure 11–D2.
Efficiency comparison between Figure 11–21 and a typical modular converter.

AN55 • TA60/D2

Selling It

One of the characteristics of a good design is that somebody wants to use it. In today's world this means it must be saleable. Doug Grant's "Analog Circuit Design for Fun and Profit" addresses a circuit specification often ignored or poorly handled by designers—is the circuit saleable? Does anyone want it; and will they select it over other alternatives? This chapter should be required reading for anyone hired into a design position.

Bob Reay describes selling another "design," namely yourself. His chapter, "A New Graduate's Guide to the Analog Interview," should be required reading for anyone trying to get hired into a design position.

The section ends with the story of the most famous timekeeper in history, John Harrison's marine chronometer. It may also be the biggest marketing nightmare in history. This is a lesson in the tenacity required for technical and economic success in the face of an almost intractable technical problem and human foibles. Harrison's stunning accomplishment combined craft, genius, and singular, uninterrupted dedication to produce a solution the world very badly wanted. His task was not, however, insulated from human failings. Imagine spending a lifetime to give the world exactly what it asked for and still needing the king of England's help to get paid!

Selling It

12. Analog Circuit Design for Fun and Profit

••

The first volume of this series of books dealt mainly with how to design analog circuits. It was an interesting collection of ideas, anecdotes, and actual descriptions of the processes used by various well-known analog circuit designers to accomplish their goals. You won't find much of that sort of thing in this chapter (although I hope it will be interesting nonetheless).

The inspiration for this chapter arose in part from a comment in the chapter of the first book submitted by Derek Bowers of Analog Devices. He admitted that some of his most elegant circuits turned out to be poor sellers, while other circuits (of which he was not particularly proud) became multi-million-dollar successes. In this chapter, I will offer a few words of advice to fledgling analog design engineers in an effort to help them distinguish between good circuits and good products. In addition, I'll alert fledgling circuit designers to a new person they will eventually encounter in their careers—the Marketeer.

Why I Wanted to Be an Engineer

As an engineering student, you probably think you have a good idea of what engineering is all about. I recall my goals when I entered engineering school in 1971. It was all so clear then. High school students with an aptitude for math and science were destined to become engineers, and I was one of them. Four years of college would be followed by a secure career in the Engineering Lab, designing circuits that would change the world. I worked a few summers as a Technician, and I knew what engineers did. They designed circuits, gave hand-drawn schematics to the drafting department to make them nice and neat, then had the Technician round up the parts and build a prototype. Then the Engineer would come back to the lab and test the prototype, and blame any shortcomings on the lousy job the Technician did building the prototype. After a few iterations, the prototype would be declared a success, the Engineer would disappear for a few days to do something in his office, then come back with a hand-sketched schematic of the next circuit. And life went on.

Then I graduated and became an Engineer. 1975 was not a good year to become an Engineer. Defense contractors had fallen on hard times, with the Vietnam War winding down. They weren't hiring Engineers. The economy was in tough shape, and the industrial companies were also hurting. Many of my fellow new Engineers were scrambling to get into a graduate school to hide until the job market got better. I was one of the lucky few that actually found a job—mostly because I had worked part-time as a Technician to pay for school, and I therefore had "experience." Just getting an interview in 1975 wasn't easy. In fact, I had already been out of school for over a month when I got a call from the university's placement office to tell me that a company had reviewed the graduating class's resumé book, and had invited me for an interview. My resumé touted some knowledge of both analog and digital circuits, and I claimed I knew which end of the soldering iron to hold. I could cobble a collection of TTL gates together to do something, and could design a circuit with an op amp in it. I even had some experience in using and testing analog-to-digital converters. Fortunately, these were important things for this position, since my grade-point average was nothing special (too many extra-curricular activities . . .). I got the job.

Then I found out what Engineering was really like.

The first day on the job, my boss handed me the manual for the then-new Intel 8080 microprocessor, and told me to read it. Every day for the first week, he'd come into my office (actually, our office—four of us shared the same office) and ask me how I was doing. He was a pretty good engineer and teacher, and I got the chance to ask him some questions about things I hadn't quite understood. It went well.

Then one day, he handed me a schematic of the 8080-based system he had just finished testing. This was my chance to see how he had designed the system's bus structure, and implemented the various sub-systems and their interfaces to the processor. It was mostly pretty straightforward stuff—all digital at this point. Then a few weeks later he came into my office and asked me to design an analog I/O interface for the system, including the signal conditioning, A/D and D/A conversion, logic interface, and various other pieces. This was the moment of truth—I was on my own for my first design.

I had a handful of specs for the instrument we were supposed to interface with—voltage levels, source impedances, bandwidths, etc. I had the specs of accuracy of the original system. I had the manufacturers' data sheets for every component imaginable. And a week or so later, I had a design done—one of those hand-drawn schematics I had worked from as a Technician, but now I was calling the shots! Then we reviewed the schematic—the boss told me he had forgotten to mention that we needed to be galvanically isolated from the instrument we were hooking into. No problem; I had used V/F conversion for the A/D, and a few opto-isolators later I had completed the revised design, including isolation, and he signed it off. I proudly marched into the lab, handed it to the Technician, and he saluted smartly on his way to build a prototype.

Then a funny thing happened. The design part stopped for a long time. There was some haggling about certain parts being no longer available. The purchasing guy complained that some of them were sole-source, and he wanted everything to be multi-sourced. So I spent some time redesigning; the basic idea stayed the same, but the schematic was revised time and time again to comply with everyone's needs. Then the software guy came over from the next office. He wanted a complete map of each I/O address, a description of each function, and the timing pauses required between operations. No problem—I wrote it all up for him over the next week or so, in between interruptions from the Tech and the purchasing guy. We met again to review what I had done, and the software guy reminded me that the last project had included some provisions for calibration and self-test. Back to the schematic—I added the required additional channels and test modes, and was finally done. The prototype had grown somewhat, and I was amazed that the Tech was still speaking to me (he'd seen all this before).

Then the boss came in and asked me to document the operation of the circuit, including a description of every component's function. The purchasing guy came in with the manufacturing guy and they asked me for a complete parts list and bill of materials, and to sign off the final schematic. After a few iterations, everything was signed off, and the product went into production. I was eager to get to the next project.

Then it got interesting. The main processor board that my boss had designed developed reliability problems—it was an obvious bug in the clock circuit, which I found by putting my finger on the pull-up resistor for the +12V clock. Half-watt resistors get hot when dissipating a whole watt. I got to fix that one. The analog input section worked fine when we used one manufacturer's V/F converter, but was noisy when we substituted an "equivalent" from another manufacturer. I tracked the problem down to a difference in the power-supply decoupling needs of the two, and conjured up a scheme that was suitable for both versions.

As production started, I was often called to determine if a component substitution was possible because one or more parts was temporarily out of stock. In some cases, the substitution had already been done, and I had to figure out why it didn't work.

A full six months later, my boss asked me to design another circuit. Think about it—almost a half year between designs. Life as an Engineer was turning out to be very different from what I had expected. At least I was getting paid.

When I was actually designing circuits, I discovered an assortment of interesting processes at work. There is recall—remembering previous circuits that may help solve the problem at hand. There is invention—defining the problem, and creating a new solution for it. There is experimentation—often, a difficult problem will require numerous tries to get to the right solution. In some cases, these processes are aided by various embodiments of design tools, from decade boxes to advanced state-of-the-art expert-system-based software. Lots of tools are available to help the designer create a solution to a problem. And each idea is weighed carefully, using all necessary processes and tools, against an endless parade of design trade-offs, to improve reliability, increase production yield, and lower costs while maintaining or improving performance.

But it never ends at the design phase. After a circuit is done, and the first units are reduced to physical hardware, it remains to determine if the thing actually solves the problem it was intended to solve. Testing, debugging, characterizing, and (often) doing it all again are part and parcel of the product development process. And lots of other authors have described their personal versions of this process in their chapters.

I occasionally design circuits at home for recreation. Most are not the same as the kind of circuits produced by my employer, but my engineering training and avocational interest in electronics motivate me to keep designing circuits from time to time. Nobody will ever buy them. Total production volume is usually one. And I get a real thrill when I see one of them work for the first time. And any engineer who has never felt the thrill of seeing his first units work perfectly first time out will probably not stay an engineer very long. In fact, the experienced engineer should feel the same sense of excitement when "it works." Often, circuits don't work the first time. After an appropriate period (hopefully a short one!) of self-flagellation, the analysis of the circuit and troubleshooting begins, usually revealing an oversight or similarly simple error. The joy of finding the error usually makes the eventual event of a working circuit anticlimactic. And building circuits at home—with no formal documentation or parts lists required—the experience is as near to pure engineering as it ever gets. When I design circuits for myself, I define, design, build, test, redesign, rebuild, and use them. Unfortunately, it doesn't work that way in the real world. Most of the time, someone else is telling you what to design. And someone else is building and testing "your" circuits. Yet someone else may redesign them. And most importantly, someone else is using your circuit, and has probably paid money to do so.

A design engineer should never lose sight of the fact that his continued gainful employment is dependent on producing circuit designs that solve a problem for which his employer will collect revenue. Circuit design for fun is best left to the home laboratory, for those engineers who still have one. Circuit design for profit is serious stuff. If you can combine the two, consider yourself lucky. Then find a second spare-time leisure pursuit having nothing to do with engineering.

I don't design circuits for a living any more. I moved from Engineering into Marketing (by way of a few years in Applications Engineering) some years back, but stayed in the electronics industry. While some marketing skills are easily transportable across industries (especially in promotion and merchandising), the product-definition part of marketing generally is most successful if the practitioner is close to the technology. I have had occasion to recruit marketing engineers from the technical ranks of our customers as well as the design and product engineering areas of our own company. Most have done well, but all have expressed great surprise at the amount of work involved in the job, compared to their previous lives in engineering (and most of them thought marketing was going to be easier!).

DILBERT reprinted by permission of UFS, Inc.

Steps in the Product Development Process

The following steps broadly outline the product development process. In all cases, the "you" refers to yourself and your colleagues in whatever enterprise employs you. Product development is seldom a single-person endeavor.

1. Concept—Find a problem that you think you can solve.
2. Feasibility—Can you really solve the problem?
3. Realization—Design and build the product.
4. Introduction—Getting the product to the customers you don't know.
5. Closure—Move on to the next problem.

Step 1. Concept—Find a Problem That You Think You Can Solve

A product is (obviously) something that is meant to be produced (manufactured, delivered to someone for use, sold, consumed; take your pick). The point is that in the present era, very few circuits are designed for recreation only. Hardware circuit hackers are still out there, including the radio amateurs, but the fact is that most circuits are designed by engineers toiling for an employer. And that employer has an obligation to its customers and shareholders to create things that solve its customers' problems, and in so doing, generate a profit. Oftentimes, these solutions take the form of innovative circuits, processes, or architectures. However, there is a weak correlation between commercial success and technical elegance or sophistication.

A product must deliver benefit to the customer; it must solve his problem. A circuit can be a part of a product, but it is never the product. A user needs to see some benefit to using your circuit over another. I recall reviewing one particular product proposal from a design engineer that detailed a novel approach to performing analog-to-digital conversion. It seemed clever enough, but as I read it, the performance claims were no better than what existed on the market already. A cost analysis indicated no improvement in cost over what existed already. Power wasn't better. No particular features seemed to be obvious—it was just another A/D converter. I just didn't see any great benefit to a customer. So I called the designer and asked him what I had missed. He replied that the architecture was novel and innovative, and there was nothing like it. We reviewed the performance he thought he could get, and a chip size estimate. After about fifteen minutes, I asked him to compare the proposed chip to another we already had in development. There wasn't any advantage obvious. Then I asked him to compare it to various academic papers. He replied that his architecture was more "creative" than various proposed schemes. But when I asked him to show me where this idea would lead (higher speed, more resolution, lower cost, added features, scalability, user features, etc.), he drew a complete blank. Even assuming device scaling or process add-ons, he (and I) couldn't think of where this would lead. I asked if the inspiration had come from a particular application or customer problem. The closest he could come was a personal-computer add-on card that he had seen once. He had no idea if the board was a big seller or not.

The project was shelved. But I suspect that one day his novel architecture (or more likely, some part of it) will be useful in solving a very different problem.

I have also had the opportunity to deal with newly hired marketing engineers. Their zeal for the perfect product often blinds them to reality, as noted in the comic strip. In defining specifications and features for a new product, there is the temptation to add every conceivable feature that any customer has ever asked for during the process of fielding requests from salespeople and customers. This leads to the frustration that engineers often have when dealing with marketeers. On the other hand, I have observed situations where the engineer has been unable to promise that a certain specification can be met, and a less-than-satisfying compromise was offered. Both parties need to spend some time analyzing which combination of features and specifications meets the requirements of the ma-

DILBERT reprinted by permission of UFS, Inc.

jority of the customers, and settle on these. "Creeping featurism" must be avoided, even if a customer calls just a week before design completion asking for one more feature, or if the designer discovers "a neat trick that could double the speed" at the last minute. Stick to the script! Last-minute changes usually result in future problems.

As difficult as it may be designing high-performance analog circuits, it's equally challenging to figure out what to design in the first place. A wonderful circuit that nobody buys is not a good product. A rather pedestrian circuit that a lot of people buy is a much better product. This is a tough concept for most of us to swallow, but it's the truth.

Making sure you understand the problem you are solving is probably harder than designing the circuit. You have to learn someone else's job and understand his problems if you are going to have any chance of solving them. Numerous techniques have evolved over the years. One very effective methodology currently in vogue is called "Voice of the Customer," or "VOC" for short. The entire VOC process is lengthy and involved, and will not be described fully here, except for the first steps, which involve customer interviewing.

I recall taking an IC designer to visit a customer in the video business. The designer had some ideas for a new A/D converter fast enough to digitize a video signal. A/D converters are generally described by their resolution (measured in bits) and speed (measured in conversions per second). We talked to the customer about his whole signal chain, from input connector to digital bus, to get a feel for the components he had used in his previous design. The A/D converter was an 8-bit part, with a certain conversion speed. As we talked, the customer began to complain that he couldn't get the resolution he wanted from the digitized image. Aha! We had discovered the problem to be solved. He needed more resolution!

I glanced at the IC designer's notes and he had definitely gotten the point—he had written "RESOLUTION" in big letters, underlined, and circled it. Then he scribbled next to it: "Only has 8 bits now—10 should be enough." Unfortunately, there is another kind of "resolution" in video; it refers to the number of pixels on the screen, and when a video engineer talks about resolution, he means the speed of the converter, not the number of bits! Having done a fair amount of reading in preparation for the visit, I picked up on the error and asked the customer for a clarification. It went something like, "How much resolution do you need, and what does that mean in terms of the A/D converter?" His response was ultimately in speed terms and we got the discussion back on track (I knew it was back on track when the IC guy wrote "RESOLUTION = SPEED!!!?!" in his notebook). It is important to understand your customer's business and language before you go on the interview.

Another time, I listened to a customer complain bitterly about an A/D converter that he claimed was outside its accuracy specs. I offered to test the device for him to verify its performance. When I tested the part, it was fine, meeting all specs on the data sheet. When I returned the unit to the customer, he insisted on demonstrating to me exactly how bad the accuracy of the converter was. I went with him into his lab, where he put the converter in the socket, and turned the system on. He then tested the system by applying a dc signal from a bench power supply to the input and displayed the digital output on a monitor. He didn't even measure the DC input—just made sure it was somewhere within the A/D converter's range, based on the reading on the front panel meter on the supply. Before I could ask how he intended to verify 12-bit accuracy with a known reference source, he showed me that the output code was very unstable, with several codes of flicker evident. This was obviously the problem—noise, not accuracy! We tried all the usual cures (changing the supply to the converter from a switcher to a linear, rerouting the grounds a bit, and adding decoupling capacitors where there hadn't been any), and each change helped. Finally, we had the output stable. A fixed input gave a steady output value, even though we hadn't checked the actual accuracy of the system (he actually had no suitable equipment for such a test anyway). But he was happy—his problem was solved. We were happy—we got the order.

The data sheet for our next A/D converter included detailed instructions on how many capacitors to use and where to locate them in the layout. It wasn't any more accurate a converter, but a lot fewer people complained about its "accuracy." And we added some tutorial information defining the various performance parameters of A/D converters, so the next customer who called complaining about accuracy would actually mean accuracy, and we would be able to diagnose and cure the problem faster.

Speaking the customer's language is critical to communicating with him. And by "language" I mean his own company jargon or slang. If you expect him to learn your terms, you'll find it a lot harder to get him to feel comfortable describing the problem he wants you to solve. And this advice applies to both engineers and marketeers attempting to interview customers.

The VOC process suggests working with a number of customers to collect images that allow you to understand their problem as they see it. This is important—satisfying the customers' needs in a way that they can

DILBERT reprinted by permission of UFS, Inc.

understand is the secret to success. The next step after collecting images of the customer's-eye view of the problem is to re-state the problem in your own language so you can figure out a solution. All engineers should spend time with customers when they are in the process of discovering a problem to solve. Too often, a visit to a customer takes the form of trying to find a problem that fits your own creative solution. This violates all known problem-solving principles, but we all do it anyway. The obvious thing to keep in mind is that solutions only exist to solve problems; without a defined problem, it is only by sheer luck that a proposed solution does the job.

Some solutions are obvious—make it faster, more accurate, cheaper, lower power—but other problems exist that can be solved without the breakthrough innovations often needed to improve one of the conventional dimensions. These can only be discovered by talking to customers and analyzing the data in a meaningful way to reveal what features or qualities of a product the customer will value. But remember—customers are in a different business than you are. It is up to you to make the effort to learn the customer's business and language in order to actually understand the problem and offer a solution!

Interviewing a prospective customer involves some preparation. You should have a reasonable list of questions you want to ask, and you should be prepared to skip around the list as the conversation wanders. I have found it extremely useful to conduct interviews in teams of two. One person asks the questions, while the other scribbles the answers as fast as possible, trying to get it all down as nearly verbatim as possible. It's important to avoid adding too much commentary or analysis here—there's plenty of time for that later. Just get the facts down. If a series of questions has been missed, the note-taker can steer the conversation back to the areas missed. When I have tried to do customer interviews solo, I have often reviewed my notes only to find phrases like "He says the biggest problem is" or "The preferred package is," where I've been unable to get it on paper fast enough, and the conversation has taken a turn to another topic too quickly. A second pair of ears and hands can help immensely.

After the interview (which should end when the customer signals to you that he's done, not when you think time is up or have another appointment), the interview team should compare notes and make sure that both have heard the same things. It is useful to re-construct the entire interview as it occurred to help the recall process. Clean up the notes as soon as possible so they can be shared and reviewed later in the process.

After you've collected several interviews, the process of analyzing the data can begin. There is a strong temptation to give more weight to the last inputs you've received, unless you've taken the time to get them all in readable form. There is also a temptation to downplay inputs that contradict your own basic assumptions. Don't do it! Always remember that your product will be more successful if it solves the customer's problem than if it fits your personal model of the way things should be.

The process used for analysis of the raw inputs can be complicated or simple. The underlying principle is always to get a customer's-eye view of what is important, and respond to it in a product definition. Commentary like, "They all say they want power consumption less than 50 milliwatts. That's ridiculous—there's a 10-horsepower motor in the system! Besides, my circuit topology takes that much just to power up the output driver," is to be avoided. Things that appear from your own perspective to be obvious contradictions like this need to be reviewed and understood, not dismissed. In the case just mentioned, you may discover that the circuit you are designing is used by the thousands in the system, and that big motor is only used to move the system into position, and powered externally. The constraint on power is probably quite real. And you should figure out how important your output driver really is in his system.

Step 2. Feasibility—Can You Really Solve the Problem (and Is It Worth Solving)?

This step follows whatever analysis tools you use to reveal the features and performance requirements of the solution you are planning. VOC, QFD, and other methods can be used, but none is a substitute for experience, judgment, and general knowledge. At this point in the process, you should feel that you understand the requirements of the customers, and the first-cut solution is probably getting clear in your mind. In fact, you may think you have enough information to actually design a circuit at this point. Resist the temptation! You are in for some surprises. At this point, don't even try to complete the design—you'll find some feature you left out, or more likely, you'll have included a feature that only one customer (probably the last one you talked to!) wanted and which sounded like an interesting design challenge. Keep it simple at this point. Don't worry too much about the cost, or even the detailed architecture inside. Take a stab at the specs and features that seem important to the customer and difficult to meet, but don't waste too much time at this point.

There are usually several alternatives to solving the customer's problem. Usually the customer won't care much about the internal architecture, so you have a lot of freedom. You should get one pretty conservative solution defined quickly, then take some time to find alternatives that are better from the standpoint of cost, power, or ease in meeting some important specification for the customer. And feel free to think "outside the box."

This last expression comes from a course I once took on innovative problem-solving. A very simple puzzle is presented—draw three parallel rows of three dots each on a piece of paper; connect all nine dots by drawing four straight lines, never lifting the writing implement from the paper. The solution to the problem was an example of "going outside the box," as shown following.

I had seen this puzzle before, and knew the trick, while others in the class were claiming it couldn't be done. I smugly told the instructor I'd be glad to show the others, since I knew the answer. Unfazed, he gave *me*

Going outside the box.

A. With four lines . . .

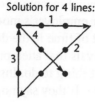

Solution for 4 lines:

the assignment of completing the puzzle by drawing only *three* lines. This put me in the same bewildered predicament as the rest of the class. After several minutes of torture, the instructor revealed the solutions—using four lines, then three, two, and even one line! (Solutions appear at the end of this chapter. . . .)

You should also talk to people that can provide assistance—other designers, applications or marketing engineers, anyone with some experience. The chances are that a circuit to do something like this has been tried before. Remember the example of the A/D architecture without a home? Perhaps this is the right problem for that solution. Don't forget to check the literature. There's no sense in re-inventing the wheel—in fact, if someone else has a patent on that particular wheel, it could get expensive. And if you come up with an idea that looks original and has benefits over previous work, consider patenting it.

If it turns out that the customer's problem does not have a solution that you can find that satisfies all the needs, there are a couple of options. One is to give up and move to the next problem. This is sometimes the best course of action. Some problems just don't have satisfactory solutions yet. File it away, keep in the back of your mind exactly what makes a solution impossible at this time, and keep your eyes open for the enabling technology. At that point, go back and see if the problem still needs a solution.

If you can't find a way to meet all the required specs, try to meet as many as you can, and try the solution out on a few willing customers. It may turn out that solving three out of four is good enough. It may be three more than anyone else has proposed!

Whether you think you're meeting some or all of the requirements, when you are closing in on the implementation, you must check to make sure you're still on course. Try describing your solution in terms of the customer's problem as you understand it. Survey methods can be used to rate individual features for importance, and "kill specs," or a series of loosely structured second-round interviews with willing customers, will work. When proposing a solution, be open to suggestions for improvement. This is not the time for defending "your" solution; after all, it isn't a solution yet—only an idea. If you are willing to make changes, customers will be willing to suggest them. And you'll find out quickly what is important that you missed, and what is superfluous. Pay attention, and bring someone with you again to take detailed notes for review soon after the visit.

At some time, you will have to decide if this problem offers enough financial incentive for you and your colleagues to spend your time (and your employer's money) solving. This is the best time, before you invest a lot of time in the detailed design. I don't advocate a detailed market analysis that attempts to prove beyond a shadow of a doubt that this is the right thing to do. Instead, ask the customers if this is the right problem to solve. If they say no, figure out the right problem to solve, and solve that one instead. If you have your heart set on solving a particular problem, make sure somebody in your company solves this customer's most important problem before someone in another company does it.

You should go through the exercise of making sure the numbers add up. If you talk to ten customers in a certain end market, and they all claim 30% market share, you have a problem. You may be able to get some data from an independent source to determine the actual shares (and thus the volume estimates for your solution), but often you will have to rely on your own estimate. And determining which of these ten customers is likely to win in his market will be based on your own feelings about their relative competence as much as any market research you will be able to do.

The failing of many product-definition processes, including VOC, is the myth that all customers are created equal, and that all customer inputs have equal weight. In many companies, a marketing department or sales department determines which customers are the ones deserving of your attention. And despite the frequent culture clashes that occur among engineering, marketing, and sales, the truth is that all three organizations need each other.

Some companies downplay the role of marketing in the product-definition process, while others recognize it for the valuable function that it can be. Even those companies that downplay its importance practice it religiously. One analog IC manufacturer has carefully chosen a group of customers it believes that it can profitably supply with circuits. It has then matched up a senior design engineer with each of these customers to learn what problems they are facing and try to figure out solutions together. Such client-based or partnership arrangements are becoming common in the industry, and represent one approach to the product definition process. If you listen carefully to what your customer is saying, you should be able to figure out what you can do for him. But the practitioners of this marketing approach will often downplay the importance of the marketing role—after all, you just need engineers talking to engineers to figure out what to do next, and it will all work out, right?

What these engineer-marketeers fail to realize is that someone picked out which customers they should get close to, and the marketing process began there, not at the product-definition step.

Whoever chooses the target customers has to think long and hard about several things. First, which companies buy enough of the sort of products we make to justify a lot of attention? Second, which of these companies are in solid financial shape? After all, a customer without money to pay the bills is probably not a customer to pursue too aggressively. Third, which customers set the standard, or are considered by their peers to be market leaders? And finally, which ones do you want to bet on? And you need to keep reviewing these questions every couple of years, because the answers to all the foregoing questions change over time.

One of the classic mistakes in the customer-selection process involved a particular component manufacturer (we'll call it "Company X," since this particular story has been handed down in the industry folklore for so long that the original company name has probably been long since lost) that chose not to extend credit to a start-up computer company started by two young engineers in a garage in Silicon Valley. That company grew to become giant Apple Computer, and certain key components in their products are never supplied by Company X.

Step 3. Realization—Design and Build the Product

Assuming you have decided to move ahead and have the commitment of all the resources you need to get the project done, this is the part where you design the circuit and develop a product. Try several approaches. Don't force a known solution into this design if it doesn't fit. Also don't try to force an innovation where none is needed. Are you aiming for the Nobel Prize or a circuit that solves a customer's problem?

The process of fine-tuning a design includes learning to tell the difference between a good circuit and a bad circuit. In most instances, the difference is obvious. One works and meets specifications (including costs!), and the other does not. Case closed. But what about the case where both circuits meet spec?

At this point, lots of questions need to be objectively answered (and some may not yet have objective answers!). Does one circuit have advantages in your manufacturing process? How about your customer's manufacturing process? Does one circuit lend itself to further improvements as technology progresses? Does one circuit have a clear path that parallels the electronics industry's unrelenting goals of faster, cheaper, lower power, smaller, more efficient? Can someone copy it easily and rob you of the profits that are rightfully yours? And most important, will one enable more profit over the long term than the other? This last one is that hardest to answer, and is left as an exercise for the reader.

And it gets messier out there in the real world. Sometimes both designs "almost" meet spec. One meets everything except the speed, while the other meets every spec except the accuracy. Now what do you do?

At this time, judgment separates the winners from the also-rans. This judgment must include common sense, experience of what has worked before and what has not, a real internal understanding of what the customer feels but is unable to express, and how the options compare with respect to all of this. Get opinions, facts, and make the call.

I won't comment too much on the actual circuit design process here. Skip to another chapter for details on how to design analog circuits. However, there is one design-related topic overlooked by many of the chapter authors in this series. It has little to do with circuit success, but everything to do with product success. It is the schedule, every engineer's enemy.

When you know what the customer needs, you will also probably need to know when he needs it. This should have a major impact on the design approach you use. The first volume of this series suggested that in the case of designing a new IC, there are risks involved in new designs, new processes, and new packages. If you are designing to a tight schedule, you should probably not try to invent anything new. The more risks you take, the more likely it becomes that you will miss your customer's schedule. And if you miss his schedule, he will miss the schedule that his customer has given him. This means that everyone loses money.

Occasionally people in sales or marketing will make a promise to a customer relating to a schedule without consulting engineering. They are trying to keep the customer interested, and figure that if they get the order, they can apply enough management pressure to make the product-development process move faster than usual. I have also observed engineers making schedule commitments to customers that can't possibly be met ("Oh, yeah . . . I can get the new design done in a week or two . . . no problem"), ignoring the fact that the design phase of a product is usually not the most time-consuming part of the development process. One-sided commitments to customers will be a problem as long as enthusiasm and emotion get in the way of rational decision-making. Aside from increasing enrollment in karate classes, nobody wins.

DILBERT reprinted by permission of UFS, Inc.

It is also a good idea to qualify your customer by asking who his customers are. If there isn't a good answer, perhaps this isn't the right customer. Remember, a customer is someone who buys products from you and sends you money, not just someone who likes your ideas and thinks he might buy something someday. The latter is more of a prospect than a customer. I've been told that if you can't write down the phone numbers of three people who will buy your product, then you don't have a product. You should try this exercise on your customer, too. If your customer has customers, try to talk to them. If you find out that all your customer's customers are planning to evaluate the new products at a particular industry event or trade show, you had better make sure that you have samples to your customer well in advance of the trade show so that he can assemble some prototypes to demonstrate. If he can't show units to his customers, both you and your customer may have to wait a year for the next show to launch a product.

While it sounds cold-hearted to focus only on customers, prospects are important, too. Never forget that. You should be responsive, courteous, and provide the support they need, and even a bit more. However, most prospects understand their place in the Grand Scheme of Things. Most of them will realize that their potential business may not represent your highest priority, and some will also become suspicious if you spend a lot of effort on their limited potential.

During the definition phase of a multi-channel D/A converter some years ago, I had determined that one potential market was numerically controlled machine tools and robots, since D/A converters are often used in position-control servomechanisms. Multi-axis motion controllers clearly needed multiple-channel D/A converters. I hit the books to find out the biggest manufacturers of machine tools and robots, and arranged a tour of them, focusing mostly on companies that were already customers of ours. The first few visits uncovered the sort of potential I had expected—on the order of a few hundred to a few thousand units each, with solid growth predicted for the next few years. Then I visited one company which was similar in size to the others (measured in annual sales), but which hadn't bought many chips from us in the previous year. In fact, they had only purchased small quantities of quite a few device types from us. This was puzzling, since they were housed in a very large building and had revenues comparable to the other companies. I presented the idea for the new product we were defining to the engineering staff. They listened attentively, made a few suggestions on certain specs that were important to them, and requested a few features. As I noted these inputs, I asked what their production volume was for the next year. They looked at each other, and after some discussion, determined that their production for the next year would be between 10 and 15 machines. I asked if they meant per week or per month, and they explained that the machines they made were very specialized, and sold into a price-insensitive market. Their production volume of 10 to 15 machines per year represented the same dollar volume as some of the other companies we had interviewed, since these machines were very big (which explained the huge building) and very expensive. They were grateful for the attention we had given them, and were happy to help us. They were also a bit surprised that we had chosen them as a target customer. However, their suggestions turned out to be useful in the product definition, and became strong selling points when we went back to the larger-volume customers.

By the time you have the first units, there are probably people waiting to try them. Some are inside your company (especially the people who have to manufacture the product in volume); some are outside your company (customers). Presumably, there has been some effort expended to develop a way to evaluate the first units to see if they meet spec. Do it quickly, and as soon as you are satisfied that the units behave as you expect, get some in the hands of someone outside the company. Try to use someone who will tell other people if he likes it, and tell only you if he doesn't like it.

Very often, your interpretation of the customer's problem and his interpretation will still be different. The customer doesn't like your product. The reason is that you didn't meet the spec that was most important to him. Perhaps he didn't tell you clearly enough (or at all). Or else you didn't understand that it was so important. It doesn't matter where the fault lies—the customer is not happy because of a failure to communicate. This is inevitable. If people always communicated clearly, there would have been far fewer wars in human history. Misunderstandings have created much more important problems than anything that may occur between you and your customer (although it doesn't feel that way when it happens). Take a deep breath and try to work it out.

Situations like this call for diplomacy and tact far beyond anything taught in engineering school. I have observed the tendency for engineers to get defensive when a customer finds a flaw in their circuit, especially if it has met the internally defined specifications. "I did my part. If it isn't good enough for the customer, that's his problem" is a fairly ridiculous statement if you think about it in the context of a supplier-customer relationship. Similarly, the marketeer who says, "We did what they told us, now they should buy it," is also ignoring the obvious fact that he didn't really understand what the customer wanted. Remember—the customer has the final say. He has the money, and if you don't keep him happy, he'll send that money to someone else. If the product doesn't meet the critical spec, get back to work and fix it!

Another problem I have observed is the case where a design works "sometimes." This is worse than a circuit that doesn't work at all. Intermittent ability to deliver a product to a customer due to use of unqualified production processes, circuit blocks, packages, or whatever, will damage a customer relationship in more ways than you can imagine. In the old days, designers got one unit working, threw the finished documentation package over the wall to manufacturing, and went on to the next thing. That's not good enough. Manufacturing and Product Assurance must be routinely involved in the product-development process. They can offer valuable insight into mistakes that others have made, and help you avoid them. And while they may often ask lots of seemingly unrelated questions about a circuit during a design review, they are trying to help.

But having discussed what can go wrong, it is equally important to mention that usually it all comes together right. You give samples to the customer on the date you promised, he tries them, and calls back to say how much he likes them. His system works exactly as he had hoped, and he looks like a hero to his management. Then he shows his system to his customers and they like it. His customers place orders, then he places orders for your product. Everyone smiles a lot.

Step 4. Introduction—Getting the Product to the Customers You Don't Know

If it's a product, there must be a customer. And at this point in the process you probably know some customers already—some are probably calling you for updates on the progress of your product, because they have decided to use it even before they have seen any units. In fact, if you don't have a first-name relationship with at least three potential customers, you ought to reconsider the whole thing. Occasionally, you'll be so far ahead of their thinking that your product will be exactly what they want even though they don't know it yet. There are some cases where this has happened—the personal computer, for example. But it happens so rarely that one of these per career is probably the limit. Without customers, you haven't designed a product—merely a circuit.

Giving a few samples to a potential customer is one way to introduce a product to the market. It works when there are only a few customers for a very specialized product. It's possible to know most of them. It gets more difficult when there are more potential customers than you can handle personally.

All customers will need help using your product. Some will need a little help, while others will need a lot of help. Still others will call you every day during the month they are designing a circuit around your product. Unless you have a lot of spare time available and need some new friends in your life, you have to create documentation adequate to allow them to use your marvelous widget without excessive hand-holding. Someone needs to write data sheets, instruction manuals, application notes, and troubleshooting guides. And that's not all. Unless you are personally going to be trained as a sales engineer, you will need to assume that other people with training in sales (yes, there is such a thing) will do the selling for you. If your product is going to be sold through a chain of distributors, you will need to provide sufficient training for them to understand your product's advantage over the competition (and how to handle situations where the competition is actually better in one respect or another). Unless you want every potential customer (or salesperson) to call you personally every time he has a question, you'll have to train other people to handle some of the questions in your place. This means time spent preparing and delivering training materials. It's difficult to fit this in while you're designing circuits.

Then there is the whole issue of external promotion to consider. There is a commonly held myth among both engineers and marketeers that derives from the "Build a better mousetrap and the world will beat a path to your door" axiom. It goes something like, "This product is so great it will sell itself." Too bad it isn't true. Here's what's wrong with that idea. The term "better" is completely subjective. If your customer hasn't been told why your product is better, he probably won't figure it out on his own.

He's probably too busy. You have to get the information to him somehow. Articles, seminars, trade shows, technical papers, newsletters—all of these are vehicles to get the information in front of the potential buyers' eyes. And all of these need careful planning and execution to optimize the return for each dollar spent. And of course, someone has to do the work.

Advertising is not as simple as it looks. A successful advertisement appears in the media that are read by the target customers, as determined both by examining the publishers' audit statements and observing what is on the desk of the customers you interview. Perhaps direct mail is a better choice. Perhaps your company has a complete suite of components for this problem—a "family" promotion of some kind may be in order. There are numerous vehicles available for product promotion—knowing them and choosing the most effective ones is the realm of the marketeer.

The goal of product promotion is to generate leads, or names of people who are interested in possibly buying your product. There are other forms of promotion, of course, aimed at establishing or enhancing a company's image so that the product promotions will remain effective. But promotion does not automatically result in revenue. Poorly planned and executed promotion plans only waste money. But an effective promotion plan can work wonders.

Even if your product is demonstrably better, the customer needs to know where the "door" to your company is located. Who does he call if he wants to buy the product? Does he know who your company is? Do the other people in his company know how to do business with your company? And lastly, if the manufacturer of the second- or even third-best mousetrap has a sales force that beats a path to your potential customers' doors, the world will have no reason to beat a path to your door, and you will not succeed. Having the world's best product simply isn't enough.

Yes, you need salespeople. Most engineers do not like salespeople. Many engineers consider circuit design a Higher Calling of some sort, and have little interest in the human interactions that enable the exchange of goods and services in a market economy. However, without these interactions, little commerce could take place.

Being on the losing end of a potential order due to a lack of relationship is frustrating. I recall one incident where after investing many months of effort, including several long-haul airplane flights, we lost a very big order at Customer A to Competitor X. I knew our product was better. I knew our price was better. I knew the overall solution cost was better. The overall system performance with our product was better. Yet we lost the order. We were all at a loss to understand why we lost the order. We had done everything right, by any measure. It took a while, but finally one of the Customer A designers told me what had happened. The order we were seeking was even more important to Competitor X than it was to us. The sales manager of X raised visibility of the impending lost business to the president of the company. The Competitor X president then phoned his old friend who was president of Customer A, and made an appointment to play golf the next weekend. Somewhere on the back nine, the issue of the new project was raised, and Mr. X asked his old friend if there was any way to use his company's product in the new project. He had heard some disturbing things about possibly losing the order. The next day, Mr. A called his engineering and purchasing managers and instructed them to use the Competitor X product. They saluted smartly, and followed orders. In this case (and there have been numerous others over the years) the human relationship outweighed the objective and fact-based decision-making processes. Losing business this way is frustrating. Getting to the point where you can win this way takes a long time, a solid track record of success, and a good sales force.

It is worth noting that most salespeople have a pretty low opinion of engineers as well. They see most engineers as unable to see the obvious importance of their customer, and can't understand why it's hard to improve the performance of a circuit by a mere factor of two by just making a minor adjustment that should take no time and entail no risk.

DILBERT reprinted by permission of UFS, Inc.

After introduction, someone must consider the management of the life-cycle of the product. Periodically reviewing the performance of a product against measurable data (sales, profits, units sold, etc.) is a necessary evil, and generally unrelated to circuit design. Long after a product has been introduced, someone (variously called a product manager, marketing engineer, or merchandising specialist, depending on the company's culture) reviews all these (and other) metrics, analyzes the cause, and undertakes corrective actions as necessary. If sales have declined, it may be that price has eroded due to new competition, a major program has ended, or some other phenomenon. It is unlikely that the manufacturing or accounting operations of your company will have visibility into the end customers, and they can only build product and report data. Someone who can examine the data and determine which course of action leads to the maximum revenue and profits must make the decisions regarding the product.

Of course, you may want to do much of this yourself. And that's fine, as long as you recognize that you will have less and less time available to design circuits. Or to learn about other kinds of circuits and systems. Consider such a decision very carefully.

Step 5. Closure—Move On to the Next Problem

While it is important to deliver circuit designs that meet certain specifications, it is not advisable to succeed once, and then rely on incremental improvements on the same idea from time to time for the rest of a career. Once you have completed the process of solving a customer's problem, it's time to declare victory and move on. Document what you did, make sure that the solution is on "autopilot," train others to understand the issues and trade-offs made, and then walk away.

You need to do new things from time to time to avoid getting stale. In the area of circuit design, doing the same things the same old way and just waiting for incremental improvements (new processes or components) can type-cast an engineer and limit his professional growth. If that's your choice, make sure you understand its implications. Most engineers I have known have looked for new ways to do things, and often find old tricks useful in solving new problems.

But where do you find new problems to solve? There are several sources of inspiration for what to do next. The best (and sadly, the most often overlooked) source of ideas for new products is your current customers. Remember the customer you designed a low-noise amplifier for last year? Perhaps he also needs a high-resolution A/D converter. Or the guy who needed a battery monitor—he might need something else vaguely analog in nature. Talk to him. But do a bit of homework yourself—find out what projects your company already offers so you don't spend a lot of time identifying a problem that others in your company are already solving. As companies grow, it becomes difficult to know what is going on in other parts of the organization; this is another place where a salesperson can be useful. He is expected to know what products his company has available now and in development to solve some customer's problem. If he hears his own customer express a similar need, he can then bring in the resources he needs to find the best available solution for his customer's problem.

I recall one visit to a customer where I had one of our design engineers with me. The customer was having a minor problem with one of our D/A converters, but had solved it by the time we got there. However, since he had the "factory guys" there, he wanted to tell us about another problem that he couldn't solve. Ever the eager marketing type, I asked him to tell me more—if my own group didn't have the solution, I could carry the message to the relevant group and get him the help he needed. The customer then launched into a lengthy dissertation on what was wrong with a particular class of IC that didn't work quite right—it was something that connected to a D/A converter, so I was curious. About five minutes into the interview, my colleague interrupted the customer to inform him that we didn't make that kind of device, and we couldn't help him. He didn't want the customer to waste his time explaining a problem we couldn't solve.

As it happened, however, another part of our company was in fact in the final design stages of a chip that was very well suited to solving the customer's problem. I had to play the diplomat and remind my colleague about the device under "secret development," and encourage the customer to keep talking. I took lots of notes, forwarded them to the appropriate group in my company, and we eventually did some very nice business with that customer.

Engineers working in high technology need to keep abreast of the latest research in their field, including new technologies. Many analog circuit designers look with disdain upon digital design; however, there are powerful techniques available in the digital domain that have performance and cost advantages over any attempt to duplicate them in the analog domain. Knowing something about them can help broaden your range of available trade-offs.

Read the journals; attend a conference or two each year, including one intended for your customers. Talk to people, especially others in your company who deal with a lot of customers. Buy things and take them apart to see how they work. Find out who is trying to solve similar problems to yours, perhaps in a different end application. The ideas you encounter may someday be useful. Learning is almost never in vain—an idea whose present worth is questionable sometimes becomes a solution to a problem in the future. And solving problems profitably is quite satisfying indeed.

And here are the solutions to the "connect-the-dots" problem . . . "draw three parallel rows of three dots each on a piece of paper; connect all nine dots by drawing four straight lines, never lifting the writing implement from the paper."

B. With three lines . . .

Solution for 3 lines:

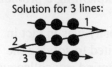

C. With two . . .

Solution for 2 lines:

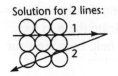

D. And finally, with one line . . .

Solution for 1 line:

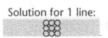

13. A New Graduate's Guide to the Analog Interview

It wasn't that long ago that armed with a couple of engineering degrees and a snappy new suit, I walked headlong into disaster: my first technical interview. The interview was with a well-known Silicon Valley integrated circuit manufacturer, and I had no idea what was in store for me. After flailing through six one-hour grueling technical sessions and my first power lunch, I remember stumbling to my car while visions of pn junctions, amplifiers, TTL gates, and flaming airplanes in a deadly tailspin swam though my brain. What went wrong?

I didn't go into the interview unprepared. I attended the "how to interview" classes held by the career placement center. The center's staff had helped me create a resumé with plenty of style and power adjectives. I was forced to watch the videotape of my practice interview in hopes that my awkward hand gestures and use of the deadly "you know" and "uh" might improve. My girlfriend (now my wife) had picked out the tie. I had five years of engineering classes and lab experience, and had spent the last two learning about analog IC design. I had torn apart my Apple II computer, designed and built my own stereo amplifier, and knew where the power-on button of a Tektronix 547 oscilloscope was located.

What went wrong? The people in the career planning office had taught me about the generic interview, my professors had taught me about analog circuit design, but it was up to me to learn how to combine the two. It took a couple of days of "on the interview training," before I finally got the hang of it, and the interviews became easier.

Now that I am sitting on the other side of the interviewing table, I find that most students still find themselves in the position I was in 10 years ago. The first interview is tough, and the last is easy. So here are some tips that I hope will make your first interview as good as your last. All it takes is a little preparation, knowing what to expect during the interview, and being able to solve a handful of basic analog circuit problems.

Preparation

Be prepared to answer this question intelligently: what do you want to do? It is surprising how many students fumble for answers when asked this question. I have actually heard students say "uh, graduate" and "get a job." Wrong. A well-thought-out answer with a dash of enthusiasm will go a long way towards getting an offer letter. As an interviewer, I would like to hear something like, "I want to join your company so I can sit at the feet of the gurus of analog integrated circuit design," but since this has yet to happen, I would settle for someone who says he has a keen interest in analog design and is willing to work hard.

All good interviewers will ask you to describe something that you have done before, so learn one circuit or system very well. It could be from a senior project, classwork, a final exam, or simply a late-night home-brew circuit hack. Have your classmates or an advisor pepper you with questions about the circuit. "What is the bandwidth? How did you compensate this node? What is the function of this transistor?" I like to ask the following question during an interview: draw me the schematic of any amplifier that you have designed and tell me about it. I then see how far the student can go in describing the circuit. The idea is to put the student at ease by having him describe a circuit that he is familiar with, while I find out how well he really understands the circuit.

If you describe a design or research project on your resumé, you better know it backward and forward. I occasionally interview a student whose resumé claims he has worked on a very challenging project, but he is unable to answer even the most basic technical questions about it. Adding a flashy project to your resumé may get you noticed, but if you are not prepared to discuss the project's technical details in depth, it is the quickest route to a rejection letter. If you don't thoroughly understand something, leave it off the resumé.

Before you go to the interview, find out what the company does. Find a data book or other literature that describes the company's products. By becoming familiar with the product line, you will be able to anticipate what technical questions you will get, and be able to ask some inspired questions. For example, when a classmate of mine was about to interview at a satellite communications company, he spent an entire day in the Stanford library reading all of the IEEE journal articles that the company's famous chief scientist had written. During the interview, my classmate was asked how he would design a certain system, so he said, "Well, at first glance I would probably do it like this . . . ," then went on to describe everything he had read in the chief scientist's articles. Of course my classmate came out of the interview looking like a genius and got the offer.

Know ahead of time what salary you want. Go to the career placement center and get a salary survey of students in your field with the same degree. It is best to know what you are worth so you can negotiate the salary you want in the beginning. Once you start working it is too late.

Prepare a set of questions that you will ask the interviewer. What is the worst and best part of his job? How does he like the company? What is the most difficult circuit he has designed? Design some questions so you get a feel for what it is like to work at that company, and whether or not you will be able to work with these people 8+ hours a day.

Finally, keep in mind that most managers think that enthusiasm, willingness to work hard, good communication skills, and amiable demeanor are much more important than the ability to solve a handful of tricky circuit problems. So when you interview, relax. Try to convey your love for analog design, your willingness to work hard, and try to stay cool. And please, remember not to call the interviewer "dude." (That actually happened more than once.)

The Interview

Most companies go through a three-step interview process. The first step is a quick on-campus interview to make sure that you are really in the electrical engineering program, you can speak in complete sentences, and you can answer some basic circuit questions. If you don't look like a complete bum, show an interest in analog design, and can recite Ohm's Law from memory, you can usually make it past this interview.

The second interview is over the phone with the hiring manager. He wants to make sure that is worth the time and effort to bring you into the plant for the final interview. The phone interview usually consists of asking what classes you took, asking you to describe the project listed on the resumé, then a series of simple circuit questions.

The third and most important interview is at the factory. The hiring manager will generally warm you up with a cup of coffee, a plant tour, and a description of the work the group is doing. Then all hell breaks loose. You will have several one-hour technical interviews with different engineers, a lunch interview where the technical staff tries to determine your compatibility with the group while you bravely try to describe pn junction theory and chew at the same time, followed by an afternoon of more technical interviews. If you have an advanced degree, you will usually be required to give a lecture to the technical staff as well.

The term "technical interview" doesn't tell the whole story; "technical grilling" is more appropriate. After the usual introductions and discussion of your career goals, etc., the grilling will begin. If the interviewer is good, he will have you describe the circuit or system listed on your resumé, which you will ace because you came prepared. Then the interviewer will pull out his favorite technical questions. These are usually designed to test your basic knowledge of circuit design, and more importantly, they allow the interviewer to evaluate your approach to solving problems that you have not seen before.

Some interviewers will have you solve the problems on paper, others on a marker board on the wall, but in either case, you will be required to think on your feet. Remember that the interviewer is looking at your approach to solving the problem and doesn't always expect you to solve it completely. When trying to solve a new problem, resist the temptation to start writing equations right away. Stop and think about what is really happening in the circuit. Try to reason out the function of different sections of the circuit and decide what parts you do and don't understand. Try to describe out loud what you are thinking. For instance, "If this node goes up, then that node goes down, so the circuit is using negative feedback." Once you understand how the circuit works, and you have a plan of attack, then you can pull out the equations.

Remember that it is always much better to say that you don't understand something than to guess. You'll never get hired if a manager thinks you are trying to b.s. your way through a problem. Rather, tell the interviewer what you do know, and what you don't understand. Tell him what you will need to know in order to solve the problem.

Try to jot down some notes about each question that you are asked. If you weren't able to solve it completely, try to finish it at home. You will be surprised at how many times the same circuit problem comes up at different interviews. When I was interviewing, I heard some questions so many times that I had to force myself to prevent the answer from sounding like a tape recording. (#1 question: What are the components of the threshold voltage for a MOS transistor?)

Make sure that you get a list of the people that interviewed you and a business card from each one. It is always a good idea to write all the interviewers thank you notes a couple of days after the interview, as it provides an easy way of reminding them of who you are and that you really want a job. Even if you don't get a job offer, they may provide valuable contacts in the future.

Sample Interview Questions

Interview questions come in all shapes and forms. I had to complete a 10-page exam for one interview. The first problem was trivial and each one got progressively harder, with the last one being mind-numbing. The interviewer used the exam to keep track of how well each university was preparing its students, and as a reference to remember each student. (Results: #1 UC Berkeley) Some companies, like Hewlett-Packard, like to ask tough questions that are not related to your field of expertise just to watch you sweat. I had this question while interviewing for a circuit design job: "You have a beaker of water with diameter x, water depth y, and you stir the water at a constant rotational velocity. How high does the water move up the sides of the beaker? I'll give you any equation you need to know." But you'll find that most questions are simple and keep appearing over and over. Here is a sample of common interview questions that I have accumulated over the years from my friends in the analog business (yes, the answers are in the back):

Q1. If you put a 0-to-5-voltstep voltage referenced to ground into the circuits shown in Figures 13–1A and 13–1B, sketch the wave forms you would expect to see at the outputs.

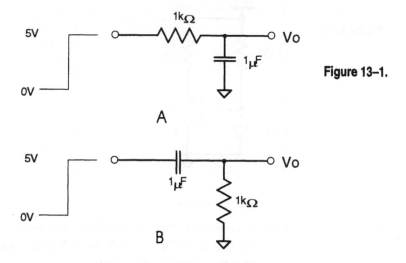

Figure 13–1.

A

B

Q2. As the base emitter voltage of the bipolar transistor Q1 in Figure 13–2 is increased from 0V, sketch the voltage at the output node.

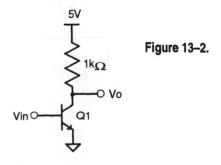

Figure 13–2.

Q3. Two loudspeakers with a passive input filter are shown in Figures 13–3A and 13–3B. Which one is the woofer, and which one is the tweeter?

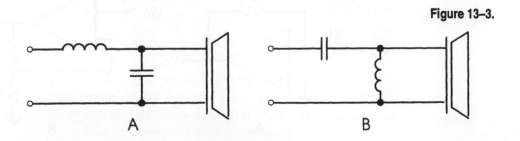

Figure 13–3.

A   B

Q4. In Figure 13–4, the diode and transistor are a matched pair. If the forward voltage of the diode is 0.7V, what is the approximate collector current in the transistor Q1?

Figure 13–4.

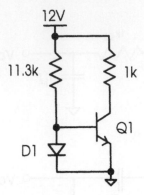

Q5. A constant-current Io is fed into the diode connected-transistor Q1 shown in Figure 13–5. What happens to the output voltage Vo as temperature is increased?

Figure 13–5.

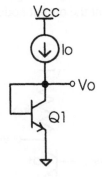

Q6. The ideal op amps of Figures 13–6A and 13–6B are connected with feedback resistors R1 and R2. What is the closed-loop DC gain of each configuration?

Figure 13–6.

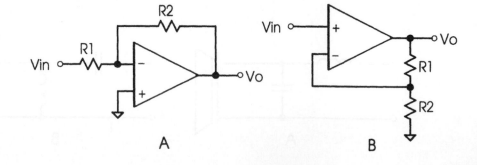

Q7. Assume that the op amps of Figures 13–6A and 13–6B have finite gain A_o. Now what is the closed-loop DC gain?

Q8. The capacitor of Figure 13–7 is connected with two ideal MOS switches. Switches T1 and T2 are alternately turned on with a frequency f_c. What is the average current flowing from node 1 to node 2? What is the equivalent impedance from node 1 to node 2?

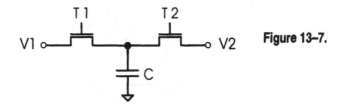

Figure 13–7.

Q9. The regulator of Figure 13–8 has an input voltage of 8V, a bias resistor R1 of 100Ω, and 10mA flowing through the 6V zener diode. Calculate the value of beta of the NPN transistor Q1 if the load current is 100mA.

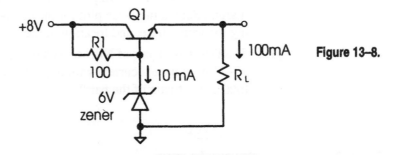

Figure 13–8.

Q10. Assume that the diode D1 of Figure 13–9 is ideal. Sketch the wave form of Vo.

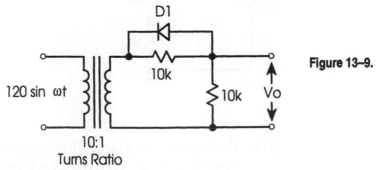

Figure 13–9.

Q11. The bipolar transistor of Figure 13–10 is biased so the voltage across R_L is 260mV. A small AC signal is applied to the input node. Qualitatively describe what the voltage at the output looks like. Calculate the AC gain.

Figure 13–10.

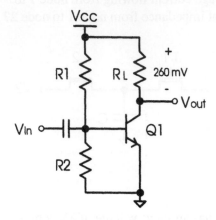

Q12. A two-pole amplifier is found to have an open-loop DC gain of 100dB, a gain-bandwidth product of 10MHz, and 45° of phase margin. Sketch the Bode plot for the open-loop amplifier, showing the gain, phase, and location of the poles.

Q13. The Darlington pair of NPN transistors Q1 and Q2 in Figure 13–11 each have a current gain of β. What is the approximate total current gain of the pair?

Figure 13–11.

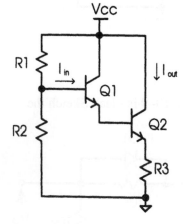

Q14. The drain current of the JFET shown in Figure 13–12 is 2.5mA when Vgs is set to –2.5V, and 2.7mA when Vgs is –2.4V. Calculate the pinch-off voltage and the drain-source saturation current.

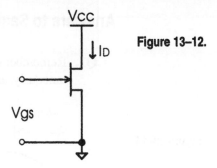

Figure 13–12.

Q15. A CMOS amplifier consisting of PMOS device Q1 and NMOS device Q2 is shown in Figure 13–13. Assuming that they both have the same gate oxide thickness, what is the approximate gain of the amplifier?

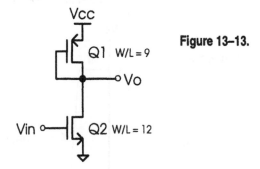

Figure 13–13.

Q16. You are probing a square wave pulse in the lab that has a rise time of 5ns and a fall time of 2ns. What is the minimum bandwidth of the oscilloscope needed to view the signal?

Q17. What is the thermal rms noise voltage of a 1k resistor at 300K?

Q18. A transistor dissipates 25 W in an ambient temperature of 25°C. Given that the thermal resistance of the transistor is 3°C/W and the maximum junction temperature is 150°C, what is the thermal resistance of the heat sink required?

Q19. Draw the equivalent circuit of an exclusive-nor gate using only inverters, nand, and nor gates. (Hey, even analog guys need to know some digital stuff.)

Q20. You are offered the following jobs; which one do you take?

a. Hacking C++ code for Windows

b. A windsurf instructor at Club Med in the Canary Islands

c. A roadie for the upcoming Rolling Stones tour

d. An analog design engineer

Answers to Sample Interview Questions

Q1. Remember that the voltage across a capacitor cannot change instantaneously, and the time constant is 1/RC, as shown in Figure 13–14.

Figure 13–14.

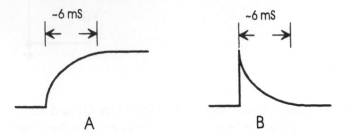

Q2. The output voltage has three distinct regions as shown in Figure 13–15: Q1 off, Q1 in the linear region, and Q1 saturated.

Figure 13–15.

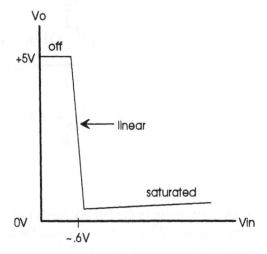

Q3. Assuming that the filter prevents high frequencies from reaching the woofer, and low frequencies from reaching the tweeter, A is the woofer, and B is the tweeter.

Q4. The current through the diode = (12 – 0.7)/11.3k = 1mA. If the diode and Q1 are a matched pair, then the circuit is a current mirror with the collector current equal to 1mA.

Q5. With a constant collector current, the output voltage will show a slope of ~ –2 mV/°C.

Q6. Figure A has an inverting gain of –R2/R1 and B has a noninverting gain of (1 + R1/R2).

Q7. Figure A has an inverting gain of 1/(1/Ao + R1/Ao – R1/R2). Figure B has a noninverting gain of (R2 + R1)/[(R2 + R1)/Ao + R2].

Q8. For every clock cycle, a small amount of charge $= C(V1 - V2)$ is transferred to and from the capacitor. Therefore, the average current is $i = q/time$ or $i = Cf_c(V1 - V2)$. The equivalent impedance is $\Delta V/i = 1/Cf_c$.

Q9. The current in the resistor is $(8 - 6)/100 = 20mA$. If the zener requires 10mA to sustain 6V, then the base current of Q1 is $20mA - 10mA = 10mA$. The transistor is then operating with a beta of $(Ie/Ib - 1) = (100mA/10mA - 1) = 9$.

Q10. With a 10:1 turns ratio, the peak voltage on the secondary side of the transformer is 12V as shown in Figure 13–16. On the positive half cycle, the diode is not conducting so the output voltage is divided in half. On the negative half cycle, the ideal diode conducts so that the full voltage appears at V_o.

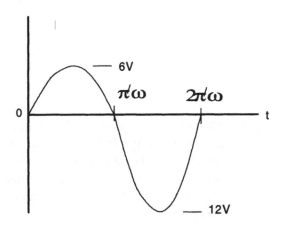

Figure 13–16.

Q11. If the input voltage is a small-signal sine wave, then the output voltage is an amplified sine wave of opposite polarity. If the output impedance of Q1 >> RL, then the gain of the circuit is to first order the g_m of Q1 times the load resistance, $A_0 = - g_m * R_L$. With $g_m = I_c/V_t$ the gain can be rewritten to $A_0 = -I_c R_L/V_t$. Recognizing that $I_c R_L = 260mV$, the equation becomes $A_0 = -260mV/V_t$ or $A_0 = -260mV/26mV = -10$.

Q12. The first pole = 100Hz, the second = 10Mhz as shown in Figure 13–17.

Q13. Current gain $= \beta (\beta + 1)$

Q14. Knowing that $I_D = I_{DSS} (1 - V_{gs}/V_p)^2$, set up simultaneous equations and solve for $I_{DSS} = 9.8mA$ and $V_p = -2.45V$.

Q15. The gain $= (g_m$ n-channel$/g_m$ p-channel$)$. Since $g_m = 2 (K'/2 * W/L * Id)^{1/2}$ and the mobility of the N-channel is approximately 3 times that of the P-channel and Id is the same for both transistors, the gain $= (3 * 12)^{1/2}/(9)^{1/2} = 12$.

Figure 13–17.

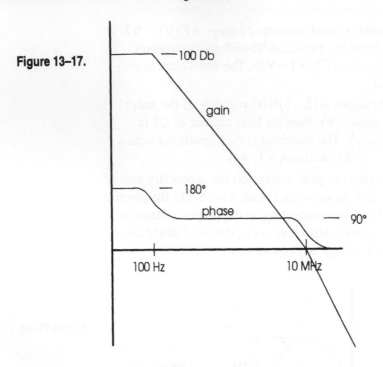

Q16. The time that it takes an RC circuit to go from 10% to 90% of its final value is $\Delta t = \ln 9 * RC$. If the bandwidth of the 'scope $BW = \frac{1}{2}\pi RC$, then the bandwidth $BW = \ln 9/(2\pi * \Delta t) = \ln 9/(2\pi * 2ns) = 174MHz$. Choose a 200MHz or faster 'scope. To reduce errors, choose a 'scope 3 times faster than the calculated value, or 600MHz.

Q17. The average noise voltage squared, $V^2 = 4kTR\,\Delta f$, so $V \sim 4nV/(Hz)^{1/2}$.

Q18. The required $\theta = (150° - 25°)/25\ W = 5°/W$. Since the package has a thermal resistance of 3°C/W, the heat sink must be a minimum of $\theta = (5°C/W - 3°C/W) = 2°C/W$.

Q19. The equation for an exclusive-or gate is $Y = ab' + ba'$. This can be rewritten as $Y = [(ab')' (ba')']'$. The logic diagram is shown in Figure 13–18.

Q20. b

Figure 13–18.

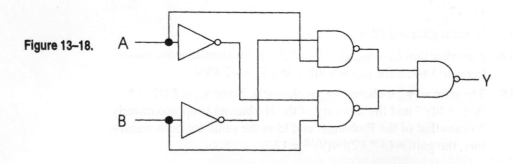

NUMBER CORRECT   RECOMMENDATION

1–5	Become a bond trader.
6–10	Buy a copy of Gray and Meyer. Memorize it.
11–15	Not bad; call up National Semiconductor.
16–19	You have a future as an analog engineer.
20	Give me a call. I know a great boardsailing spot where we can sail and discuss job opportunities.

14. John Harrison's "Ticking Box"*

There was never a shortage of inventive genius in England, and many fertile minds were directed towards the problem of finding longitude at sea. In 1687 two proposals were made by an unknown inventor which were novel, to say the least. He had discovered that a glass filled to the brim with water would run over at the instant of new and full moon, so that the longitude could be determined with precision at least twice a month. His second method was far superior to the first, he thought, and involved the use of a popular nostrum concocted by Sir Kenelm Digby called the "powder of sympathy." This miraculous healer cured open wounds of all kinds, but unlike ordinary and inferior brands of medicine, the powder of sympathy was applied, not to the wound but to the weapon that inflicted it. Digby used to describe how he made one of his patients jump sympathetically merely by putting a dressing he had taken from the patient's wound into a basin containing some of his curative powder. The inventor who suggested using Digby's powder as an aid to navigation proposed that before sailing every ship should be furnished with a wounded dog. A reliable observer on shore, equipped with a standard clock and a bandage from the dog's wound, would do the rest. Every hour, on the dot, he would immerse the dog's bandage in a solution of the powder of sympathy and the dog on shipboard would yelp the hour.

Another serious proposal was made in 1714 by William Whiston, a clergyman, and Humphrey Ditton, a mathematician. These men suggested that a number of lightships be anchored in the principal shipping lanes at regular intervals across the Atlantic ocean. The lightships would fire at regular intervals a star shell timed to explode at 6440 feet. Sea captains could easily calculate their distance from the nearest lightship merely by timing the interval between the flash and the report. This system would be especially convenient in the North Atlantic, they pointed out, where the depth never exceeded 300 fathoms! For obvious reasons, the proposal of Whiston and Ditton was not carried out, but they started something. Their plan was published, and thanks to the publicity it received in various periodicals, a petition was submitted to Parliament on March 25, 1714, by "several Captains of Her Majesty's Ships, Merchants of London, and

---

*Reprinted from "The Story of Maps"

Commanders of Merchantmen," setting forth the great importance of finding the longitude and praying that a public reward be offered for some practicable method of doing it. Not only the petition but the proposal of Whiston and Ditton were referred to a committee, who in turn consulted a number of eminent scientists including Newton and Halley.

That same year Newton prepared a statement which he read to the committee. He said, "That, for determining the Longitude at Sea, there have been several Projects, true in the Theory, but difficult to execute." Newton did not favor the use of the eclipses of the satellites of Jupiter, and as for the scheme proposed by Whiston and Ditton, he pointed out that it was rather a method of "keeping an Account of the Longitude at Sea, than for finding it, if at any time it should be lost." Among the methods that are difficult to execute, he went on, "One is, by a Watch to keep time exactly: But, by reason of the Motion of a Ship, the Variation of Heat and Cold, Wet and Dry, and the Difference of Gravity in Different Latitudes, such a Watch hath not yet been made." That was the trouble: such a watch had not been made.

The idea of transporting a timekeeper for the purpose of finding longitude was not new, and the futility of the scheme was just as old. To the ancients it was just a dream. When Gemma Frisius suggested it in 1530 there were mechanical clocks, but they were a fairly new invention, and crudely built, which made the idea improbable if not impossible. The idea of transporting "some true Horologie or Watch, apt to be carried in journeying, which by an Astrolabe is to be rectified . . ." was again stated by Blundeville in 1622, but still there was no watch which was "true" in the sense of being accurate enough to use for determining longitude. If a timekeeper was the answer, it would have to be very accurate indeed. According to Picard's value, a degree of longitude was equal to about sixty-eight miles at the equator, or four minutes by the clock. One minute of time meant seventeen miles—towards or away from danger. And if on a six weeks' voyage a navigator wanted to get his longitude within half a degree (thirty-four miles) the rate of his timekeeper must not gain or lose more than two minutes in forty-two days, or three seconds a day.

Fortified by these calculations, which spelled the impossible, and the report of the committee, Parliament passed a bill (1714) "for providing a publick reward for such person or persons as shall discover the Longitude." It was the largest reward ever offered, and stated that for any practical invention the following sum would be paid:

£10,000 for any device that would determine the longitude within 1 degree.

£15,000 for any device that would determine the longitude within 40 minutes.

£20,000 for any device that would determine the longitude within 30 minutes (2 minutes of time or 34 miles).

As though aware of the absurdity of their terms, Parliament authorized the formation of a permanent commission—the Board of Longitude—and empowered it to pay one half of any of the above rewards as soon as a majority of its members were satisfied that any proposed method was practicable and useful, and that it would give security to ships within eighty miles of danger, meaning land. The other half of any reward would be paid as soon as a ship using the device should sail from Britain to a port in the West Indies without erring in her longitude more than the amounts specified. Moreover, the Board was authorized to grant a smaller reward for a less accurate method, provided it was practicable, and to spend a sum not to exceed £2000 on experiments which might lead to a useful invention.

For fifty years this handsome reward* stood untouched, a prize for the impossible, the butt of English humorists and satirists. Magazines and newspapers used it as a stock cliché. The Board of Longitude failed to see the joke. Day in and day out they were hounded by fools and charlatans; the perpetual motion lads and the geniuses who could quarter a circle and trisect an angle. To handle the flood of crackpots, they employed a secretary who handed out stereotyped replies to stereotyped proposals. The members of the Board met three times a year at the Admiralty, contributing their services and their time to the Crown. They took their responsibilities seriously and frequently called in consultants to help them appraise a promising invention. They were generous with grants-in-aid to struggling inventors with sound ideas, but what they demanded was results. Neither the Board nor any one else knew exactly what they were looking for, but what everyone knew was that the longitude problem had stopped the best minds in Europe, including Newton, Halley, Huygens, von Leibnitz and all the rest. It was solved, finally, by a ticking machine in a box, the invention of an uneducated Yorkshire carpenter named John Harrison. The device was the marine chronometer.

Early clocks fell into two general classes: nonportable timekeepers driven by a falling weight, and portable timekeepers such as table clocks and crude watches, driven by a coiled spring. Gemma Frisius suggested the latter for use at sea, but with reservations. Knowing the unreliable temperament of spring-driven timekeepers, he admitted that sand and water clocks would have to be carried along to check the error of a spring-driven machine. In Spain, during the reign of Philip II, clocks were solicited which would run exactly twenty-four hours a day, and many different kinds had been invented. According to Alonso de Santa Cruz there were "some with wheels, chains and weights of steel: some with chains of catgut and steel: others using sand, as in sandglasses: others with water in place of sand, and designed after many different fashions:

---

*Editor's note: The prize was equal to about 6 million 1994 dollars.

others again with vases or large glasses filled with quicksilver: and, lastly, some, the most ingenious of all, driven by the force of the wind, which moves a weight and thereby the chain of the clock, or which are moved by the flame of a wick saturated with oil: and all of them adjusted to measure twenty-four hours exactly."

Robert Hooke became interested in the development of portable timekeepers for use at sea about the time Huygens perfected the pendulum clock. One of the most versatile scientists and inventors of all time, Hooke was one of those rare mechanical geniuses who was equally clever with a pen. After studying the faults of current timekeepers and the possibility of building a more accurate one, he slyly wrote a summary of his investigations, intimating that he was completely baffled and discouraged. "All I could obtain," he said, "was a Catalogue of Difficulties, first in the doing of it, secondly in the bringing of it into publick use, thirdly, in making advantage of it. Difficulties were proposed from the alteration of Climates, Airs, heats and colds, temperature of Springs, the nature of Vibrations, the wearing of Materials, the motion of the Ship, and divers others." Even if a reliable timekeeper were possible, he concluded, "it would be difficult to bring it to use, for Sea-men know their way already to any Port. . . ." As for the rewards: "the Praemium for the Longitude," there never was any such thing, he retorted scornfully. "No King or State would pay a farthing for it."

In spite of his pretended despondency, Hooke nevertheless lectured in 1664 on the subject of applying springs to the balance of a watch in order to render its vibrations more uniform, and demonstrated, with models, twenty different ways of doing it. At the same time he confessed that he had one or two other methods up his sleeve which he hoped to cash in on at some future date. Like many scientists of the time, Hooke expressed the principle of his balance spring in a Latin anagram; roughly: Ut tensio, sic vis, "as the tension is, so is the force," or, "the force exerted by a spring is directly proportional to the extent to which it is tensioned."

The first timekeeper designed specifically for use at sea was made by Christian Huygens in 1660. The escapement was controlled by a pendulum instead of a spring balance, and like many of the clocks that followed, it proved useless except in a flat calm. Its rate was unpredictable; when tossed around by the sea it either ran in jerks or stopped altogether. The length of the pendulum varied with changes of temperature, and the rate of going changed in different latitudes, for some mysterious reason not yet determined. But by 1715 every physical principle and mechanical part that would have to be incorporated in an accurate timekeeper was understood by watchmakers. All that remained was to bridge the gap between a good clock and one that was nearly perfect. It was that half degree of longitude, that two minutes of time, which meant the difference between conquest and failure, the difference between £20,000 and just another timekeeper.

One of the biggest hurdles between watchmakers and the prize money was the weather: temperature and humidity. A few men included barometric pressure. Without a doubt, changes in the weather did things to clocks and watches, and many suggestions were forthcoming as to how this principal source of trouble could be overcome. Stephen Plank and William Palmer, watchmakers, proposed keeping a timekeeper close to a fire, thus obviating errors due to change in temperature. Plank suggested keeping a watch in a brass box over a stove which would always be hot. He claimed to have a secret process for keeping the temperature of the fire uniform. Jeremy Thacker, inventor and watchmaker, published a book on the subject of the longitude, in which he made some caustic remarks about the efforts of his contemporaries. He suggested that one of his colleagues, who wanted to test his clock at sea, should first arrange to have two consecutive Junes equally hot at every hour of every day. Another colleague, referred to as Mr. Br . . . e, was dubbed the Corrector of the Moon's Motion. In a more serious vein, Thacker made several sage observations regarding the physical laws with which watchmakers were struggling. He verified experimentally that a coiled spring loses strength when heated and gains it when cooled. He kept his own clock under a kind of bell jar connected with an exhaust pump, so that it could be run in a partial vacuum. He also devised an auxiliary spring which kept the clock going while the mainspring was being wound. Both springs were wound outside the bell by means of rods passed through stuffing boxes, so that neither the vacuum nor the clock mechanism would have to be disturbed. In spite of these and other devices, watchmakers remained in the dark and their problems remained unsolved until John Harrison went to work on the physical laws behind them. After that they did not seem so difficult.

Harrison was born at Foulby in the parish of Wragby, Yorkshire, in May, 1693. He was the son of a carpenter and joiner in the service of Sir Rowland Winn of Nostell Priory. John was the oldest son in a large family. When he was six years old he contracted smallpox, and while convalescing spent hours watching the mechanism and listening to the ticking of a watch laid on his pillow. When his family moved to Barrow in Lincolnshire, John was seven years old. There he learned his father's trade and worked with him for several years. Occasionally he earned a little extra by surveying and measuring land, but he was much more interested in mechanics, and spent his evenings studying Nicholas Saunderson's published lectures on mathematics and physics. These he copied out in longhand including all the diagrams. He also studied the mechanism of clocks and watches, how to repair them and how they might be improved. In 1715, when he was twenty-two, he built his first grandfather clock or "regulator." The only remarkable feature of the machine was that all the wheels except the escape wheel were made of oak, with the teeth, carved separately, set into a groove in the rim.

Many of the mechanical faults in the clocks and watches that Harrison saw around him were caused by the expansion and contraction of the metals used in their construction. Pendulums, for example, were usually made of an iron or steel rod with a lead bob fastened at the end. In winter the rod contracted and the clock went fast, and in summer the rod expanded, making the clock lose time. Harrison made his first important contribution to clockmaking by developing the "gridiron" pendulum, so named because of its appearance. Brass and steel, he knew, expand for a given increase in temperature in the ratio of about three to two (100 to 62). He therefore built a pendulum with nine alternating steel and brass rods, so pinned together that expansion or contraction caused by variation in the temperature was eliminated, the unlike rods counteracting each other.

The accuracy of a clock is no greater than the efficiency of its escapement, the piece which releases for a second, more or less, the driving power, such as a suspended weight or a coiled mainspring. One day Harrison was called out to repair a steeple clock that refused to run. After looking it over he discovered that all it needed was some oil on the pallets of the escapement. He oiled the mechanism and soon after went to work on a design for an escapement that would not need oiling. The result was an ingenious "grasshopper" escapement that was very nearly frictionless and also noiseless. However, it was extremely delicate, unnecessarily so, and was easily upset by dust or unnecessary oil. These two improved parts alone were almost enough to revolutionize the clockmaking industry. One of the first two grandfather clocks he built that were equipped with his improved pendulum and grasshopper escapement did not gain or lose more than a second a month during a period of fourteen years.

Harrison was twenty-one years old when Parliament posted the £20,000 reward for a reliable method of determining longitude at sea. He had not finished his first clock, and it is doubtful whether he seriously aspired to winning such a fortune, but certainly no young inventor ever had such a fabulous goal to shoot at, or such limited competition. Yet Harrison never hurried his work, even after it must have been apparent to him that the prize was almost within his reach. On the contrary, his real goal was the perfection of his marine timekeeper as a precision instrument and a thing of beauty. The monetary reward, therefore, was a foregone conclusion.

His first two fine grandfather clocks were completed by 1726, when he was thirty-three years old, and in 1728 he went to London, carrying with him full-scale models of his gridiron pendulum and grasshopper escapement, and working drawings of a marine clock he hoped to build if he could get some financial assistance from the Board of Longitude. He called on Edmund Halley, Astronomer Royal, who was also a member of the Board. Halley advised him not to depend on the Board of Longitude, but to talk things over with George Graham, England's leading horologist. Harrison called on Graham at ten o'clock one morning, and together they talked pendulums, escapements, remontoires and springs until eight o'clock in the evening, when Harrison departed a happy man. Graham had advised him to build his clock first and then apply to the Board of

Longitude. He had also offered to loan Harrison the money to build it with, and would not listen to any talk about interest or security of any kind. Harrison went home to Barrow and spent the next seven years building his first marine timekeeper, his "Number One," as it was later called.

In addition to heat and cold, the archenemies of all watchmakers, he concentrated on eliminating friction, or cutting it down to a bare minimum, on every moving part, and devised many ingenious ways of doing it; some of them radical departures from accepted watchmaking practice. Instead of using a pendulum, which would be impractical at sea, Harrison designed two huge balances weighing about five pounds each, that were connected by wires running over brass arcs so that their motions were always opposed. Thus any effect on one produced by the motion of the ship would be counteracted by the other. The "grasshopper" escapement was modified and simplified and two mainsprings on separate drums were installed. The clock was finished in 1735.

There was nothing beautiful or graceful about Harrison's Number One. It weighed seventy-two pounds and looked like nothing but an awkward, unwieldy piece of machinery. However, everyone who saw it and studied its mechanism declared it a masterpiece of ingenuity, and its performance certainly belied its appearance. Harrison mounted its case in gimbals and for a while tested it unofficially on a barge in the Humber River. Then he took it to London where he enjoyed his first brief triumph. Five members of the Royal Society examined the clock, studied its mechanism and then presented Harrison with a certificate stating that the principles of this timekeeper promised a sufficient degree of accuracy to meet the requirements set forth in the Act of Queen Anne. This historic document, which opened for Harrison the door to the Board of Longitude, was signed by Halley, Smith, Bradley, Machin and Graham.

On the strength of the certificate, Harrison applied to the Board of Longitude for a trial at sea, and in 1736 he was sent to Lisbon in H.M.S. Centurion, Captain Proctor. In his possession was a note from Sir Charles Wager, First Lord of the Admiralty, asking Proctor to see that every courtesy be given the bearer, who was said by those who knew him best to be "a very ingenious and sober man." Harrison was given the run of the ship, and his timekeeper was placed in the Captain's cabin where he could make observations and wind his clock without interruption. Proctor was courteous but skeptical. "The difficulty of measuring Time truly," he wrote, "where so many unequal Shocks and Motions stand in Opposition to it, gives me concern for the honest Man, and makes me feel he has attempted Impossibilities."

No record of the clock's going on the outward voyage is known, but after the return trip, made in H.M.S. Oxford, Robert Man, Harrison was given a certificate signed by the master (that is, navigator) stating: "When we made the land, the said land, according to my reckoning (and others), ought to have been the Start; but before we knew what land it was, John

Harrison declared to me and the rest of the ship's company, that according to his observations with his machine, it ought to be the Lizard—the which, indeed, it was found to be, his observation showing the ship to be more west than my reckoning, above one degree and twenty-six miles." It was an impressive report in spite of its simplicity, and yet the voyage to Lisbon and return was made in practically a north and south direction; one that would hardly demonstrate the best qualities of the clock in the most dramatic fashion. It should be noted, however, that even on this well-worn trade route it was not considered a scandal that the ship's navigator should make an error of 90 miles in his landfall.

On June 20, 1737, Harrison made his first bow to the mighty Board of Longitude. According to the official minutes, "Mr. John Harrison produced a new invented machine, in the nature of clockwork, whereby he proposes to keep time at sea with more exactness than by any other instrument or method hitherto contrived . . . and proposes to make another machine of smaller dimensions within the space of two years, whereby he will endeavour to correct some defects which he hath found in that already prepared, so as to render the same more perfect . . ." The Board voted him £500 to help defray expenses, one half to be paid at once and the other half when he completed the second clock and delivered same into the hands of one of His Majesty's ship's captains.

Harrison's Number Two contained several minor mechanical improvements and this time all the wheels were made of brass instead of wood. In some respects it was even more cumbersome than Number One, and it weighed one hundred and three pounds. Its case and gimbal suspension weighed another sixty-two pounds. Number Two was finished in 1739, but instead of turning it over to a sea captain appointed by the Board to receive it, Harrison tested it for nearly two years under conditions of "great heat and motion." Number Two was never sent to sea because by the time it was ready, England was at war with Spain and the Admiralty had no desire to give the Spaniards an opportunity to capture it.

In January, 1741, Harrison wrote the Board that he had begun work on a third clock which promised to be far superior to the first two. They voted him another £500. Harrison struggled with it for several months, but seems to have miscalculated the "moment of inertia" of its balances. He thought he could get it going by the first of August, 1741, and have it ready for a sea trial two years later. But after five years the Board learned "that it does not go well, at present, as he expected it would, yet he plainly perceived the Cause of its present Imperfection to lye in a certain part [the balances] which, being of a different form from the corresponding part in the other machines, had never been tried before." Harrison had made a few improvements in the parts of Number Three and had incorporated in it the same antifriction devices he had used on Number Two, but the clock was still bulky and its parts were far from delicate; the machine weighed sixty-six pounds and its case and gimbals another thirty-five.

Harrison was again feeling the pinch, even though the Board had given him several advances to keep him going, for in 1746, when he reported on Number Three, he laid before the Board an impressive testimonial signed by twelve members of the Royal Society including the President, Martin Folkes, Bradley, Graham, Halley and Cavendish, attesting the importance and practical value of his inventions in the solution of the longitude problem. Presumably this gesture was made to insure the financial support of the Board of Longitude. However, the Board needed no prodding. Three years later, acting on its own volition, the Royal Society awarded Harrison the Copley medal, the highest honor it could bestow. His modesty, perseverance and skill made them forget, at least for a time, the total lack of academic background which was so highly revered by that august body.

Convinced that Number Three would never satisfy him, Harrison proposed to start work on two more timekeepers, even before Number Three was given a trial at sea. One would be pocketsize and the other slightly larger. The Board approved the project and Harrison went ahead. Abandoning the idea of a pocketsize chronometer, Harrison decided to concentrate his efforts on a slightly larger clock, which could be adapted to the intricate mechanism he had designed without sacrificing accuracy. In 1757 he began work on Number Four, a machine which "by reason alike of its beauty, its accuracy, and its historical interest, must take pride of place as the most famous chronometer that ever has been or ever will be made." It was finished in 1759.

Number Four resembled an enormous "pair-case" watch about five inches in diameter, complete with pendant, as though it were to be worn. The dial was white enamel with an ornamental design in black. The hour and minute hands were of blued steel and the second hand was polished. Instead of a gimbal suspension, which Harrison had come to distrust, he used only a soft cushion in a plain box to support the clock. An adjustable outer box was fitted with a divided arc so that the timekeeper could be kept in the same position (with the pendant always slightly above the horizontal) regardless of the lie of the ship. When it was finished, Number Four was not adjusted for more than this one position, and on its first voyage it had to be carefully tended. The watch beat five to the second and ran for thirty hours without rewinding. The pivot holes were jeweled to the third wheel with rubies and the end stones were diamonds. Engraved in the top-plate were the words "John Harrison & Son, A.D. 1759." Cunningly concealed from prying eyes beneath the plate was a mechanism such as the world had never seen; every pinion and bearing, each spring and wheel was the end product of careful planning, precise measurement and exquisite craftsmanship. Into the mechanism had gone "fifty years of self-denial, unremitting toil, and ceaseless concentration." To Harrison, whose singleness of purpose had made it possible for him to achieve the impossible, Number Four was a satisfactory climax to a lifetime of effort. He was proud of this timekeeper, and in a rare burst of eloquence he wrote, "I think I may make bold to say, that there is neither any other Mechanical or Mathematical thing in the World that is more beautiful or curious in texture than this my watch or Time-keeper for the Longitude . . . and I heartily thank Almighty God that I have lived so long, as in some measure to complete it."

After checking and adjusting Number Four with his pendulum clock for nearly two years, Harrison reported to the Board of Longitude, in March 1761, that Number Four was as good as Number Three and that its performance greatly exceeded his expectations. He asked for a trial at sea. His request was granted, and in April, 1761, William Harrison, his son and right-hand man, took Number Three to Portsmouth. The father arrived a short time later with Number Four. There were numerous delays at Portsmouth, and it was October before passage was finally arranged for young Harrison aboard H.M.S. Deptford, Dudley Digges, bound for Jamaica. John Harrison, who was then sixty-eight years old, decided not to attempt the long sea voyage himself; and he also decided to stake everything on the performance of Number Four, instead of sending both Three and Four along. The Deptford finally sailed from Spithead with a convoy, November 18, 1761, after first touching at Portland and Plymouth. The sea trial was on.

Number Four had been placed in a case with four locks, and the four keys were given to William Harrison, Governor Lyttleton of Jamaica, who was taking passage on the Deptford, Captain Digges, and his first lieutenant. All four had to be present in order to open the case, even for winding. The Board of Longitude had further arranged to have the longitude of Jamaica determined de novo before the trial, by a series of observations of the satellites of Jupiter, but because of the lateness of the season it was decided to accept the best previously established reckoning. Local time at Portsmouth and at Jamaica was to be determined by taking equal altitudes of the sun, and the difference compared with the time indicated by Harrison's timekeeper.

As usual, the first scheduled port of call on the run to Jamaica was Madeira. On this particular voyage, all hands aboard the Deptford were anxious to make the island on the first approach. To William Harrison it meant the first crucial test of Number Four; to Captain Digges it meant a test of his dead reckoning against a mechanical device in which he had no confidence; but the ship's company had more than a scientific interest in the proceedings. They were afraid of missing Madeira altogether, "the consequence of which, would have been Inconvenient." To the horror of all hands, it was found that the beer had spoiled, over a thousand gallons of it, and the people had already been reduced to drinking water. Nine days out from Plymouth the ship's longitude, by dead reckoning, was 13° 50' west of Greenwich, but according to Number Four and William Harrison it was 15° 19' W. Captain Digges naturally favored his dead reckoning calculations, but Harrison stoutly maintained that Number Four was right and that if Madeira were properly marked on the chart they would sight it the next day. Although Digges offered to bet Harrison five to one that he was wrong, he held his course, and the following morning at 6 A.M. the lookout sighted Porto Santo, the northeastern island of the Madeira group, dead ahead.

The Deptford's officers were greatly impressed by Harrison's uncanny predictions throughout the voyage. They were even more impressed when they arrived at Jamaica three days before H.M.S. Beaver, which had sailed for Jamaica ten days before them. Number Four was promptly taken ashore and checked. After allowing for its rate of going (2⅔ seconds per day losing at Portsmouth), it was found to be 5 seconds slow, an error in longitude of 1¼' only, or 1¼ nautical miles.

The official trial ended at Jamaica. Arrangements were made for William Harrison to make the return voyage in the Merlin, sloop, and in a burst of enthusiasm Captain Digges placed his order for the first Harrison-built chronometer which should be offered for sale. The passage back to England was a severe test for Number Four. The weather was extremely rough and the timekeeper, still carefully tended by Harrison, had to be moved to the poop, the only dry place on the ship, where it was pounded unmercifully and "received a number of violent shocks." However, when it was again checked at Portsmouth, its total error for the five months' voyage, through heat and cold, fair weather and foul (after allowing for its rate of going), was only $1^m 53\frac{1}{2}°$, or an error in longitude of 28½' (28½ nautical miles). This was safely within the limit of half a degree specified in the Act of Queen Anne. John Harrison and son had won the fabulous reward of £20,000.

The sea trial had ended, but the trials of John Harrison had just begun. Now for the first time, at the age of sixty-nine, Harrison began to feel the lack of an academic background. He was a simple man; he did not know the language of diplomacy, the gentle art of innuendo and evasion. He had mastered the longitude but he did not know how to cope with the Royal Society or the Board of Longitude. He had won the reward and all he wanted now was his money. The money was not immediately forthcoming.

Neither the Board of Longitude nor the scientists who served it as consultants were at any time guilty of dishonesty in their dealings with Harrison; they were only human. £20,000 was a tremendous fortune, and it was one thing to dole out living expenses to a watchmaker in amounts not exceeding £500 so that he might contribute something or other to the general cause. But it was another thing to hand over £20,000 in a lump sum to one man, and a man of humble birth at that. It was most extraordinary. Moreover, there were men on the Board and members of the Royal Society who had designs on the reward themselves or at least a cut of it. James Bradley and Johann Tobias Mayer had both worked long and hard on the compilation of accurate lunar tables. Mayer's widow was paid £3000 for his contribution to the cause of longitude, and in 1761 Bradley told Harrison that he and Mayer would have shared £10,000 of the prize money between them if it had not been for his blasted watch. Halley had struggled long and manfully on the solution of the longitude by compass variation, and was not in a position to ignore any part of £20,000. The Reverend Nevil Maskelyne, Astronomer Royal, and compiler of the Nautical Almanac, was an obstinate and uncompromising apostle of "lunar distances" or "lunars" for finding the longitude, and had closed his mind to any other method whatsoever. He loved neither Harrison nor his watch. In view of these and other unnamed aspirants, it was inevitable that the Board should decide that the amazing performance of Harrison's timekeeper was a fluke. They had never been allowed to examine the mechanism, and they pointed out that if a gross of watches were carried to Jamaica under the same conditions, one out of the lot might perform equally well—at least for one trip. They accordingly refused to give Harrison a certificate stating that he had met the requirements of the Act until his timekeeper was given a further trial, or trials. Meanwhile, they did agree to give him the sum of £2500 as an interim reward, since his machine had proved to be a rather useful contraption, though mysterious beyond words. An Act of Parliament (February, 1763) enabling him to receive £5000 as soon as he disclosed the secret of his invention, was completely nullified by the absurdly rigid conditions set up by the Board. He was finally granted a new trial at sea.

The rules laid down for the new trial were elaborate and exacting. The difference in longitude between Portsmouth and Jamaica was to be determined de novo by observations of Jupiter's satellites. Number Four was to be rated at Greenwich before sailing, but Harrison balked, saying "that he did not chuse to part with it out of his hands till he shall have reaped some advantage from it." However, he agreed to send his own rating, sealed, to the Secretary of the Admiralty before the trial began. After endless delays the trial was arranged to take place between Portsmouth and Barbados, instead of Jamaica, and William Harrison embarked on February 14, 1764, in H.M.S. Tartar, Sir John Lindsay, at the Nore. The Tartar proceeded to Portsmouth, where Harrison checked the rate of Number Four with a regulator installed there in a temporary observatory. On March 28, 1764, the Tartar sailed from Portsmouth and the second trial was on.

It was the same story all over again. On April 18, twenty-one days out, Harrison took two altitudes of the sun and announced to Sir John that they were forty-three miles east of Porto Santo. Sir John accordingly steered a direct course for it, and at one o'clock the next morning the island was sighted, "which exactly agreed with the Distance mentioned above." They arrived at Barbados May 13, "Mr. Harrison all along in the Voyage declaring how far he was distant from that Island, according to the best settled longitude thereof. The Day before they made it, he declared the Distance: and Sir John sailed in Consequence of this Declaration, till Eleven at Night, which proving dark he thought proper to lay by. Mr. Harrison then declaring they were no more than eight or nine

Miles from the Land, which accordingly at Day Break they saw from that Distance."

When Harrison went ashore with Number Four he discovered that none other than Maskelyne and an assistant, Green, had been sent ahead to check the longitude of Barbados by observing Jupiter's satellites. Moreover, Maskelyne had been orating loudly on the superiority of his own method of finding longitude, namely by lunar distances. When Harrison heard what had been going on he objected strenuously, pointing out to Sir John that Maskelyne was not only an interested party but an active and avid competitor, and should not have anything to do with the trials. A compromise was arranged, but, as it turned out, Maskelyne was suddenly indisposed and unable to make the observations.

After comparing the data obtained by observation with Harrison's chronometer, Number Four showed an error of 38.4 seconds over a period of seven weeks, or 9.6 miles of longitude (at the equator) between Portsmouth and Barbados. And when the clock was again checked at Portsmouth, after 156 days, elapsed time, it showed, after allowing for its rate of going, a total gain of only 54 seconds of time. If further allowance were made for changes of rate caused by variations in temperature, information posted beforehand by Harrison, the rate of Number Four would have been reduced to an error of 15 seconds of loss in 5 months, or less than 1/10 of a second a day.

The evidence in favor of Harrison's chronometer was overwhelming, and could no longer be ignored or set aside. But the Board of Longitude was not through. In a Resolution of February 9, 1765, they were unanimously of the opinion that "the said timekeeper has kept its time with sufficient correctness, without losing its longitude in the voyage from Portsmouth to Barbados beyond the nearest limit required by the Act 12th of Queen Anne, but even considerably within the same." Now, they said, all Harrison had to do was demonstrate the mechanism of his clock and explain the construction of it, "by Means whereof other such Timekeepers might be framed, of sufficient Correctness to find the Longitude at Sea. . . ." In order to get the first £10,000 Harrison had to submit, on oath, complete working drawings of Number Four; explain and demonstrate the operation of each part, including the process of tempering the springs; and finally, hand over to the Board his first three timekeepers as well as Number Four.

Any foreigner would have acknowledged defeat at this juncture, but not Harrison, who was an Englishman and a Yorkshireman to boot. "I cannot help thinking," he wrote the Board, after hearing their harsh terms, "but I am extremely ill used by gentlemen who I might have expected different treatment from. . . . It must be owned that my case is very hard, but I hope I am the first, and for my country's sake, shall be the last that suffers by pinning my faith on an English Act of Parliament." The case of "Longitude Harrison" began to be aired publicly, and several of his friends launched an impromptu publicity campaign against the Board and against

Parliament. The Board finally softened their terms and Harrison reluctantly took his clock apart at his home for the edification of a committee of six, nominated by the Board; three of them, Thomas Mudge, William Matthews and Larcum Kendall, were watchmakers. Harrison then received a certificate from the Board (October 28, 1765) entitling him to £7500, or the balance due him on the first half of the reward. The second half did not come so easily.

Number Four was now in the hands of the Board of Longitude, held in trust for the benefit of the people of England. As such, it was carefully guarded against prying eyes and tampering, even by members of the Board. However, that learned body did its humble best. First they set out to publicize its mechanism as widely as possible. Unable to take the thing apart themselves, they had to depend on Harrison's own drawings, and these were redrawn and carefully engraved. What was supposed to be a full textual description was written by the Reverend Nevil Maskelyne and printed in book form with illustrations appended: The Principles of Mr. Harrison's Time-Keeper, with Plates of the Same. London, 1767. Actually the book was harmless enough, because no human being could have even begun to reproduce the clock from Maskelyne's description. To Harrison it was just another bitter pill to swallow. "They have since published all my Drawings," he wrote, "without giving me the last Moiety of the Reward, or even paying me and my Son for Our Time at a rate as common Mechanicks; an Instance of such Cruelty and Injustice as I believe never existed in a learned and civilised Nation before." Other galling experiences followed.

With great pomp and ceremony Number Four was carried to the Royal Observatory at Greenwich. There it was scheduled to undergo a prolonged and exhaustive series of trials under the direction of the Astronomer Royal, the Reverend Nevil Maskelyne. It cannot be said that Maskelyne shirked his duty, although he was handicapped by the fact that the timekeeper was always kept locked in its case, and he could not even wind it except in the presence of an officer detailed by the Governor of Greenwich to witness the performance. Number Four, after all, was a £10,000 timekeeper. The tests went on for two months. Maskelyne tried the watch in various positions for which it was not adjusted, dial up and dial down. Then for ten months it was tested in a horizontal position, dial up. The Board published a full account of the results with a preface written by Maskelyne, in which he gave it as his studied opinion "That Mr. Harrison's Watch cannot be depended upon to keep the Longitude within a Degree, in a West-India Voyage of six weeks, nor to keep the Longitude within a Half a Degree for more than a Fortnight, and then it must be kept in a Place where the Thermometer is always some Degrees above freezing." (There was still £10,000 prize money outstanding.)

The Board of Longitude next commissioned Larcum Kendall, watchmaker, to make a duplicate of Number Four. They also advised Harrison that he must make Number Five and Number Six and have them tried at sea, intimating that otherwise he would not be entitled to the other half of the reward. When Harrison asked if he might use Number Four for a short time to help him build two copies of it, he was told that Kendall needed it to work from and that it would be impossible. Harrison did the best he could, while the Board laid plans for an exhaustive series of tests for Number Five and Number Six. They spoke of sending them to Hudson's Bay and of letting them toss and pitch in the Downs for a month or two as well as sending them out to the West Indies.

After three years (1767–1770) Number Five was finished. In 1771, just as the Harrisons were finishing the last adjustments on the clock, they heard that Captain Cook was preparing for a second exploring cruise, and that the Board was planning to send Kendall's duplicate of Number Four along with him. Harrison pleaded with them to send Number Four and Number Five instead, telling them he was willing to stake his claim to the balance of the reward on their performance, or to submit "to any mode of trial, by men not already proved partial, which shall be definite in its nature." The man was now more than ever anxious to settle the business once and for all. But it was not so to be. He was told that the Board did not see fit to send Number Four out of the kingdom, nor did they see any reason for departing from the manner of trial already decided upon.

John Harrison was now seventy-eight years old. His eyes were failing and his skilled hands were not as steady as they were, but his heart was strong and there was still a lot of fight left in him. Among his powerful friends and admirers was His Majesty King George the Third, who had granted Harrison and his son an audience after the historic voyage of the Tartar. Harrison now sought the protection of his king, and "Farmer George," after hearing the case from start to finish, lost his patience. "By God, Harrison, I'll see you righted," he roared. And he did. Number Five was tried at His Majesty's private observatory at Kew. The king attended the daily checking of the clock's performance, and had the pleasure of watching the operation of a timekeeper whose total error over a ten week's period was 4½ seconds.

Harrison submitted a memorial to the Board of Longitude, November 28, 1772, describing in detail the circumstances and results of the trial at Kew. In return, the Board passed a resolution to the effect that they were not the slightest bit interested; that they saw no reason to alter the manner of trial they had already proposed and that no regard would be paid for a trial made under any other conditions. In desperation Harrison decided to play his last card—the king. Backed by His Majesty's personal interest in the proceedings, Harrison presented a petition to the House of Commons with weight behind it. It was heralded as follows: "The Lord North, by His Majesty's Command, acquainted the House that His Majesty, having been informed of the Contents of the said Petition, recommended it to the Consideration of the House." Fox was present to give the petition his full support, and the king was willing, if necessary, to appear at the Bar of the House under an inferior title and testify in Harrison's behalf. At the same time, Harrison circulated a broadside, The Case of Mr. John Harrison, stating his claims to the second half of the reward.

The Board of Longitude began to squirm. Public indignation was mounting rapidly and the Speaker of the House informed the Board that consideration of the petition would be deferred until they had an opportunity to revise their proceedings in regard to Mr. Harrison. Seven Admiralty clerks were put to work copying out all of the Board's resolutions concerning Harrison. While they worked day and night to finish the job, the Board made one last desperate effort. They summoned William Harrison to appear before them; but the hour was late. They put him through a catechism and tried to make him consent to new trials and new conditions. Harrison stood fast, refusing to consent to anything they might propose. Meanwhile a money bill was drawn up by Parliament in record time; the king gave it the nod and it was passed. The Harrisons had won their fight.

Guidance and Commentary

The book concludes with six chapters offering guidance and commentary. Eric Swanson, long a proponent of "Moore's Law," explains why he feels this law dominates all design approaches. John Willison advises on all sorts of things in a highly efficient and far-ranging editorial. Jim Williams explains why a laboratory in your home can be an invaluable intellectual and economic investment, and provides details on how to assemble a lab.

In an especially memorable essay, Barrie Gilbert discusses how to promote innovation in the IC business. The chapter is an elegant answer to a world full of managerial "methods" which purport to systematize innovation.

Carl Nelson discusses combining loose thinking with strict conformance to mother nature's laws to produce good circuits. Art Delagrange expresses similar views, with examples drawn from a lifetime of design work.

15. Moore's Law

···

Call me a heretic, but in the late 1970s, long before I'd heard of Philbrick, Widlar, or Gilbert, I learned about Moore's Law. Gordon Moore came down to a VLSI conference at Caltech armed with a "moon curve" somewhat like that shown in Figure 15–1. His message was simple: memory density increases fourfold every three years. Run the linear-year versus log-density curve out for a couple of decades, and you reach levels of integration so fabulous that you might as well be at the moon.

Moore also claimed that increases in memory density trickle down to less significant areas like microprocessors, and he challenged the design community to try to figure out what on earth to do with all those extra transistors. Fifteen years later, the minicomputer is dead, Moore's Intel is very big, and, just like clockwork, memories are a thousand times denser. The analog-oriented readers of this book may appreciate the following memory aids. Chip complexity increases 4X every three years, 12dB every three years, 4dB/year, 40dB/decade. Integration, it seems, is a second-order high-pass filter.

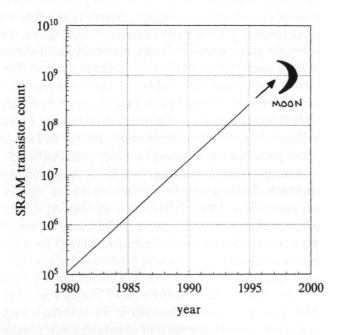

Figure 15–1.
A moon curve.

For me, learning Moore's Law before moving on to analog circuits proved very helpful. Moore's Law gives digital designers a drive to obsolete the past, and to do so quickly. A 64K DRAM designer knew better than to rest on his laurels, lest Moore's Law run him over. Young digital designers know that their first chips must whip the old guys' chips. In contrast, the analog design community takes the view that its new kids may approach, with time, the greatness of the old guys. That view is wrong; our expectations for young designers should be set much higher.

Fortunately, a good-sized piece of the analog IC business now follows Moore's Law. State-of-the-art mixed-signal circuits crossed the VLSI threshold of 10000 transistors around 1985. A decade later we're at the million-transistor level. Has the analog design business fundamentally changed? Are we reduced to button-pushing automatons? Is elegance gone?

The next three sections attempt to answer such questions. First, we take a look at some of the competition between brute-force integration and design elegance. Elegance lives on, but brute force must be respected! Next, we look at a few of the interesting subcircuits of analog CMOS. True to the analog tradition, state-of-the-art subcircuits are born in the brain or in the lab, not on the workstation, never synthesized. Finally, we'll look at elegance at a different level, how analog circuits with thousands of transistors can have legitimate elegance of their own.

Brute Force vs. Elegance

The evolution of analog-to-digital converters bears an interesting relationship to Moore's Law. Figure 15–2 plots the dynamic range of state-of-the-art analog-to-digital converters as a function of their sampling frequency. Over many decades of ADC speed, the world's best converter performance falls on a reasonably well-defined line. The line doesn't stand still over the years; it moves upward. The rate of improvement has remained remarkably constant since the first monolithic converters appeared in the mid 1970s. Transistors get faster and more numerous. CMOS technology rises to compete with bipolar. Power supply and signal voltages decrease. New architectures emerge and are perfected. And converter performance improves by a very predictable 2dB/year.

Analog-to-digital converters are noisy by analog signal processing standards. Today's world-class converters have input-referred noise spectral densities of $100nV/\sqrt{Hz}$ or so. Perhaps ADC evolution will stop when converters reach the noise performance of, say, 50Ω resistors, but we won't reach such quasi-fundamental limits for a generation! Also, no matter how noisy analog-to-digital converters may be, they represent the only path from the analog world to decent-quality memory.

What's the tie-in to Moore's Law? The Law gives us a 4X increase in ADC complexity every three years. We can take it easy for the next three years and simply integrate four of today's converters on a common sub-

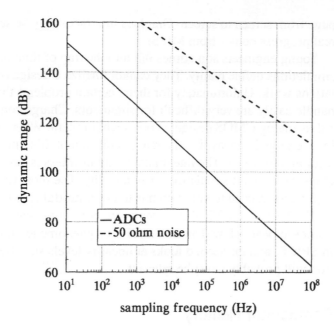

Figure 15–2.
Monolithic ADC
performance 1993.

strate. We'll connect the same analog signal to all four converter inputs and design some simple logic to add the four digital outputs every sampling period. If each converter's noise is dominated by thermal noise (uncorrelated from converter to converter), we get a 6dB improvement in dynamic range. The Moore's Law increase in integration underpins ADC improvement of 2dB/year!

To my knowledge, brute-force replication of ADCs has never yielded a converter of world-class performance. World-class converters exploit design cleverness to achieve lower manufacturing costs than brute-force alternatives. Yet the brute-force option serves as a competitive reminder to clever engineers—they had better not take too long to perfect their cleverness!

Born in the Lab

Certainly, the complexity of analog VLSI demands computer circuit simulation. Circuit simulation hasn't changed much over the years. Perhaps the only significant progress in this area comes from better computers. Solving big, nonlinear differential equations is still, after all, solving big, nonlinear differential equations. CAD tool vendors, ever-notorious for overhyping their products, claim that improved graphics interfaces translate into huge productivity increases, but in practice these interfaces add little. Now-obsolete batch-mode tools forced designers to think before simulation, and thinking is a very healthy thing. Today's 17-inch workstation monitors cannot display enough detail, nowhere near as much as yesterday's quarter-inch-thick printouts, and engineers must constantly page from screen to screen. Graphics interfaces may be sexy and fun, but real progress comes from MIPS.

Young engineers sometimes fall into the trap of thinking that computer simulations define reality. They cannot finalize a design before the simulations work. Unfortunately for them, certain problems that real circuits handle easily are very difficult for simulators. Charge conservation, critical to analog CMOS design, is one such problem. When your simulator loses charge in a way that the real circuit cannot, it's time to discount the error and move on. The integrated circuit business is paid to ship real chips, not to have simulations match reality. The most valuable design experience related to simulation is to be comfortable with its limitations!

The best analog VLSI subcircuits are born in the lab. Two of my favorites appear below. The first involves a phenomenon not even modeled in SPICE, and the second looks at linearity levels so extreme that simulation will probably never be relevant.

Comparator Memory

I've always been amused that the most accurate low-speed analog-to-digital converters are built from solid-state electronics' most miserable low-frequency devices. Actually, that overstates the case. GaAs MESFETs are even worse than silicon MOSFETs, but CMOS devices are pretty bad. Start with poor device matching, maybe 10X worse than bipolar. Add in huge doses of 1/f noise. Complete the recipe with the power supply, temperature, and impedance sensitivities of charge injection. Small wonder the old bipolar companies thought they needed biCMOS to do anything useful! They were wrong; few state-of-the-art converters ever use biCMOS. Moore's Law is enough.

Dave Welland's self-calibrated CS5016 contains wonderful architecture. The 5016 is a 16-bit, 50kHz, successive-approximation converter whose architecture dates back to 1984, building on early self-calibration work done by Dave Hodges and Hae-Seung Lee at Berkeley. All of these folks recognized the fact that, given enough transistors, an analog-to-digital converter could figure out all by itself how to divide up a reference voltage into precisely equal pieces. The principle may be obvious, but the death is in the details. Noise constantly creeps in to corrupt the measurement of all those little pieces, and the effects of noise must be removed just right.

Once the DAC inside the ADC is properly calibrated, the comparator is all that's left. Everyone knows that 1/f noise in the comparator is bad news. While CMOS 1/f noise is bad, it's always eliminated by either autozeroing or chopping, and by now it's axiomatic in the design business that one of those two techniques can be counted on for any analog VLSI application. The 5016 autozeroes its comparator in a way we'll describe later.

A less-appreciated requirement for the comparator inside an SAR ADC is that it had better be memoryless. For some analog inputs, the successive-approximation algorithm requires the comparator to make its most sensitive decision in the approximation cycle immediately following huge comparator overdrive. If the comparator has any memory at all, missing codes can result.

Sure enough, when the 5016 silicon debugging reached Dave's original design intent, we began to see the telltale fingerprints of comparator memory. And not just your basic thermal-induced memory. These memory symptoms disappeared at high temperature and were far worse at low temperature. Slowing down the chip's master clock provided alarmingly little benefit. Something strange was happening, and it surely wasn't modeled in SPICE.

While Dave worked to characterize the problem on the 5016 probe station, I decided to build the NMOS differential amplifier shown in Figure 15–3. This decision was fortunate, because the ancient discrete MOSFETs I used (3N169s) had about 5 times the memory of Orbit Semiconductor's 3μm MOSFETs. Simple Siliconix transmission gates moved the differential pair from huge overdrive to zero overdrive, and the results are shown in Figures 15–4 and 15–5. The MOSFET which carries the most current during overdrive experiences a temporary threshold voltage increase. When the differential pair is returned to balance at the input, the output exhibits a nasty, long recovery tail. Soak the differential pair in

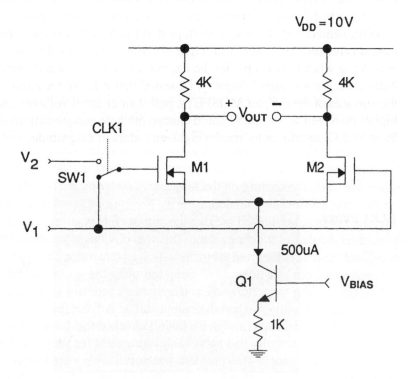

Figure 15–3.
Memory test circuit.

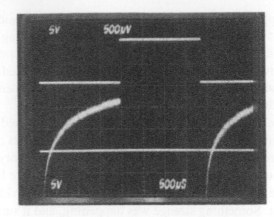

CLK1 5V/div

V_{OUT} 500µV/div

Figure 15–4.
Memory of negative overdrive ($V_1 =$ 5V, $V_2 = 3$V).

the overdriven state for 2× longer, and the lifetime of the recovery tail increases by 1.5×. Increase the magnitude of the overdrive beyond the point where all of the diffpair current is switched, and the shape of the tail stops changing.

I figured that some sort of equilibrium between fixed charges and mobile charges at the Si-SiO$_2$ interface had to be involved. At high temperatures this equilibrium could be re-established in much less time than one comparison cycle, but at low temperatures recovery could be slow. Ever-longer "soak times" would allow ever-slower fixed-charge states to participate in the threshold shift. Since the fixed-charge population could never significantly reduce the mobile carrier population in the channel, saturation of the memory effect would occur once all of the mobile carriers were switched during overdrive. The story seemed to hang together.

Well, venture-capital-funded startups don't survive based on stories, they survive based on fixes. You might not believe me if I told you what I was doing when I came up with the fix, but rest assured I wasn't pushing buttons on a workstation! Anyway, I guessed that if I could accumulate the surfaces of the diffpair MOSFETs, pull their channel voltages much higher than their gates, then I could change the hole concentrations at their Si-SiO$_2$ interfaces by maybe eighteen orders of magnitude, and

Figure 15–5.
Memory of positive overdrive ($V_1 = 5$V, $V_2 = 7$V).

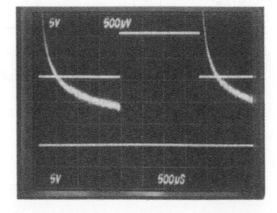

CLK1 5V/div

V_{OUT} 500µV/div

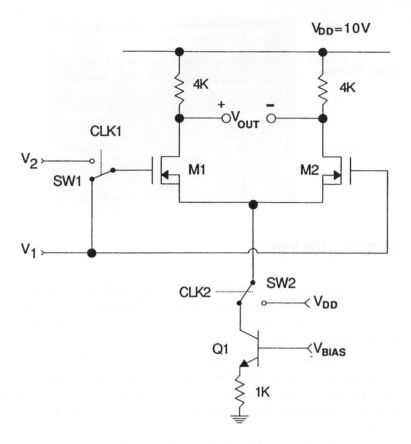

Figure 15–6.
Erasure test circuit.

erase that memory. The next day I built the circuit of Figure 15–6 and
used it to produce the photographs of Figures 15–7 and 15–8. Only a few
nanoseconds of flush time were required to completely erase the differential pair's memory. It may seem like a strange thing to do to a precision
comparator, but we've shipped a lot of product with flushed MOSFETs.

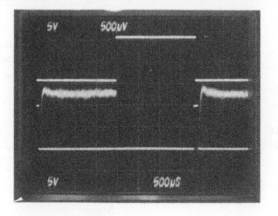

Figure 15–7.
Erased negative
overdrive ($V_1 = 5V$,
$V_2 = 3V$).

CLK1 5V/div

V_{OUT} 500µV/div

CLK2 5V/div

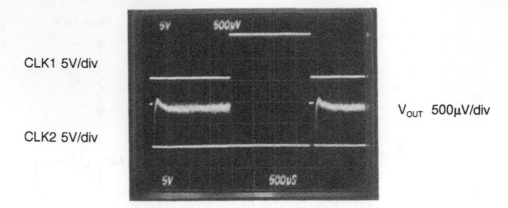

CLK1 5V/div

CLK2 5V/div

V_{OUT} 500µV/div

Figure 15–8.
Erased positive
overdrive (V_1 = 5V,
V_2 = 7V).

Sampling the Input

When I interviewed at Crystal Semiconductor in 1984, I thought that self-calibration was the greatest thing since sliced bread and that customers would love it. Wrong. Customers loved the accuracy of calibrated parts, but they hated to calibrate to get it. If I had been managing the 5016 development, I wouldn't have paid much attention to post-calibration temperature and power-supply shifts. Now I know better. The true quality of a self-calibrated design is best measured by how stable its performance is without recalibration. Fortunately, Dave Welland was way ahead of me there.

A conceptual view of the 5016 sampling path appears in Figure 15–9. The comparator input stage is used as part of an op amp during signal acquisition. This neatly autozeroes the stage's 1/f noise when we use it during conversion. Dave recognized that the charge injection of the sampling switch adds error charge to the signal charge, and he minimized the error in an ingenious way. Rather than using a conventional transmission gate, Dave built a tri-state output stage into the closed-loop sampling

Figure 15–9.
CS5016 sampling
architecture.

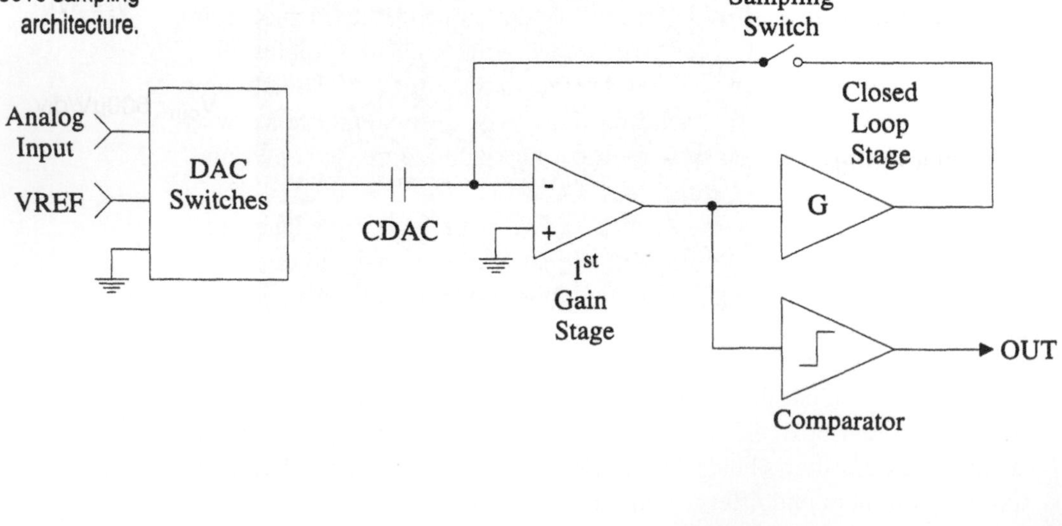

Sampling
Switch

Closed
Loop
Stage

Analog
Input

VREF

DAC
Switches

CDAC

−
+
1st
Gain
Stage

G

Comparator

OUT

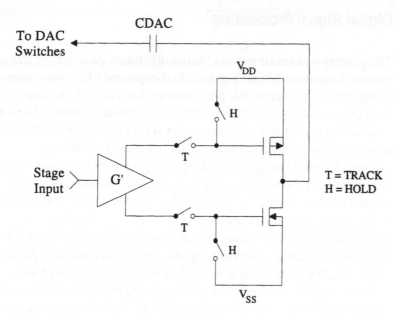

To DAC Switches

CDAC

V_DD

H

T

Stage Input

G'

T = TRACK
H = HOLD

T

H

V_SS

Figure 15–10.
CS5016 sampling switch detail.

path. A simplified schematic is shown in Figure 15–10. The gate voltage swings on the sampling devices are much smaller than those associated with transmission gates, and the swings can be designed to be power-supply independent. Furthermore, at shutoff, most of the channel charge of the active-region sampling MOSFETs flows out their sources and away from CDAC. The results speak for themselves: the 5016's offset shifts with temperature by less than .02LSB/°C.

More recently, the progress in open-loop sampling has been amazing. Just about every commercial delta-sigma converter samples its analog input with simple capacitor-transmission-gate structures. These open-loop sample-holds help achieve overall converter linearities now surpassing 120dB.[1] Optimization of sampling circuit performance only involves a few dozen transistors, and we're getting better all the time! Charge injection from the MOSFET is getting pretty well understood, but newcomers should be forewarned that little of this understanding comes from the Sparcstation!

Elegance at a Higher Level

The real elegance of analog VLSI circuits occurs beyond the subcircuit level. Successful analog VLSI architectures trivialize all but a few of a chip's analog subcircuits. Successful architectures are minimally analog. Successful architects know digital signal processing. Analog VLSI circuits may be complex beasts, but when an architecture is really clean, you know. Your competition knows, too, a few years down the road!

Digital Signal Processing

The journey was hectic but fun. Active-RC filters gave way to integrated switched-capacitor filters. The switched-capacitor filters were noisy compared to their active-RC predecessors, but you can't be expected to work miracles with a mere ~1000pF of total capacitance! Fortunately, switched-capacitor filter characteristics stayed put, allowing them to dominate sensitivity-critical, noise-tolerant telecom applications.

Once Moore's Law gave us cheap enough digital transistors and cheap enough data converters, the switched-capacitor technology was doomed. Digital filters of wonderful precision have no sensitivity and no drift. Their noise performance can be perfectly simulated, and you can ask all the essential questions before silicon tapes out. Digital filters can be tested with digital test vectors to quality levels unheard of in the analog world. Analog chauvinists take heed: don't fight if you can't win.

Fortunately for us dinosaurs, analog design experience helps produce better digital filters. Optimizing datapath wordwidths is very similar to optimizing kT/C noise. Quantizing filter coefficients is similar to quantizing signals. Elegant, high-performance analog filter designs will always be difficult to put into production, but once an elegant DSP design is done, it's out of design forever. You can't beat that!

The power of digital signal processing is never more apparent than when you're dealing with adaptive signal processing. Adaptive digital filters, be they echo cancelers or data channel equalizers, are simply more intelligent, more interesting, than their fixed-function predecessors. Take the time to understand them, and you'll be hooked.

Noise and Robustness

Just about everyone who transmits digital bits tries to send those bits as far as possible, as fast as possible, down the cheapest possible medium, until recovery of those bits is an analog problem. Digital data detection is one of the best long-term analog businesses there is. I'll add that nobody is too enthusiastic about sending a clock alongside the data. Thus, timing recovery is another of the great long-term analog businesses.

Data detection and timing recovery circuits were among the first to embrace Moore's Law. Adaptive equalizers routinely clean up the frequency responses of real-world data channels. Maximum-likelihood receivers figure out the most likely transmitted sequence despite a channel's additive noise. Error-correcting codes operate on the detected bit sequence, often providing further orders-of-magnitude improvement in system bit error rates. All of these techniques, originally perfected for board-level designs, are found in abundance on today's AVLSI chips.

Traditional analog signal processing carries with it a certain hopelessness with respect to noise. As analog processing complexity grows, additive noise sources grow in number, and system performance fades away. Modern digital communication system designers must be forgiven if they look at the weapons of the analog world and see Stone Age technology in the Iron Age. Throwing rocks may still get the job done, but the weapons of Moore's Law are elegant at a higher level.

The Elegance of Digital Audio

Audio data converters represent highly refined combinations of analog and digital signal processing. Converter performance now surpasses what reasonable people can claim to hear, and still it improves by 2dB/year. Can this possibly make sense?

We all know that the great musical performances of our parents' generation are irrevocably lost. Lost by the limitations of recording electronics of that era. Lost thanks to the deterioration of analog storage media. Lost for a variety of technological reasons, but lost. Digital recording and storage now rule. The great sounds of our generation will never be lost. The job is not finished. Combine a small array of microphones, very good analog-to-digital converters, and digital signal processing, and the results can be magic. We'll be able to listen in new ways, to hear the contributions of individuals to the sounds of their orchestras.

Reference

1. Don Kerth et al., "A 120dB Linear Switched-Capacitor Delta-Sigma Modulator," *ISSCC Digest of Technical Papers* (February 1994).

16. Analog Circuit Design

Analog circuit design involves a strange mix of intuition, experience, analysis, and luck. One of the nicest things about the job is that you feel very smart when you make something work very well.

A necessary condition for being a good analog designer is that you know about 57 important facts. If you know these 57 important facts, and know them well enough that they become part of your working intuition, you may become a good analog circuit designer.

Undoubtedly, there is an organized way to present these facts to the "interested student." The facts could be prioritized, or they could be alphabetized, or derived from first principles. The priority could be assigned, with the most important facts coming first on the list (for emphasis) or last on the list (for suspense). Some day, when I have a lot of time, I'm going to put the list in the right order.

It is difficult to present all the facts in such a way that will make sense to everyone. Sometimes I think that the only way to do this is by working examples over a twenty-year period. But we don't have time for that, and so as a poor alternative, I'm just going to write the facts down as they occur to me. There is a very good chance that you have run into most of these facts already. If so, just take heart that someone else has been tormented by the same problems.

Here's the list of things you should know:

0. If you even look at another engineer's approach to solving an analog circuit design problem before you solve the problem yourself, you greatly reduce the chance that you will do something creative.

1. Capacitors and resistors have parasitic inductance. A good rule of thumb is 4nH for a leaded component, and about 0.4nH for a surface-mount chip component. This means that a 100pF leaded capacitor will have a self-resonance at 250MHz. This can be just great, if you are using the part to bypass a 250MHz signal, but might be a nuisance otherwise.

2. If you don't want a transistor with a high bandwidth to oscillate in a circuit, place lossy components in at least two out of its three leads. A 33Ω resistor in the base and collector leads will usually do the trick without degrading performance. Ferrite beads in the leads work well to fix the same problem.

3. If you are probing a circuit with a dc voltmeter and the readings are not making any sense (for example, if there is a large offset at the input to an op amp, but the output is not pinned) suspect that something is oscillating.

4. Op amps will often oscillate when driving capacitive loads. A good way to think about this problem is that the low-pass filter formed by the output resistance of the op amp together with the capacitance of the load is adding enough phase shift (taken together with the phase shift through the op amp) that your negative feedback has become positive feedback.

5. The base-emitter voltage (V_{be}) of a small signal transistor is about 0.65V and drops by about 2mV/°C. Yes, the V_{be} goes down as the temperature goes up.

6. The Johnson noise of a resistor is about $0.13nV/\sqrt{Hz\sqrt{\Omega}}$. So, multiply 0.13nV by the square root of the resistance value (in Ohms) to find the noise in a 1Hz bandwidth. Then multiply by the square root of your bandwidth (in Hertz) to find the total noise voltage. This is the rms noise voltage: you can expect about 5–6 times the rms value in a peak-to-peak measurement.

Example: a 1kΩ resistor has about $4.1nV/\sqrt{Hz}$, or about 41m Vrms in a 100MHz bandwidth, which would look like about 0.2mV peak-to-peak on a 100MHz 'scope. Note that the Johnson noise voltage goes up with the square root of the resistance.

7. The Johnson noise current of a resistor is equal to the Johnson noise voltage divided by the resistance. (Thanks to Professor Ohm.) Note that the Johnson noise current goes down as the resistance goes up.

8. The impedance looking into the emitter of a transistor at room temperature is 26Ω divided by the emitter current in mA.

9. All amplifiers are differential, i.e., they are referenced to a "ground" somewhere. Single-ended designs just ignore that fact, and pretend (sometimes to a good approximation) that the signal ground is the same as the ground that is used for the feedback network or for the non-inverting input to the op amp.

10. A typical metal film resistor has a temperature coefficient of about 100 ppm/°C. Tempcos about 10× better are available at reasonable cost, but you will pay a lot for tempcos around a few ppm.

11. The input noise voltage of a very quiet op amp is $1nV/\sqrt{Hz}$. But there are a lot of op amps around with $20nV/\sqrt{Hz}$ of input noise.

Also, watch out for input noise current: multiply the input noise current by the source impedance of the networks connected to the op amp's inputs to determine which noise source is most important, and select your op amps accordingly. Generally speaking, op amps with bipolar front-ends have lower voltage noise and higher current noise than op amps with FET front-ends.

12. Be aware that using an LC circuit as a power supply filter can actually multiply the power supply noise at the resonant frequency of the filter. A choke is an inductor with a very low Q to avoid just this problem.

13. Use comparators for comparing, and op amps for amplifying, and don't even think about mixing the two.

14. Ceramic capacitors with any dielectric other than NPO should be used only for bypass applications. For example, Z5U dielectrics exhibit a capacitance change of 50% between 25°C and 80°C, and X7R dielectrics change their capacity by about 1%/V between 0 and 5V. Imagine the distortion!

15. An N-channel enhancement-mode FET is a part that needs a positive voltage on the gate relative to the source to conduct from drain to source.

16. Small-signal JFETs are often characterized by extremely low gate currents, and so work very well as low-leakage diodes (connect the drain and source together). Use them in log current-to-voltage converters and for low-leakage input protection.

17. If you want to low-pass filter a signal, use a Bessel (or phase linear) filter for the least overshoot in the time domain, and use a Cauer (or elliptic) filter for the fastest rolloff in the frequency domain. The rise time for a Bessel-filtered signal will be .35 divided by the 3dB bandwidth of the filter. Good 'scope front-ends behave like Bessel filters, and so a 350MHz 'scope will exhibit a 1.0ns rise time for an infinitely fast input step.

18. A decibel (dB) is always 10 times the log of the ratio of two powers. Period. Sometimes the power is proportional to the square of the voltage or current. In these cases you may want to use a formula with a twenty in it, but I didn't want to confuse anybody here.

19. At low frequencies, the current in the collector of a transistor is in phase with the current applied to the base. At high frequencies, the collector current lags by 90°. You will not understand any high-frequency oscillator circuits until you appreciate this simple fact.

20. The most common glass-epoxy PCB material (FR-4) has a dielectric constant of about 4.3. To build a trace with a characteristic impedance of 100Ω, use a trace width of about 0.4 times the thickness of the FR-4 with a ground plane on the other side. To make a 50Ω trace, you will need a trace width about 2.0 times the thickness of the FR-4.

21. If you need a programmable dynamic current source, find out about operational transconductance amplifiers. NSC makes a nice one called the LM13600. Most of the problem is figuring out when you need a programmable dynamic current source.

22. An 5V relay coil can be driven very nicely by a CMOS output with an emitter follower. Usually 5V relays have a "must make" specification of 3.5V, so this configuration will save power and does not require any flyback components.

23. A typical thermocouple potential is 30μV/°C. If you care about a few hundred microvolts in a circuit, you will need to take care: route all your signals differentially, along the same path, and avoid temperature gradients. DPDT latching relays work well for multiplexing signals in these applications as they do not heat up, thus avoiding large temperature gradients, which could generate offsets even when the signals are routed differentially.

24. You *should* be bothered by a design which looks messy, cluttered, or indirect. This uncomfortable feeling is one of the few indications you have to know that there is a better way.

25. If you have not already done so, buy 100 pieces of each 5%, ¼W carbon film resistor value and arrange them in some nice slide-out plastic drawers. When you are feeling extravagant, do the same for the 1% metal film types.

26. Avoid drawing any current from the wiper of a potentiometer. The resistance of the wiper contact will cause problems (local heating, noise, offsets, etc.) if you do.

27. Most digital phase detectors have a deadband, i.e., the analog output does not change over a small range near where the two inputs are coincident. This often-ignored fact has helped to create some very noisy PLLs.

28. The phase noise of a phase-locked VCO will be at least 6dB worse than the phase noise of the divided reference for each octave between the comparison frequency and the VCO output frequency. Hint: avoid low-comparison frequencies.

29. For very low distortion, the drains (or collectors, as the case may be) of a differential amplifier's front-end should be bootstrapped to the source (or emitter) so that the voltages on the part are not modulated by the input signal.

30. If your design uses a $3 op amp, and if you are going to be making a thousand of them, realize that you have just spent $3000. Are you smart enough to figure out how to use a $.30 op amp instead? If you think you are, then the return on your time is pretty good.

31. Often, the Q of an LC tank circuit is dominated by losses in the inductor, which are modeled by a series resistance, R. The Q of such a part is given by $Q = \omega L/R$.

At the resonant frequency, $f = 1/2\pi\sqrt{LC}$, the reactance of the L and C cancel each other. At this frequency, the impedance of a series LC circuit is just R, and the impedance across a parallel LC tank is Q^2R.

32. Leakage currents get a factor of 2 worse for every 10°C increase in temperature.

33. When the inputs to most JFET op amps exceed the common-mode range for the part, the output may reverse polarity. This artifact will haunt the designers of these parts for the rest of their lives, as it should. In the meantime, you need to be very careful when designing circuits with these parts; a benign-looking unity follower inside a feedback loop can cause the loop to lock up forever if the common-mode input to the op amps is exceeded.

34. Understand the difference between "make-before-break" and "break-before-make" when you specify switches.

35. Three-terminal voltage regulators in TO220 packages are wonderful parts and you should use a lot of them. They are cheap, rugged, thermally protected, and very versatile. Besides their recommended use as voltage regulators, they may be used in heater circuits, battery chargers, or virtually any place where you would like a protected power transistor.

36. If you need to make a really fast edge, like under 100pS, use a step recovery diode. To generate a fast edge, you start by passing a current in the forward direction, then quickly (in under a few nanoseconds) reverse the current through the diode. Like most diodes, the SRD will conduct current in the reverse direction for a time called the reverse recovery time, and then it will stop conducting very abruptly (a "step" recovery). The transition time can be as short as 35pS, and this will be the rise time of the current step into your load.

Well, there you have it. These are the first 37 of the 57 facts you must know to become an analog circuit designer. I have either misplaced, forgotten, or have yet to learn the 20 missing items. If you find any, would you let me know? Happy hunting!

17. There's No Place Like Home

What's your choice for the single best aid to an interesting and productive circuit design career? A PhD? An IQ of 250? A CAD workstation? Getting a paper into the Solid State Circuit Conference? Befriending the boss? I suppose all of these are of some value, but none even comes close to something else. In fact, their combined benefit isn't worth a fraction of the something else. This something else even has potential economic rewards. What is this wondrous thing that outshines all the other candidates? It is, simply, a laboratory in your home. The enormous productivity advantage provided by a home lab is unmatched by anything I am familiar with. As for economic benefits, no stock tip, no real estate deal, no raise, no nothing can match the long term investment yield a home lab can produce. The laboratory is, after all, an investment in yourself. It is an almost unfair advantage.

The magic of a home lab is that it effectively creates time. Over the last 20 years I estimate that about 90% of my work output has occurred in a home lab. The ability to grab a few hours here and there combined with occasional marathon 5–20 hour sessions produces a huge accumulated time benefit. Perhaps more importantly, the time generated is highly leveraged. An hour in the lab at home is worth a day at work.

A lot of work time is spent on unplanned and parasitic activities. Phone calls, interruptions, meetings, and just plain gossiping eat up obscene amounts of time. While these events may ultimately contribute towards good circuits, they do so in a very oblique way. Worse yet, they rob psychological momentum, breaking up design time into chunks instead of allowing continuous periods of concentration. When I'm at work I do my job. When I'm at home in the lab is where the boss and stockholders get what they paid for. It sounds absurd, but I have sat in meetings praying for 6 o'clock to come so I can go home and get to work. The uninterrupted time in a home lab permits persistence, one of the most powerful tools a designer has.

I favor long, uninterrupted lab sessions of at least 5 to 10 hours, but family time won't always allow this. However, I can almost always get in two to four hours per day. Few things can match the convenience and efficiency of getting an idea while washing dishes or putting my son to sleep and being able to breadboard it *now*. The easy and instant availability of lab time makes even small amounts of time practical. Because no

one else uses your lab, everything is undisturbed and just as you left it after the last session. Nothing is missing or broken,[1] and all test equipment is familiar. You can get right to work.

Measured over months, these small sessions produce spectacular gains in work output. The less frequent but more lengthy sessions contribute still more.

Analog circuits have some peculiar and highly desirable characteristics which are in concert with all this. They are small in scale. An analog design is almost always easily and quickly built on a small piece of copper-clad board. This board is readily shuttled between home and work, permitting continuous design activity at both locations.[2] A second useful characteristic is that most analog circuit development does not require the most sophisticated or modern test equipment. This, combined with test equipment's extremely rapid depreciation rate, has broad implications for home lab financing. The ready availability of high-quality used test equipment is the key to an affordable home lab. Clearly, serious circuit design requires high performance instrumentation. The saving grace is that this equipment can be five, twenty, or even thirty years old and still easily meet measurement requirements. The fundamental measurement perfor-

---

1. It is illuminating to consider that the average lifetime of an oscilloscope probe in a corporate lab is about a year. The company money and time lost due to this is incalculable. In 20 years of maintaining a home lab I have never broken a probe or lost its accessories. When personal money and time are at risk, things just seem to last longer.
2. An extreme variant related to this is reported by Steve Pietkiewicz of Linear Technology Corporation. Faced with a one-week business trip, he packed a complete portable lab and built and debugged a 15-bit A-D converter in hotel rooms.

mance of test equipment has not really changed much. Modern equipment simplifies the measurement process, offers computational capability, lower parts count, smaller size, and cost advantages (for new purchases). It is still vastly more expensive than used instrumentation. A Tektronix 454 150MHz portable oscilloscope is freely available on the surplus market for about $150.00. A new oscilloscope of equivalent capability costs at least ten times this price.

Older equipment offers another subtle economic advantage. It is far easier to repair than modern instruments. Discrete circuitry and standard-product ICs ease servicing and parts replacement problems. Contemporary processor-driven instruments are difficult to fix because their software control is "invisible," often convoluted, and almost impervious to standard troubleshooting techniques. Accurate diagnosis based on symptoms is extremely difficult. Special test equipment and fixtures are usually required. Additionally, the widespread usage of custom ICs presents a formidable barrier to home repair. Manufacturers will, of course, service their products, but costs are too high for home lab budgets. Modern, computationally based equipment using custom ICs makes perfect sense in a corporate setting where economic realities are very different. The time and dollar costs associated with using and maintaining older equipment in an industrial setting are prohibitive. This is diametrically opposed to home lab economics, and a prime reason why test equipment depreciates so rapidly.

The particular requirements of analog design combined with this set of anomalies sets guidelines for home lab purchases.[3] In general, instruments designed between about 1965 and 1980 meet most of the discussed criteria. Everybody has their own opinions and prejudices about instruments. Here are some of mine.

Oscilloscopes

The oscilloscope is probably the most important instrument in the analog laboratory. Tektronix oscilloscopes manufactured between 1964 and 1969 are my favorites. Brilliantly conceived and stunning in their execution, they define excellence. These instruments were designed and manufactured under unique circumstances. It is unlikely that test equipment will ever again be built to such uncompromising standards. Types 547 and 556 are magnificent machines, built to last forever, easily maintained, and almost a privilege to own. The widely available plug-in vertical amplifiers provide broad measurement capability. The 1A4 four-trace and 1A5 and 1A7A differential plug-ins are particularly useful. A 547 equipped with a

---

3. An excellent publication for instrument shopping is "Nuts and Volts," headquartered in Corona, California. Telephone 800/783-4624.

1A4 plug-in provides extensive triggering and display capability. The dual beam 556, equipped with two vertical plug-ins, is an oscilloscope driver's dream. These instruments can be purchased for less than the price of a dinner for two in San Francisco.[4] Their primary disadvantages are size and 50MHz bandwidth, although sampling plug-ins go out to 1GHz.

The Tektronix 453 and 454 portables extend bandwidth to 150MHz while cutting size down. The trade-off is lack of plug-in capability. The later (1972) Tektronix 485 portable has 350MHz bandwidth but uses custom ICs, is not nearly as ruggedly built, and is very difficult to repair. Similarly, Tektronix 7000 series plug-in instruments (1970s and 80s) feature very high performance but have custom ICs and are not as well constructed as earlier types. They are also harder to fix. The price-risk-performance ratio is, however, becoming almost irresistible. A 500MHz 7904 with plug-in amplifiers brings only $1000.00 today, and the price will continue to drop.

Sampling 'scopes and plug-ins attain bandwidths into the GHz range at low cost. The Tektronix 661, equipped with a 4S2 plug-in, has 3.9GHz bandwidth, but costs under $100.00. The high bandwidths, sensitivity, and overload immunity of sampling instruments are attractive, but their wideband sections are tricky to maintain.

Other 'scopes worthy of mention include the Hewlett-Packard 180 series, featuring small size, plug-in capability and 250MHz bandwidth. HP also built the portable 1725A, a 275MHz instrument with many good attributes. Both of these instruments utilize custom ICs and hybrids, raising the maintenance cost risk factor.

Related to oscilloscopes are curve tracers. No analog lab is complete without one of these. The Tektronix 575 is an excellent choice. It is the same size as older Tektronix lab 'scopes and is indispensable for device characterization. The more modern 576 is fully solid state, and has extended capabilities and more features. A 576 is still reasonably expensive (»$1500.00). I winced when I finally bought one, but the pain fades quickly with use. A 575 is adequate; the 576 is the one you really want.

Oscilloscopes require probes. There are so many kinds of probes and they are all so wonderful! I am a hopeless probe freak. It's too embarrassing to print how many probes I own. A good guideline is to purchase only high quality, name brand probes. There are a lot of subtleties involved in probe design and construction, particularly at high frequencies. Many off-brand types give very poor results. You will need a variety of 1× and 10× passive probes, as well as differential, high voltage, and other types. 50Ω systems utilize special probes, which give exceptionally clean results at very high frequency.

---

4. It is highly likely that Tektronix instruments manufactured between 1964 and 1969 would have appreciated at the same rate as, say, the Mercedes-Benz 300 SL . . . if oscilloscopes were cars. They meet every criterion for collectible status except one; there is no market. As such, for the few aberrants interested, they are surely the world's greatest bargain.

Active probes are also a necessity. This category includes FET probes and current probes. FET probes provide low-capacitive loading at high frequency. The 230MHz Tektronix P-6045 is noteworthy because it is easy to repair compared to other FET probes. A special type of FET probe is the differential probe. These devices are basically two matched FET probes contained within a common probe housing. This probe literally brings the advantages of a differential input oscilloscope to the circuit board. The Tektronix P6046 is excellent, and usually quite cheap because nobody knows what it is. Make sure it works when you buy it, because these probes are extraordinarily tricky to trim up for CMRR after repair. Finally, there are clip-on current probes. These are really a must, and the one to have is the DC-50MHz Tektronix P-6042. They are not difficult to fix, but the Hall effect-based sensor in the head is expensive. AC only clip-on probes are not as versatile, but are still useful. Tektronix has several versions, and the type 131 and 134 amplifiers extend probe capability and eliminate scale factor calculations. The Hewlett-Packard 428, essentially a DC only clip-on probe, features high accuracy over a 50mA to 10 amp range.

Power Supplies

There are never enough power supplies. For analog work, supplies should be metered, linear regulators with fully adjustable voltage output and current limiting. The HP 6216 is small and serves well. At higher currents (i.e., to 10 amps) the Lambda LK series are excellent. These SCR pre-regulated linear regulators are reasonably compact, very rugged, and handle any load I have ever seen without introducing odd dynamics. The SCR pre-regulator permits high power over a wide output voltage range with the low noise characteristics of a linear regulator.

Signal Sources

A lab needs a variety of signal sources. The Hewlett-Packard 200 series sine wave oscillators are excellent, cheap, and easily repaired. The later versions are solid state, and quite small. At high frequencies the HP 8601A sweep generator is a superb instrument, with fully settable and leveled output to 100MHz. The small size, high performance, and versatility make this a very desirable instrument. It does, however, have a couple of custom hybrid circuits, raising the cost-to-repair risk factor.

Function generators are sometimes useful, and the old Wavetek 100 series are easily found and repaired. Pulse generators are a must; the Datapulse 101 is my favorite. It is compact, fast, and has a full complement of features. It has fully discrete construction and is easy to maintain. For high power output the HP214A is excellent, although not small.

Voltmeters

DVMs are an area where I'm willing to risk on processor-driven equipment. The reason is that the cost is so low. The Fluke handheld DVMs are so cheap and work so well they are irresistible. There are some exceptionally good values in older DVMs too. The 5½ digit Fluke 8800A is an excellent choice, although lacking current ranges. The 4½ digit HP3465 is also quite good, and has current ranges. Another older DVM worthy of mention is the Data Precision 245–248 series. These full featured 4½ digit meters are very small, and usually sell for next to nothing. Their construction is acceptable, although their compactness sometimes makes repair challenging.

AC wideband true RMS voltmeters utilize thermal converters. These are special purpose instruments, but when you must measure RMS they are indispensable. The metered Hewlett-Packard 3400A has been made for years, and is easy to get. This instrument gives good accuracy to 10MHz. All 3400s look the same, but the design has been periodically updated. If possible, avoid the photochopped version in favor of the later models. The HP3403C goes out to 100MHz, has higher accuracy, and an autoranging digital display. This is an exotic, highly desirable machine. It is also harder to find, more expensive, and not trivial to repair.

Miscellaneous Instruments

There are literally dozens of other instruments I have found useful and practical to own. Tektronix plug-in spectrum analyzers make sense once you commit to a 'scope mainframe. Types 1L5, 1L10, and 1L20 cover a wide frequency range, but are harder to use than modern instruments. Distortion analyzers are also useful. The HP334A is very good, and has about a .01% distortion floor. The HP339A goes down to about .002%, and has a built in low distortion oscillator. It is also considerably more expensive. Both are "auto-nulling" types, which saves much knob twiddling. Frequency counters are sometimes required, and the little HP5300 series are very good general purpose units. The old 5245L is larger, but the extensive line of plug-ins makes this a very versatile instrument. Occasionally, a chart recorder makes sense, and the HP7000A (XY) and HP680 (strip) are excellent. The 7000A has particularly well thought out input amplifiers and sweep capabilities. Other instruments finding occasional use are a variable voltage reference (the Fluke 332 is huge, but there is no substitute when you need it) and a picoammeter. Kiethley picoammeters (e.g., type 610) are relatively hard to find, but read into the femtoampere range. "Diddle boxes" for both resistance and capacitance are very useful. These break down into precision and non-precision types. General Radio and ESI built excellent precision types (e.g., G.R. 1400 Series), but many have been abused . . . look (and smell) inside before you buy. Non-precision types by EICO and Heathkit are everywhere, and cost essentially nothing. The precision variable air capacitors built by General Radio (types 722D and the later 1422D) are particularly applicable for transducer simulation. They are also worth buying just to look at; it is hard to believe human beings could build anything so beautiful.

Oscilloscope cameras are needed to document displays. Modern data recording techniques are relegating 'scope cameras to almost antique status, which has happily depressed their price. My work involves a significant amount of waveform documentation, so I have quite a bit of specialized camera equipment. The Tektronix C-30 is a good general purpose camera which fits, via adapters, a wide variety of oscilloscopes. It is probably the best choice for occasional work. The Tektronix C27 and C12 are larger cameras, designed for plug-in 'scopes. Their size is somewhat compensated by their ease of use. However, I do not recommend them unless you do a lot of photographic documentation, or require highly repeatable results.

Finally, cables, connectors, and adapters are a must have. You need a wide variety of BNC, banana jack, and other terminator, connector, adapter, and cable hardware. This stuff is not cheap; in fact it is outrageously expensive, but there is no choice. You can't work without it and the people who make it know it.

No discussion of a home laboratory is complete without comment on its location. You will spend many hours in this lab; it should be as comfortable and pleasant a place as possible. The use of space, lighting, and furnishings should be quite carefully considered. My lab is in a large

Figure 17–2.
You will spend many hours in this lab. It should be a comfortable and pleasant place.

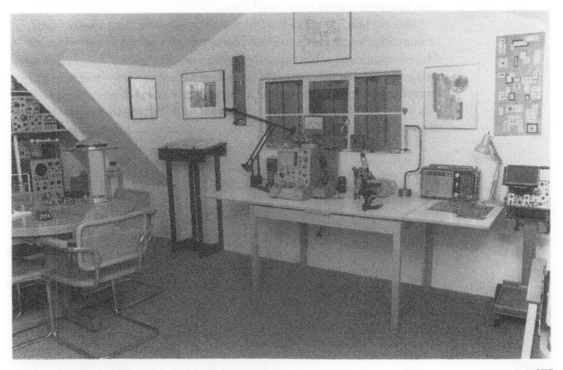

Figure 17–3.
Maintaining lab organization is painful, but increases time efficiency.

room on the second floor, overlooking a very quiet park. It is a bright, colorful room. Some of my favorite pictures and art are on the walls, and I try to keep the place fairly clean. In short, I do what I can to promote an environment conducive to working.

Over the last 20 years I have found a home lab the best career friend imaginable. It provides a time efficiency advantage that is almost unfair. More importantly, it has insured that my vocation and hobby remain happily and completely mixed. That room on the second floor maintains my enthusiasm. Engineering looks like as good a career choice at 45 as it did at 8 years old. To get that from a room full of old equipment has got to be the world's best bargain.

Figure 17–4.
It's convenient to
be able to write up
lab results as they
occur.

18. It Starts with Tomorrow

Fostering Innovation in the Chip Biz

What are the roots of innovation? How does it actually happen in the microelectronics industry today? How can it be fostered and enhanced? These are questions of considerable interest to managers. It is suggested here that innovation is a very personal process, beginning with a strong interest in tomorrow's needs and the visualization of *significantly different* solutions. Modern management methods aimed at enhancing the rate and quality of innovative product design may fail because they depend too much on what are essentially *algorithmic* approaches to group improvement, with diminished emphasis on the need to recognize, encourage and support the singular vision. A recurrent theme in today's corporations is that new product concepts must be firmly—perhaps even exclusively—based on marketing data acquired through listening to the "Voice of the Customer." While recognizing the critical role and immense value of market research, this view is thought to be an incomplete and inadequate characterization of the fundamental challenge, which requires a stronger emphasis on the role of *anticipation* in product innovation, stemming both from a broad *general knowledge* of the traditional marketplace and a high-spirited sense of tomorrow's needs before these are articulated.

> "I do not think there is any thrill that can go through the human heart like that felt by the inventor as he sees some creation of the brain unfolding to success . . . Such emotions make a man forget food, sleep, friends, love, everything . . ."
>
> —NIKOLA TESLA, 1896

Innovation! A high-spirited word, much in evidence in today's technological world. Though not a unique twentieth-century phenomenon, the relentless introduction of ever more innovative products has become a particularly evident, in fact, a *characteristic* aspect of modern techno-cultures. In some industries, where product life-cycles are short, achieving a high rate of innovation is a key strategic objective. Nowhere is this dependence on focused, purposeful innovation of the highest quality more apparent than in the microelectronics business. But what are the fundamental sources of the innovative spark? What distinguishes innovative products from effective and adequate—but totally predictable—follow-ons and spin-offs? What separates creative flair from routine incrementalism? What can be done to encourage innovation and elevate the output of finely wrought, ground-breaking, universally acclaimed products in a modern high-tech company?

This essay expresses my personal views about how innovation really happens, in contrast to how it might be thought to happen, in a modern company. These opinions are based on forty years of plying my trade as a designer of electronic devices, circuits, and systems. Longevity of service brings no guarantees of wisdom. However, it may perhaps help one to see the field with a reasonably broad perspective and to address contemporary issues of managing innovation possessed of some familiarity with the overall challenge. Because it is a personal position, it may be useful to begin with a sketch of my early years.

I lost my father in World War II, a blow I've never been able to quantify. He was a classical pianist; had he lived, I no doubt would have pursued my own love of serious music full-time. Instead, I settled on radio and chemistry as hobbies, partly influenced by my much-senior brother, who had already made some receivers. I had an upstairs experiments room, fastidiously organized, in which I built many receivers and transmitters, using brass-and-ebony components from the earliest days of radio, screwed down on to soft-wood bases—the quintessential breadboard! Usually, there was a further board screwed and angle-bracketed to provide a front panel, on which I mounted such things as input and output binding terminals, switches for range-changing, and rugged multi-vaned variable capacitors for tuning. I had begun with the obligatory 'crystal set,' and progressed through one-valvers, several TRF receivers, and a seven-valve superhet that I designed from the ground up. Having no electricity in my home (it was lit by gas mantles, and heated in the winter by coal fire in just one room), all of my early experiments were battery powered, giving me a taste for the low-power challenge inherent in the design of contemporary integrated circuits!

With the cessation of hostilities, a plethora of government-surplus electronics equipment hit the market at incredible prices. Using money earned from a newspaper route, I purchased as much as I could afford. My bounty included exquisite VHF receivers, enigmatic IFF systems (*sans* detonator), A-scan and PPI radar 'indicators units' (one of which had *two* CRTs!), and some beautiful electromechanical servosystems, whose oscillatory hunting in search of a steady state was mesmerizing. This stuff was deeply alluring and *bristling* with possibilities. With these war-spared parts, I built my first clearly remembered oscilloscopes and a TV receiver, in 1949–50. It was the only TV on the block, and on it my family and I watched the Grand Coronation of Elizabeth II.

These were profoundly joyous and fulfilling days of discovery. I recall the thrill of 'inventing' the super-regenerative receiver, the cross-coupled multivibrator (with triodes, of course, not transistors), voltage regulators, pentode timebase generators, pulse-width modulation, AVC and AFC, electronic musical generators, and a good deal more, all out of a free-wheeling 'what if?' approach to my hobby. I was hardly deflated to later learn that others had beaten me to the tape, often by several decades! The urge was always to pursue an *original* design, from ground zero, and to try to understand the fundamentals. I occasionally bought magazines like *Practical Wireless*, but I couldn't imagine actually building something from those pages! It was the same with model aircraft and boats: I'd much rather find out what worked and what didn't by direct experience (read failure) than build to somebody else's plans. Copying, even with the prospect of achieving superior performance, was no fun *at all*.

I started my first job on a brisk, leaf-shedding autumn morning in September 1954, at the Signals Research and Development Establishment (SRDE). The labs were a rambling group of low wooden buildings, carefully secreted among trees and bristling with exotic antennas, perched atop chalk cliffs overlooking the English Channel. Oddly, it didn't seem like *work*, at all: I was actually getting paid for cheerfully pursuing what I had passionately enjoyed doing since single-digit years, but with *immensely* augmented resources (the British Government!). The point-contact transistor was one of the new playthings in my lab. Six years later, at Mullard, I designed an all-transistorized sampling oscilloscope, and emigrated to the USA in 1964, to pursue 'scope design at Tektronix, in Oregon.

There, during the late sixties, I was given considerable latitude—even encouragement—to develop novel semiconductor devices and circuits at Tek. My chosen emphasis was on high-frequency nonlinear circuits. Out of this period came monolithic mixers and multipliers, and the discovery of the generalized 'translinear-principle.' In 1972, back in England for a while, I worked as a Group Leader at the Plessey Research Labs, on optical character recognition systems (using what nowadays would probably be called 'neural network techniques,' but which I just called adaptive signal processing), optical holographic memories and various communications ICs. I was also writing a lot of software at that time, for simulating three-dimensional current-transport behavior in various 'super-integrated' semiconductor structures, including carrier-domain multipliers, magnetometers, and a type of merged logic like I^2L.

My relationship with Analog Devices goes back 22 years, fully half of my working life. While still with Plessey, I was contacted by Ray Stata. We discussed the idea of working for Analog Devices. I was unable to leave England at that time because my mother was seriously ill, so we worked out a deal, the result of which was that I 're-engineered' the two bedrooms on the top floor of my three-story house in Dorset, on the south coast of England, one into a well-equipped electronics lab (including an early-production Tektronix 7000-series 'scope I had helped design), the other into a library-quiet, carpeted, and cork-walled office, equipped with a large desk, overlooking Poole Harbor, and an even-larger drawing board. During this happy sojourn I designed several 'firsts'—the first complete IC multiplier designed for laser-trimming, the first monolithic RMS-DC converter, the first monolithic V/F converter with 0.01% linearity, the first dual two-quadrant multiplier, all of which I laid out myself, using mylar, pencils, and many, many erasers, the sort wrapped in a spiral paper sheath.

The formal VOC emphasis was decades away. Yet these products were not conceived with total disregard for practical utility or market potency. Nor was the importance of listening to the customer an alien idea. Clearing out some old files recently, I was amused to find my thoughts about this in a memo written following a brainstorming session we had in November of 1975:

"I would be in favor of having one engineer who spends <u>more than half</u> his time traveling around the country collecting in-depth information, and whose responsibilities were to ensure that our current development program constantly matches the mood of the marketplace . . . [and] be alert for important new opportunities. He would not function primarily as a salesman . . . [but] would carry a constantly-updated portfolio of applications material and would offer to work with the customer on particular requirements. This type of <u>professional</u> link is far more beneficial (both to us and the customer) than [excessive emphasis on the deliberations of product selection committees] and anyway will probably become essential as our products become more sophisticated." [Original underlining]

The Wellsprings of Innovation

I've always been interested in the process of innovation and its traveling companion, creativity. I'm curious about why and how it arises in one's own work and how it might be fostered in one's co-workers. At root, innovation is a matter of matching the needs of the market—in all of its many facets and dimensions—to the ideas, materials, tools, and other constructive means at our disposal. Something, perhaps, that might best be entrusted to a team. There is no question that the sheer scale and complexity of many modern IC projects demand the use of teams, and that good team-building skills are an essential requirement of the effective engineering manager. Yet it seems to me that innovation remains a highly individual, at times even lonely, quest, and that enhancing one's own innovative productivity—both in terms of quantity and quality—must always be a *personal*, not a group or corporate, challenge.

Undoubtedly, the innovative spirit can be seriously hampered by a lack-luster infrastructure, run by senior managers who have their minds on higher things, and by executives who view their corporation as little more than a contract-winning and revenue-generating machine, to be optimized up by frequent rebuilding and generous oiling with business-school dogmas. Conversely, the often frail, tenuous groping toward individually distinguished performance on the part of young designers can be transformed by a supportive corporation, one that has many erstwhile engineers at the top, which recognizes latent talent, and which is willing to take a gamble on the individual. I have worked under both regimes, and can truthfully say that at the Tektronix of the sixties and at Analog Devices throughout its history, their top executives succeeded in fostering engineering excellence through the massive support of competent technical contributors, and the thoughtful, attentive consideration and encouragement of the idiosyncratic visions of such people.

Innovative urges originate within the individual, and can be either quenched or fanned into a blaze by corporate attitudes. But where do the *ideas* come from in the first place? I like to say that "Innovation Starts with Tomorrow." It is the "Start of the Art"—the *new* art that will one day become commonplace, even classic. Prowling at the boundary between the present and the future, the innovator never ceases to peer through the cracks and holes in the construction fence for telltale signs of new opportunities, as our world changes day by day. Innovation consists of this persistent, vigilant *boundary watch* followed by a *creative response* to what is seen. Essential precursors to innovation are a prolonged study of a certain class of problems, a thorough familiarity with the field of application, and total immersion in the *personal* challenge of making a significant contribution to the state of the art.

But is this enough? Many authors have grappled with the enigma of creativity. Some believe that it happens when normally disparate frames of reference suddenly merge in a moment of insight. For example, Arthur Koestler writes[1]

"... a familiar and unnoticed phenomenon ... is suddenly perceived at an unfamiliar and significant angle. Discovery often means simply the uncovering of something which has always been there but was hidden from the eye by the blinkers of habit."

Instances of this type of discovery come to mind: Watt and the steam kettle (probably apocryphal); Fleming and penicillin; Archimedes and his tub; etc. But others, including myself, reject the widely held idea that *radically creative* concepts can arise from a methodical, conscious, logical process. P. B. Medawar, who won the Nobel Prize for Medicine in 1960, believes[2] that it is a matter of "hypothetico-deduction." He states that hypothesis generation is

"... a creative act in the sense that it is the invention of *a possible world*, or a possible fragment of the world; experiments are then

1. Arthur Koestler, *The Act of Creation: A Study of the Conscious and Unconscious in Science and Art* (New York: Dell Publishing Co., 1967), 108.
2. P. B. Medawar, *The Art of the Soluble* (London: Methuen & Co. Ltd., 1967), 89.

done to find out whether the imagined world is, to a good enough approximation, the real one." [Italics mine]

According to Medawar, the creative process begins with an act of imagination, more like an act of faith, without a strong factual basis; the testing of the hypothesis that follows, on the other hand, requires deduction, a quite different activity.[3] Others, like Edward de Bono,[4] believe in the similar notion of "lateral thinking." In this scenario, one consciously jumps out of the familiar boundaries into a what-if world where the rules are different, establishes a workable structure which is self-consistent within this temporary frame of reference, then seeks to re-establish connections with the 'real world.' I find this matches my mode of working very closely.

I've got my own theory about the sources of the creative spark. I begin by noting that the most well-known aspect of the 'creative moment' is that it is mercurial and elusive. I suspect that human free will and creativity both have a *thermal* basis. Minds are an epiphenomenon of their physical substratum, the brain, which is diffused with thermal noise.[5] In particular, large aggregations of neurons are subject to statistical fluctuations at their outputs, and almost certainly exhibit a chaotic aspect, in the formal sense. That is, a small inclination on the part of just a single neuron to fire too soon, without 'the right reason,' can trigger an avalanche in coupled neurons in the group, whose states may cluster around a neural 'strange attractor.' This microcosm gets presented to our consciousness (a few milliseconds later) for consideration; we interpret it as that inexplicable, but very welcome, revelation.

When this happens in its milder, everyday forms (such as choosing what to select from a lunch menu) we simply call it free will; when it happens while we're thinking about a problem (or maybe *not* thinking about the problem), and culminates in that felicitous, euphoric, amusing "Aha!" moment, then we give the name 'creativity' to this cerebral sparkling, and call the outcome a Startling New Idea. What we each end up doing with these serendipitous sparks depends on our mood, on our orientation to opportunity, and on the strength we can draw from our internal 'databases.' For the innovator, these databases (roughly equivalent to experience) would include such things as general and specific market knowledge, and familiarity with relevant technologies, and of what has been successfully done already ('prior art') in the form of circuit topologies, IC products, and complete systems. Allowing these sparks full rein to control the immediate outcome, by inviting interaction with these databases and by suspending judgment, is essential to the creative process.

---

3. "The Reith Lectures Are Discussed," *The Listener* (published by the British Broadcasting Corporation), (January 11 1968): 41.
4. See, for example, "de Bono's Thinking Course," *Facts on File Publications* (1982).
5. Viewed as an electrochemical entity, the neuron could be said to exhibit the ionic noise of a chemical reaction; but this, too, ultimately has a thermal basis.

Clearly, whatever is going on in the attic, we are not deterministic state machines, that, like computers, always deliver the same response for the same stimuli. Nor are all our conclusions reached dianoetically. It is for this reason that even the most advanced computers are so utterly boring and lacking in creative sparkle. On the other hand, even inexpensive home computers today are very, very good at retaining huge archives of knowledge, accessible within milliseconds, and are very, very good at carrying out difficult calculations, of the sort that 'radio engineers' up until the '70s would have to do by hand, or with the help of a 'slip stick' (for years, the primary icon of the engineering professions), wastefully consuming a large fraction of their working day.

Computers, *in this very limited sense*, may have better 'experience' on which to draw than we have. But they are *rule-based*, and don't have our probabilistic sparkle (because we don't allow them to). The present symbiosis between *unruly* human minds and cool-headed digital computers, who live by the rule, and who can reliably provide us, their users, with instant access to vast amounts of knowledge, has already transformed the process of innovation, although in a rather predictable fashion. An even stronger symbiosis will result, I believe, with the *eventual* installation of non-determinism in neural network-based thinking systems, perhaps in the next decade. This courageous step is destined to radically alter the way we will innovate in the future, with quite unpredictable consequences. I find this a fascinating prospect.

The philosopher Jean-Francois Lyotard comments[6]

"In what we call thinking the mind isn't 'directed' but suspended. You don't give it rules. You teach it to receive. You don't clear the ground to build unobstructed: you make a little clearing where the penumbra of an almost-given will be able to enter and modify its contour."

This reference to the negative effects of 'direction' and 'rules' is telling. There is a tension that arises in a corporate environment between the need to have structure in certain cases and the need to leave other things unstructured. Innovation does not thrive in a rule-rich context; on the other hand, it can be significantly enhanced in a tool-rich context, particularly if these tools provide access to a large body of knowledge and allow one to play uncountable 'what-if' games with this knowledge. Such tools have proven time and again to provide profound and *completely unexpected* insights: the new world of fractals was unknown and probably unknowable without computers, and the same can be said of chaos theory, whole-body imaging, molecular engineering, and much

---

6. J-F Lyotard, "The Inhuman: Reflections on Time," tr. G. Bennington and R. Bowlby (Stanford: Stanford University Press, 1991): 19.

else. Imaginative use of computers is nowadays almost synonymous with innovation, at least, in the way they open up our minds to visualizing new possibilities. It remains up to us, though, to turn fragile, promising ideas into robust, marketable products, which is the true essence of innovation.

The precise moment when a new concept, or 'art,' first bears fruit is especially significant. Although always about ideas and personal insights, innovation is not so much about *knowing how* (that is, 'know-how') as about *actually making things happen*. It frequently involves recourse to the use of *markedly unusual and unexpected methods* in achieving its objectives. And although we generally use the term 'innovation' in connection with *practical* advances, theory may play an important role in the process, and one can also innovate in a purely theoretical direction (for example, Norbert Wiener's seminal statistical theory of communication). But there always has to be a tangible product, usable by many others, as output.

I've stressed that innovation—invention—is largely a matter of one's personal perceptions of, and responses to, one's surroundings. It arises out of seeing the myriad opportunities that abound in a utilitarian culture in an ever-fresh and bold new light. Opinions may differ, but I believe it's about being *convinced* first, of the validity of that singular vision, a bold assurance arising in equal measure from first, experience of what works combined with a firm grasp of the current needs of the market; and second, an awareness of the necessity to continually channel that vision into profitable realities. In managing our self-image, it's *okay* to appropriate to oneself such terms as product champion, conceptualizer, mentor, inventor, or master of the art, if we feel that truly describes what we are and what we do. It would, of course, be immodest to make those claims *publicly* about oneself. Nevertheless, these are the kinds of 'good words' strongly motivated achievers might well choose to describe their aspirations in private moments. It's *okay* to feel proud of one's best achievements, if they've proven market-worthy.

Invention thrives in a multi-disciplinary mind. During the course of developing a new IC product, the well-equipped innovator will, over the course of a typical day, need to take on the mind of circuit designer (concerned with basic concepts and structure), technical writer (explaining one's work to other team members, or thinking about how the applications are going to be presented), semiconductor device specialist (during transistor design for critical cells), marketeer (maintaining a focus on the needs of the customer, firming up the formal specs, continually verifying fitness of use, etc.), test engineer (in preparation for production), and accountant (watching die size, yield, cost).

Innovative design is far removed from the serial, step-by-step process that is sometimes suggested. It is an extremely iterative, exploring, yearning, and discovering process. Dissatisfaction abounds at every turn. Revisions occur with stunning regularity. One's attention is constantly readjusting, at every level. For example, during the design phase we need to address numerous minor yet essential aspects of our circuit's behavior (say, stability with a nasty load); a few minutes later we may be attending to, say, optimizing a device geometry; sometimes tinkering at the input boundary, sometimes at output; checking temperature behavior again, then AC, then transient, then beta sensitivity, operation at the supply limits, and round and round again. Out of all this, the final solution gradually comes into sharper focus.

Even while primarily in the 'design' mode, it may be hard to think about the product entirely as a challenge in circuit refinement. One needs to frequently pop out of the schematic view of 'reality' and briefly review, say, applications issues, trying on the shoes of the user again, to see how well they fit; a moment later, plunging back into schematicland, scrutinizing the horizon one more time for things that maybe aren't yet quite right; or challenging oneself again about the justification for the overall architecture, or the need to reconsider the practicality of the present chip structure, the roughed-out layout, and so on. The dynamics of 'getting it all together' involve a lot of judgment, art, trial and error, and are far removed from the popular image of the nerdy engineer, methodically pursuing serial, rule-based, 'scientific,' forward-pushing progress. Only the neophyte engineer remains in the same mode for hours, or even weeks, at a time.

We need better tools. Faster simulation is always in demand. Certain aspects of circuit behavior—such as finding the 1dB compression point of a mixer—need a lot of CPU time, and have to be performed in the background, although it is often difficult to take the next step until such a result is available. The simulation of other behavioral aspects may simply not be possible at all, or to the required accuracy, and one is left to devise ingenious analytical methods to solve such problems in the classical way, using circuit theory! Though a well-understood challenge, the need for very rapid turn-around in a simulation context is rarely viewed as an *essential* aspect of the ergonomics of innovation. Fast machines don't just provide quick answers; they are better able to run beside us in a partnership. But *we* should be the gating factor; it should be *our* wits that limit the rate of progress, not those of an unresponsive machine.

Modeling the Market

Innovation in a large corporation depends on a lot more than our willingness to put the best of our personal insights and creative talents to work. We need to establish and maintain firm anchor-points in the marketplace, always the final arbiter of success and failure for the serious product designer. While our primary focus must be on the *technical* issues relating to the systems, components, and technologies that we and competing companies each develop, we must also thoroughly understand the *dynamics and psychology* of our particular industry. Further, we must understand not only our current markets intimately, but go well beyond: we must constantly *anticipate* the future needs of these markets. As innovators, we must be neither overly timid, nor cocksure, about that challenge.

If timid, we might fall into the trap of modeling this 'world of the market' as a fortress of rationality, where there are compelling reasons why our customers' present solutions are not just satisfactory, but intimidatingly superior; a place where all the good ideas that relate to some particular business have long ago been figured out, and around which an impenetrable wall has been built. Such an apologetic approach to our domain of opportunity would be unwise. The truth is that the majority of users of advanced components and systems are daily managing to scrape by with *barely adequate* solutions to their technical problems, and are constantly on the lookout for more competitive, more reliable, more powerful alternatives. They crave to be *advised* by their vendors. With an eye on leadership, we shouldn't let them down.

Yet we cannot afford to be too confident about our prowess to serve the world of the market. In a 'good company,' with a trail of successes behind it, one may occasionally hear scornful comments about one's competitors. The innovative spirit has no place for either derision or complacency. One's view of the market and of one's competitors needs at all times to be *honest, focused, realistic, and balanced*. In this outward-embracing view of our world, we must also include advanced ideas coming out of academia, the ideas of others in industry, as expressed in the professional journals and at conferences and workshops, and the commentaries of journalists writing about our business in the trade books and financial-world newspapers. In short, we need to be effective *gatekeepers*, balancing self-motivated innovation against careful reflection of external factors, the eager anticipation of the challenge against thoughtful assimilation.

In our field of microelectronic product design, one innovator was both a legend and an enigma: this was Bob Widlar, who died in 1991.[7] Those who knew him recalled that he was very hard to relate to. Bob Dobkin, who worked alongside Widlar, said: "Widlar knew it all, he knew he knew it all and nobody else knew anything." He was a maverick and a nonconformist, with many stubborn ideas of his own. Yet he did amazing things with silicon, and introduced many firsts: "He pioneered the three-terminal voltage regulator, on-chip power devices, the bandgap voltage regulator, super-beta transistors and a full bag of clever and interesting circuit and device techniques," said Jim Solomon.

"One thing that everyone should know: Bob was concerned with all aspects of his craft (or art), *including* 'marketing,' in the true sense of understanding the economics and systems applications of his products," observes Analog's Lew Counts, who believes we would do well to replace

---

7. I am grateful to Lew Counts for reminding me of the tribute written by Jim Solomon in the August 1991 issue of the *IEEE Journal of Solid-State Circuits*, vol. 26, no. 8, pp. 1087–1088.

some of the organizational paperwork currently on our desks with Bob's seminal articles. Although we may be disappointed to find nothing explicit in them about his motivation for the development of a new IC, nevertheless his market orientation and his *overall grasp of the possibilities of the medium* were an ever-present and tangible aspect of his work. One wonders whether Bob Widlar would fare as well in modern corporations as he did at National Semi in the '60s, or what his reaction would be to some of the prevalent "improvement" methodologies.

To a large extent, today's innovators rely on existing markets to guide their thinking. To what extent has it always been that way? How were innovators of long ago motivated? Was their way of creating new products applicable to us today, locked as we are in a co-dependent embrace with the customer? We engineers are inheritors of the spirit of a long lineage of innovators, and the beneficiaries of their energies. The golden braid of knowledge technologies[8] can be threaded through The Savvy Sumerians (c. 3,000 BCE), Archimedes (287–212 BCE), Lenny da Vinci (1452–1519), Jack Gutenberg (?–1468), Bill Gilbert (1544–1603), Humpie Davy (1769–1830), Mike Faraday (1791–1867), Charlie Babbage (1791–1871), Sam Morse (1791–1872), Wern von Siemens (1816–1982), Jim Maxwell (1831–1979), Tom Edison (1847–1931), Hank Hertz (1857–1894), Chuck Steinmetz (1865–1923), Nick Tesla (1856–1943), Guggi Marconi (1874–1937), Norbert Wiener (1894–1964), Ed Armstrong (1890–1982), and many more.

The key feature of the work of these giants of technology, and dozens more like them, is that *they didn't wait to be told to innovate.* What they did stemmed from a fundamental urge to produce solutions that significantly challenged the norms, and could even transform the world. The Sumerians' insight that the physical tokens[9] used to keep track of financial transactions (and much else) could be replaced by distinctive marks on soft clay tablets, which were later transformed into records of archival quality by exposure to the noonday sun, was innovation springing from great independence of mind. (Who would have been the 'customer' for *writing*? The very thought is laughable.)

Most of us feel (justifiably) that we cannot aspire to the greatness of such inventors, particularly in our limited, highly specific domain of virus-scale electronics. Nevertheless, it is proper—and not immodest, in my view—to seek to emulate their example. Like them, we need to have a clear conception of what would be *useful*; to always be ready to *propose solutions* without first needing to be asked; to be *confident* and passionately *committed* to one's vocation; to maintain a high level of *concentration*; to feel *resourceful, capable, well-equipped, determined*, to never cease devising a string of *self-imposed challenges* for solution; to practice

---

8. Which is what electronics is all about, in the final analysis.
9. See "Before Writing: Vol 1, From Counting to Cuneiform," by Denise Schmandt-Besserat, University of Texas Press (1992) for an enlightening account of the precursors of writing.

persistence. It is out of these attitudes—the 'state of the heart'—that the best innovation wells forth.

Listening to Voices

From the tastefully furnished executive wings of modern corporations we often hear that, unequivocally, the process of innovation must begin with the customer. It is said that product concepts must be founded directly on one's intimate understanding of each customer's needs, garnered through countless hours of listening attentively to what we are told is the Right Thing To Do.

There is no denying the importance of paying close attention to our customers' needs—particularly when that means *a large group of customers with similar needs*. But is this the motive-force of innovation? Sometimes. An earnest and sincere involvement with the customer is frequently important—even essential—right from the start, and skillful probing may well lead to valuable and unexpected insights, which need our careful assessment before a new development is started. Usually, though, the *real* 'start of the art' is quite fuzzy. Many valuable product ideas reach right back into our early fragmentary awareness of the general needs of an emerging market, with roots in our broad knowledge of practices and trends. The final design will invariably be based as much on our own stored ideas about critical requirements and specifications for products in the customer's field of business, and techniques to address those needs, which we've painstakingly garnered over a long period of time, as it is on the customer's voice, attended to for a few short hours.

Often, that costly trip to the customer is not so much to fuel the innovative process as to establish the realism and scale of the *business* opportunity, the volumes, pricing and schedule, information on which to base decisions about multiple-project resourcing. Although we will be very attentive to what the customer may tell us about technical matters, it is unusual that something is learned about the *function* or *specifications* that is completely unfamiliar.

This is particularly true of mature generic products (such as most amplifiers) and of many application-specific ICs (ASICs) and other special-purpose products that address well-developed markets. The success of new types of product, and user-specific ICs (USICs), having hitherto unavailable (or even unattainable) functions, depends very heavily on meeting an 'external' set of requirements down to the letter, and obviously requires much more careful listening. On occasions, though, even these don't get completely defined by a customer, but rather by a lengthy process of sifting through the special requirements of a system specified in *general* terms, often by reference to operational standards in the public domain. Numerous case histories point to this lesson.

In one such case history, the customer—a major computer manufacturer—knew in broad terms what was wanted (an encoder to convert a computer image into a television image), but had practically no idea how it should be done. In fact, engineers at this company approached Analog Devices because they knew that, somehow, *modulation* was involved, and that we were the undisputed leader in analog *multipliers*. Ironically, although these *can* perform modulation, they would have been a poor choice in implementing this function. Thus, we had been the customer's first choice on the basis of our known leadership in an irrelevant field! On the other hand, many of our other signal-processing skills were very relevant, and of course, we would say they made a good choice!

In the months that followed, it took an enormous amount of research to find out what *really* needed to be done. We drew on standard knowledge about television, consulted the relevant standards, and talked with specialists in the industry. Bit by bit, pixel by pixel, the design incorporated the best of all this knowledge, and has since become a very successful product. But the success of this product cannot, in all honesty, be attributed to any fine insights we learned from the one original customer. We got into the TV encoder business because, much earlier, we had developed certain products (in this case, analog multipliers and mixers) out of an awareness of their *general* utility, and because, once we had sized up the opportunity by a few trips down Customer Lane, we then *independently researched the subject to really internalize the challenge.*

During the early days of Analog Devices this pattern used to be fairly common: we'd demonstrate competence in some field, by having made a *unilateral* decision to add a novel function to the catalog (sometimes on the whim of a solitary product champion), without a clear voice from the marketplace, then later would discover that these generic competencies aroused the attention of new customers for ASICs and USICs. The task of picking winners was later entrusted to a small committee, which met, sporadically and infrequently, in pizza parlors, private homes, and Chinese restaurants. Our batting average was probably no better than what might have been achieved by giving the product champions full rein. Still, it worked; the catalog swelled, and the stock climbed. Innovation was happening apace. And not just in product design, but in new processes, new packaging techniques, new testing methods.

I strongly believe that seeding the market with well-conceived, *anticipatory* generics, the '70s paradigm,' if you like, remains a very serviceable strategy for a microelectronics company; such here-and-now products will probably be of more value to one's customers than a quadrivium of questionnaires and a plethora of promises. On the other hand, it would be foolish to overlook the profound importance of developing and strengthening one's relationships with key customers; without them the most daringly innovative product would be so much artfully coordinated sand.

I'm not advocating a mindless return to the methods that happened to work well in an earlier age. It is a matter of *emphasis*. It's about maintaining a balanced outlook, about the optimal use of resources and about managing risk. Innovation is always risky; but deliberately putting the brakes on free-spirited innovation is not without risk either. Which would you rather do: (1) Bet on a few people who have proven to be well-rounded, resourceful, and intimately familiar with the techniques and systems used in any market sector; instruct them to spend a lot of time in mentoring less-experienced employees and in encouraging them to sift through the numerous items of professional and trade literature (including standards documents) in constantly broadening their own awareness of the field; give them all the most powerful CAD tools available, and a lot of freedom to be creative . . . or . . . (2) Encourage your designers to become more effective communicators, to write and later review questionnaires, carry out statistical analysis on the replies, generate product recommendations, and then form teams to act on them to the letter? Both classes of activity are important. But if you were forced to choose between scenario (1) and (2), which do you think would be the more potent strategy?

Case Histories in Communications Products

Analog Devices' involvement in the radio world took a big step forward several years ago (although we didn't realize it at the time) when a Japanese customer requested a quotation on a special multi-stage log-amp. This request had arisen out of the customer's evaluation of our AD640. Here was another product that was not the result of a market definition process. When it went into production, *not a single customer had yet been identified*. It was the world's first five-stage log-amp, and the only log-amp to use laser trimming to provide very exact calibration. I personally felt it was just a good idea to add log-amps to the growing repertoire of wideband nonlinear circuits in the catalog, and to continue to pursue such products of general value.

What this customer wanted seemed preposterous: twice the dynamic range of the AD640, single- (rather than dual-) supply operation, at about one-tenth the power, having new and very tight phase requirements, and various other extra features, all in a much smaller package and (of course) at some small fraction of the cost of two AD640s. I vividly recall standing by the fax machine just outside Paul Brokaw's office, reading with much amusement the request that had minutes before come in from our Tokyo office, wondering what kind of fools they must think we were to even *consider* bidding on such a thing. After all, we were a high-class outfit: we didn't make jelly beans for the masses.

But the seed (or was it bean!) was planted, and the technical challenge took root. I couldn't put it aside. During a lot of nocturnal sims (when our time sharing VAX-780 was more open-minded) I became excited by the possibility of actually meeting both the performance and the cost objectives. At some point, we decided to "Just Do It," and eventually, out of that customer's one-page request came the nine-stage AD606. I dispensed with the laser-trimming used for the AD640, instead pared the design down to accurate essentials, found new ways to extend the dynamic range and meet the phase skew requirements, threw out one supply, and whittled 40 pins down to 16.

However, even though strongly based on the one customer's request (which had been articulated as little more than a general desire to combine the function of two AD640s in a single low-power, low-cost chip), and even though *we were listening hard* for every scrap of guidance they could provide us, the actual specifications for the AD606 were once again very hard to elicit from the systems engineers for whom we were specifically designing the part. They, it seemed, knew less than we did about what needed to be done. So, where these specifications were missing, we interpolated and extrapolated with our best guesses as to what an *ideal* receiver would do in the circumstances, adding one or two innovative features that weren't in the original request.

The subsequent learning process surrounding the AD606 project—about the systems in which the part was to be used, as well as the accumulated know-how of designing, specifying, and testing such parts—became a major team effort that substantially furthered our capabilities in RF receiver circuits for digital phone systems, and opened doors to new opportunities. In developing it, we gained invaluable experience and learned much that was later to help us advance the state of the art in multi-stage log-amps into newer, even stranger territories.

Before long, the same customer clamored for more function (the addition of a UHF mixer), an even lower supply voltage (2.7V min), even lower power (20mW), and, of course, all for the same low price! Now we were *really* listening to the voice of the customer, because, in spite of the tight margins, the business opportunity looked like a good one. But that C-voice was still weak. We were not really being given a *performance definition* for an ASIC, so much as being asked to add general new capabilities, while lowering the supply voltage and power. Yet again, we were forced to do a lot of *independent system research* to produce a design, for which the number AD607 was assigned.

As it turned out, this design deviated in important ways from the original expectations[10] of the customer (a mixer plus log-amp) in that it relied on an overly innovative approach[11] in order to address some new dynamic range issues and circumvent the technical limitations of the all-NPN process we had been using for the AD640 and AD606. It used a very fast-acting AGC loop with accurate linear-in-dB gain control to implement a log-amp in an unusual manner. This time, the customer didn't believe our approach would work, mainly, I believe, because no one had ever made log-amps in this way before.

---

10. Not from detailed performance requirements, though, even less a suggested architecture. Neither of these were provided.

11. Timing is all-important; product concepts can be neither too advanced nor too pedestrian.

So the 'AD607' was put on a back burner, even though the forward-looking concept was felt to have general market potential in cellular phone systems. After some re-thinking we eventually developed the AD608, which was just what the customer wanted, although it required us to use XCFB, an advanced, and as yet unproven, IC process.[12] The risks were weighed; XFCB won. The AD607 was later redesigned on this process for use in GSM digital phone systems, in which it promises to provide a highly effective solution.

It's very important to understand that in this, and many similar case histories, the barely audible Voice of the Customer quickly gave way to the much louder and more authoritative Voice of the Committee that wrote the GSM standards, and to the Voice of the Consultants that we hired to help us more rapidly progress into this new field. (You notice how they're all VOCs?) Thus, in pursuing an innovative approach to product development, a *wide range of voices* are often being heeded, including those all-important ones that sound from inside, the Voice of Conviction and the Voice of Commonsense.

We need to be careful in connection with the last of these voices, though. Comfortable common sense can be the nepenthe that smothers innovation. All of us are inclined at times to view things in the same old fading light, particularly if the accepted solution seems "only common-sensical." (I am bound to think of the lemming-like use of op-amps where voltage gain is needed. Op-amps are *far* from the best choice in many applications. How alluring is their promise of 'infinite gain,' but how far from the truth! Still, they remain the *commonsense* choice for thousands of users, and new products are ignored.)

Often, there are situations where we need to pay close attention not so much to what the customer may say to us, but to what is *really the problem that is in need of a solution*. Thus, only a few years back, the commonsense way to boil a pan of water was to add heat directly, either by dissipating a kilowatt or two in an electric resistor, or by the oxidation of some energy-rich material (gas, oil, wood, whatever). Few would have been so crazy as to have suggested the use of a peculiar vacuum tube called a magnetron. In fact, it's pretty certain that no one actually working in the kitchen (the Customer, in this case) would have *ever* thought about the need for a different approach to something as prosaic as heating food. Yet, the overnight success of the inexpensive microwave oven is just one of innumerable examples of products which owe their genesis to a *truly innovative* approach to the marketplace—one that *foresees* an opportunity before it is articulated, or even which sees a way of *generating* a need where there currently isn't one. Out of the introduction of the microwave oven came a totally new, co-dependent industry, that of instant meals.

---

12. XFCB, for "Extra Fast Complementary Bipolar," a DI process bringing long-anticipated benefits to low-voltage, low-power circuitry. See later comments on the genesis of this IC process.

Time and again we find that innovation has meant that someone (often, literally one person) saw a bold new way of achieving a commonplace task, and had heeded the Voice of Courage and proceeded even without the slightest hint from the marketplace of its utility. This relentless and self-eclipsing search for 'a better way' is the hallmark of the innovative engineer. Thus it is unavoidably true that the innovator is frequently the iconoclast: not content with merely making a useful contribution to advance the state of the art, he or she seeks to *redefine* that art, to *restart* the art all over again from a totally different perspective, often obsoleting last year's best ideas in the process.

History through Dark Glasses

Come with me on a journey into pseudohistory. It is a chilly winter's evening in November, 1878. A young man of thirty has recently finished reading a book about how to be a successful marketeer. It was called *"YOURS IS THE MARKET,"* subtitled *"How to Find Out What People Really Need and Thereby Become Rich and Famous."* Although it was actually written by an inscrutable Japanese sage in Kyoto, it had recently become popular through the best-selling translation and Americanization by a famous Harvard professor with the improbable name of Yucan Sellum. This book proclaimed that "*. . . the first step to a successful product is thorough market research,*" and having taken this very much to heart, Tom had set out to systematically poll the residents of Menlo Park, New Jersey, to find out what they *Really Needed.*

He was getting a little tired, first because he'd walked many miles, but also because the responses were all so boringly and predictably similar, and he felt he'd amassed plenty enough information to comprise a statistically-valid sample set. He decided, though, that he'd complete a round-100 inquiries: "That surely will tell me *exactly* what People Really Need," he thought to himself. (In fact, he was subconsciously recalling Prof. Sellum's words: "*It is obvious that the more people to whom you talk, the more likely it is that you will find out exactly what the People Really Need. By the time you have interviewed one hundred people, it is only obvious that the probability is close to 100% that you'll know precisely what is marketable.*")

He knocked on the 99th door, and started the algorithm. "Good evening, sir, my name's Tom Edison, and I am interested to know what you might find inconvenient or inadequate about the present way you light your home. Is there perchance some improvement that you'd like to see on the market?" "I dunno who you are, young man," growled the homeowner, "but yes, I can think of a couple of things. First, if you can invent a stronger, brighter gas mantle, people will beat a path to your door. Those durned things are *always* breaking! And second, if you can invent a way that causes leaking gas pipes to be self-healing, you'll quickly find yourself off these streets. You can write that down. Here! Take this quarter and buy yourself dinner: you look starved!"

Tom was a little discouraged. Though he *was* hungry, he didn't need charity. Years ago, as a twenty-three-year-old, back in his Newark days, he'd made $40,000 from the unsolicited invention of the Universal Stock Printer for Western Union, and had developed several derivatives of the Morse telegraph. He'd also breathed new life into Bell's telephone by the invention of the more powerful carbon microphone, and he'd invented that phonograph thingy, too. It is said he was writing about 400 patent disclosures a year.

No one had wanted the phonograph, of course, nor the improved telephone, come to that, but Thomas Alva Edison had a pretty keen eye for what innovation was all about, and could readily shrug off the myopic naysayers. He used to declare that he was a "commercial inventor" who worked for the "silver dollar." What he meant by that was that he consciously directed his studies to devices that could satisfy *real needs* and thereby come into widespread popular use. But, all that was before his conversion by Prof. Sellum; ah, those heady days were the *old* way of doing things, he now sadly realized.

As he plodded the streets, he felt just a mite resentful. When it came to home lighting, he would have really welcomed an opportunity to promote his *current* ideas. Nevertheless, with the noble Professor's words emblazoned across his forehead, Tom went resolutely up the seven steps to the final door, and oscillated the brass knocker. Sharp echoes resounded from within the chilly and austere interior.

While waiting, he thought: "Hmmm . . . I could fix things so that the touch of a little button on this door would melodiously ring a bell in the living room, and an annunciator panel would show which door was involved . . ." He became excited as numerous elaborations of the idea coursed through his lively consciousness. Then he quickly corrected himself. "Nah, no one's ever asked for *that*, so it's probably not a good idea."

As he was reflecting on the senselessness of *even thinking* about ignoring the Harvard Professor's sound advice, and actually inventing and marketing something that no one had asked for, the door abruptly swung open, and a stern, ruddy-faced matron of ample proportions confronted him. "YES!?" she hissed.

"Good evening, ma'am, my name's Thomas Edison, and I'm interested in knowing what you find inconvenient or perhaps inadequate about the present way you light your home. Is there some improvement that you'd like to see marketed?" "Boy, there's *nothing* in the *slightest* wrong with the lighting in *my* home. We use *oil* lamps, the same as *all* of us in this family do, and have done for generations. Now, if you can find a way to make our oil-lamps burn *twice as bright* and *twice as long* from one filling, *that* would be something you could sell. But since you can't, be *off* with you, and find something *better* to do with your life!" The sound of the heavy black door being slammed in his face convinced him that he'd listened to enough voices for one night.

When Edison got back to his lab, he sank down into his favorite old leather chair, and with a sigh of the sort only a marketeer knows, he ran his fingers through his prematurely graying hair. All the rest of his guys had gone home by this late hour. It was already quite dark. He reached over and flipped a switch. Instantly, the desk was flooded with a warm yellowish light, emanating from a glass bottle connected by a couple of coiled wires to a generator spinning[13] somewhere in the basement, whence drifted the distinctive whiff of ozone emanating from sparking commutators.

On the desk were the patent disclosures for his new tungsten lamp, alongside hundreds of pages of notes on numerous other kinds of filaments with which he had experimented. On top of all these was the good Professor's best-selling and popular guide to success, heavily dog-eared and yellow-highlighted with Tom's fluorescein-filled fountain-pen ("Another 'bright' idea of mine," he'd quipped). From this seminal work, he had learned about a new way to success: *Listen to the Voice of the Customer.*

Reaching into the deep pockets of his trench coat, Edison wearily pulled out his spiral-bound reporter's pad, and reviewed the day's research. The message was clear. Of the 83 that had actually voiced some definite opinion, the customers had noted two key improvements needed in their home lighting systems: better gas mantles, and higher-efficiency wicks for their oil lamps. "Too bad nobody ever asked me if I had any ideas of my own," he sighed, ruefully recalling Sellum's strong advice that the VOC process must be conducted "*with decorum*" and "*in such a way that . . . one only elicits those facts which the customer freely wishes to impart to the researcher.*" (Chapter 13, Para. 13, page 1313).

Thomas Alva Edison opened one of his large oak filing cabinets, and tossed in all the tungsten-filament papers, heaving another great sigh. Maybe *someday* he'd find a use for all that work. He then took a sharp pencil and a clean sheet of paper, and wrote:

"Trip Report, 18th November, 1878. Spent all day doing a VOC in Menlo. Spoke with 100 people re lighting improvements; got good info. from 83. . . . Action Item: Write Product Development Proposal re Improvements to Gas Mantles and Oil-Lamp Wicks. Do before Monday exec. council mtng. Call a KJ to consider weaknesses in present methods of mnfng mantles. Memo: be sure Monica obtained an adequate supply of Post-It™ pads."

Innovating in the Nineties

Of course, Edison didn't work that way or write such rubbish. So far as we know, he never pounded the streets looking for ideas; as far as we know, he never conducted market surveys; he certainly didn't spend his

---

13. However, not humming. Edison was fixated on DC, and jealously blinded to the value of AC power.

time generating product proposals. But he did have a flair for knowing what was marketable.[14] We probably can't pursue invention in precisely the same free-wheeling fashion that Edison did. In certain important ways, our world is different. But the boisterous entrepreneurial spirit which he and other long-dead pioneers exhibited can still be a source of inspiration to us today. The basic challenge remains essentially the same: thoroughly master your technologies; become intimately familiar with the needs of the market in the broadest possible terms; respond to these, but spend only the minimum necessary time, while pursuing new solutions in readiness for that moment when the market opportunities that you saw on the far horizon come into full view of everybody.

Still, what *is* it about our world, and the way we innovate nowadays, that has changed so much? Why *can't* we still turn out product ideas as profusely as Edison did? Why, when eavesdropping on cocktail-time conversations, do we technical people chuckle (or maybe sigh) at hearing someone use such embarrassingly *old-fashioned* terms as 'Inventor' and 'Genius'? First, we must acknowledge that men like Gauss, Henry, Ampere, Weber—and Edison—were extraordinarily gifted, possessed of relentless energy and self-assurance. They were born into a time when the enormous scope of opportunities for electrical and magnetic devices had yet to be fully understood and their potency in everyday life demonstrated. Arguably, it's easy to be a pioneer when numerous untried and exciting ideas surround you like so many low-hanging plums.

But is this the correct explanation of their success? Are we not today "born into" a world where the latent potency of global personal communication systems, enabled by spacecraft and satellites, by cheap multi-million transistor DSPs and high-performing analog ICs, is poised to transform our lives far beyond what we witness today? This is a world in which sub-micron CMOS, 100GHz silicon germanium heterojunction transistors, optical signal-processing, neural networks, nanomachines, MCMs, and MMICs are all waiting to be exploited by the eager innovator. Is it not true that a modern IC company, with its broad range of technologies and wide applicability, can be equally a springboard to unimagined new conquests? I very much doubt whether it's much harder to be a technical pioneer today than it was at the turn of the century.

Of course, Edison was not inspired by a mythical Prof. Sellum specializing in cute organizational methods. Rather, he devoured the published works of another remarkable innovator, Michael Faraday, himself burning with the red-hot zeal of an adventurer and world-class discoverer. Faraday worked at the fringe. Indeed, when we study the history of the great inventors, we find that they were often fired by ideas which, in their day,

---

14. Usually, anyway; but in defending his empire of DC generators and distribution systems, he even used mendacious disinformation slurs to impede Tesla's promotion of AC as a better choice than his own.

were right at the ambiguous leading edge—really more like a soft slope—of some 'new paradigm.'

Edison was no different in this respect. Many of the ideas he later turned to practical advantage were first conceived, but only tenuously exploited, in less market-oriented Europe. He owed a great, although rarely noted, debt to a Serbian of unequivocally greater genius, Nikola Tesla, who worked for Edison for a while.[15] Incidentally, Tesla points to another necessary quality of the innovator: long hours. His were 10:30 a.m. to 5:00 a.m. the following morning, with a brief break for a ritualistic dinner, every evening, in the Palm Room of the Waldolf-Astoria hotel. The interplay between these two innovators makes a fascinating study. Edison was the eternal pragmatist who disliked Tesla for being an egghead; he prided himself on "knowing the things that would not work," and approached his work by a tenacious and tedious process of elimination. Of this "empirical dragnet," Tesla would later say, amusedly:

> "If Edison had a needle to find in a haystack, he would proceed at once with the diligence of the bee to examine straw after straw until he found the object of his search. I was a sorry witness of such doings, knowing that a little theory and calculation would have saved him ninety percent of his labor."[16]

But Tesla was also a touchy and difficult man for others to work with. He expected the same long hours from his technicians as he himself put into his work. For these men, electrical engineering was a vast, unexplored frontier, bristling with opportunities to innovate precisely because there was yet essentially no electrical industry. Delivered into this vacuum, basic inventions could have a dramatic impact; competition would come only much later.

These circumstances are not unique to any age. It's only the details that differ as time passes our way. Sure, there are plenty of light bulbs and electric motors already, and plenty of op-amps and microprocessors. The chief question for the contemporary innovator in microelectronics is: what are *there not* plenty of? That was the essence of Edison's quest, and he accordingly imagined, then innovated, ingenious and eminently practical electrical, mechanical, and electromechanical devices, with profit unashamedly in mind.

Through the nervously-flashing retinas of his own eyes, Tesla looked out on the same world and had startlingly different visions of the future,

---

15. Tesla introduced him to the wonders of alternating current. Edison treated him very badly, even cheating him out of $50,000 after he successfully completed a project with which Edison challenged him, and didn't think he'd achieve. As noted earlier, Edison later launched smear campaigns when it looked like Tesla's visionary ideas about AC power systems threatened the commercial empire based on DC.

16. Quoted from *Tesla, Man Out of Time*, by Margaret Cheney, Barnes & Noble (1993), p 32.

including even radio and radar, VTOL aircraft, robotics, and much else. But in some respects his approach was similar: like Edison, he was possessed of a lot of personal energy and self-assurance; he knew of his unique talents. Above all, he had a *strong sense of mission* and what some might regard as a fanatical single-mindedness (see opening quote). Even today, the best innovation, in my view, springs from *owning the subject*, and pursuing an individual pilgrimage, toward destinations which are largely of one's own making. We aren't making the best products just because some customer suggested them to us, or even assured us of big orders, but because we have a passion to bring some art, in which we have a *large personal investment*, to the pinnacle of perfection.

Opportunity, Imagination, Anticipation

When we look at the world intersected by the time-slice given about equally to each one of us, what do we see? Opportunities! Not fewer (because "all the good inventions have already been made" and "all the practical needs of the market are already being satisfied by a huge industry"), but *many more*, precisely because of the massive infrastructure that now exists.

Think about how hard it would have been in Faraday's time to wind a solenoid. Where would he have obtained a few hundred feet of enameled copper wire? Not from the local Radio Shack. Undaunted, he *imagined* his way forward. Today, making a solenoid is literally child's play. Indeed, many of today's kids are doing things with technology that would baffle Faraday. Thus empowered by the infrastructure, our level of innovation can be so much more potent; we can do great things with the technical resources at our disposal. While Faraday may have spent a week or a month or a year getting the materials together and then winding a coil or two, we just order what we need from the Allied catalog.

So the 'innovating-in-a-vacuum-was-easy' theory doesn't make a lot of sense; it couldn't have been any easier *because* there was no infrastructure: it was probably a lot harder. Today, we are beset on all sides by *astounding* technology waiting to be put to innovative use. And just like Faraday, Edison, and Tesla, and all those other pioneers, we need to *anticipate* the imminent need for this or that new component—from what we know of the market's current needs, and based on what we know about our technologies, whether primitive or advanced—and to anticipate its value and realize its potential before everybody else does. These aspects of innovation are timeless, and they are not strongly susceptible to methodological enhancement by clinical studies of innovation in the *Harvard Business Review* (though they make interesting reading).

Still, we haven't answered the question about how our world is different from earlier times. Might it be the high complexity and sophistication of modern technological projects? Faraday's solenoids, Edison's filament lamps, carbon microphones, DC motors and dynamos, and Tesla's super-coils and induction motors, though revolutionary, seem in retrospect quite simple, almost naive. Perhaps that's part of it. But underlying even the most complex of modern devices, circuits, and systems, there are always just a few simple ideas. For example, there is today a strong market need for exceptionally-low-noise amplifiers, in medical ultrasound equipment, in analytical instruments, and in many communication systems. The principles of low-noise design have not altered for decades; this is not at all a matter of complexity, but of sound engineering practice based on a clear understanding of cosmic clockwork. Yet here, as in so many other situations, opportunities for innovative solutions remain.

Even complex microprocessors make use of conceptually simple high-level logical entities, the details of which become quite secondary in executing a design, which, furthermore, is often only evolutionary, based on a large existing knowledge base. Architects of megamillion-transistor memories are no more innovative than those advancing the state of the art in the underlying cells that are used in such memories. The complexity argument seems to be a red herring.

Maybe today's markets differ in that they are mature: they are already well served by many effective solutions, offered by numerous competing companies.

There can be no doubt that it is easier to innovate when there are simply no existing solutions, and no one else in the field with whom to compete. "Edison had it easy!" you might say; "Bring him back into *these* times and see just how well his genius would serve him!" I've often wondered about that. The modernist's view of the world, and an awareness of the seductive power of myths, leads one to realize that the great figures of history were probably not in any essential way much different from you or me. The notion that the era of Great Innovation and Pioneering is past could be enervating. Certainly, Edison would be a very *different* figure in today's world, but we can only speculate about whether he'd achieve more, or less.

What Lies Ahead?

I believe we are right at the edge of a massive thrust forward into the age of what I like to call 'Epitronics,' by which I mean electronics in the service of knowledge. Such systems are electronic *only* because electronics provides cheap, miniature, and very fast substrata for the realization of knowledge systems, not because of any *essentially-electrical* aspect of the function of these systems. The term *epi*tronic points to this 'floating-above' aspect of complex data-handling systems: what they are transcends what they are built from. Today, general-purpose computers are the most obvious 'knowledge engines'; their internal representation is entirely in the form of dimensionless logical symbols; the fact that their processing elements happen to be electrically-responding gates is only incidental; computers are in no philosophically important way electronic; they belong to the class of epitronic systems.

Communication channels, by comparison, handle *knowledge in transit*, that is, information. (Knowledge accumulates when information flows; thus these are an integral/derivative pair, like voltage and charge.) Communications systems are physical—they are 'more Newtonian' in that they are *essentially* electrical, and involve signal representations that have profoundly significant dimensions, such as voltage, current, energy, charge, and time, present in components that also have dimensional attributes, such as resistance, capacitance, and inductance, and have *fundamentally* temperature-dependent behavior. These differences in the way we utilize electronics may someday lead to two quite separate fields of endeavor. Even now, there are hundreds of computer architects in the world who know little or nothing about how circuits work, nor do they need to. But the situation is different for the communications system designer, who invariably does need to be acquainted with both digital and analog signal-processing techniques,[17] and very fluent in at least one.

There are other ways in which our times differ from those of the last century. For one, corporations have to be concerned about their obligation to the investment community and the appearance of the financials in the quarterly report. As a consequence, there is much less room for taking risks. Taken to an extreme, the minimization of risk requires a retreat into the safe harbor of incrementalism. In the heyday of the late 19th century, this was not such a critical issue governing business decisions. In a modern microelectronics culture, we tend to encourage fishing in safe waters, rather than undertaking bold journeys out onto the high seas in search of uncharted territories and islands of opportunity.

For another, Edison, and pioneers like him throughout history, were rarely seeking just 'better solutions' (such as stronger gas mantles or long-life wicks); rather, they were bent on finding *radically different ways* to address widespread unserviced needs. Indeed, the word 'innovate' embodies the essential idea of introducing something 'new,' not just 'improved.' Unavoidably, so much of modern microelectronic engineering *is* derivative: the lower-power op-amp; the quad op-amp; the faster-settling op-amp . . . all doubtless serving real needs, but all based on the same traditional approach to feedback amplifier design.

We need to continually challenge ourselves, by asking such questions as: How might this function be approached if the system constraints were altered? What lies beyond the op-amp as the next 'universal' amplifier cell? How about a microprocessor which is internally massively-parallel,

---

17. It is interesting to note that the scorn poured on 'old-fashioned' analog approaches is nowadays confined to the pages of the *Wall Street Journal* and trade books. The job market has recently woken up to the fact that experienced analog engineers are in very short supply, which ought to have been foreseen.

and may use millions of transistors, but which has just three pins (VPOS, GND, and DATA-CONTROL-I/O) and sells for a dollar in flip-chip form? From what we know about physics, engineering, and the fundamental limitations to realizability, how might 30GHz monolithic transceivers be structured and fabricated by the end of the next decade?

It's unlikely we will be able to fully answer such questions, but it helps to think *a lot* about the far future, which each of us is having a small but significant part in creating. In 1945, when the domain of electrical devices was already quite mature, but *electronics* was still a brand-new word, Arthur C. Clarke, a normally modest Englishman, envisaged a totally new way of deploying electronic technologies—in a global network of satellites in geosynchronous orbits. He even sketched out highly innovative details of implementation, along with other visionary concepts, in his large output of published works.[18] When he made these suggestions, few would have foreseen the critical importance of communications satellites in every corner of modern human life.

There is also the difference of *project scale*. Today's projects are often team efforts, requiring the coordination of many people, often with a significant range of disciplines. But, one may wonder, was it so very different in Edison's time? He was, for example, the 'Team Leader' behind the construction of the generating station on Pearl Street, and for the wiring of a few hundred mansions in New York City which this station served. One does not need to know all the details to be fairly certain this was an interdisciplinary task of considerable magnitude and daring.

The operative word in this case was not 'team' (of *course* a lot of people were needed to carry out Edison's vision) but '*leader.*' The image of an admired team manager, orchestrating great clusters of dedicated manpower, is not supported by the pages of history. He seems to have been able to put together groups of technicians whose members worked well together, and then set them in motion, but *he* wanted the public acknowledgment for the achievement. He is to be credited in the way he anticipated emergent needs, understood the potential of his own ideas, and then steered others to actually create the reality, but he was far removed from the modern concept of the democratic, team-building engineering manager.

Leadership in Innovation

Let's briefly address the tension arising between 'leading' and 'responding to' the market, which my Edison parody lamely seeks to illuminate. Suppose one reads an ad with the slogan: "National Maxilinear of Texas—Your Leading Supplier in Microelectronics, Responding to Every Need of the Marketplace!" or some such jingle. I don't think that is *quite*

18. See, for example, "Extraterrestrial Relays," *Wireless World* (October 1945).

a contradiction, but it comes pretty close: to my mind, such a hypothetical reference to 'leading' would be weakened by the subsequent reference to 'responding.' Surely, leadership must involve going *ahead* of the pack, stealthily and methodically seeking *new paths*, taking the *risk* that the road ahead may be littered with unseen dangers. This doesn't require genius. It's leaders, not geniuses, who fight for and claim new territories; the settlers, with their gilt-framed "Home Sweet Home," rocking chairs, and Wedgewood chinaware, come later.

Edison certainly took risks in connection with his pioneering inventions, but he did not seem to have regarded himself as a genius. Nor did he need to be, in order to be a strong leader. By contrast, someone such as Albert Einstein probably *was* a genius, but he didn't possess Edison's innovative powers, in the sense that he left no practical invention as a legacy, and I think few would describe Einstein as a leader. Edison could also conceptualize, but in a nuts-and-bolts sort of way, for he bypassed much theory—even scorned it—and set about immediately turning his ideas into *tangible products* for which nobody had yet expressed the slightest interest, with the full expectation of quickly demonstrating their *practical* value.

History provides abundant lessons of people who forged entire new industries out of a singular vision, often one whose potential was totally unappreciated by contemporaries. Thus, even though of obvious value today, there was no clamor *from the public at large* for the printing press, the telephone, photography, vacuum tubes and the cathode-ray tube, the superhet receiver, tape recording, the transistor, the plain-paper copier, digital watches, pocket calculators, the Sony "Walkman," the CD player, or countless other examples. Each of these were the outcome of a stubborn conviction, often on the part of *just one person*, that some idea or another had *intrinsic utility* and could *generate* whole new markets, not merely serve the measurable market.

We noted earlier that Edison's method was "to innovate devices that could satisfy real needs and thereby come into widespread popular use." It was necessarily based on a *strong sense of what those needs were*—or would be! This paradigm, it seems to me, is the essence of leadership, which, as an obvious—even tautological—matter of definition, means leading, not following; anticipating, not merely responding. Two examples, gleaned from idle breakfast-time reading, of leadership-inspired innovations for which absolutely no prior market existed, are worth quoting here. The first is the invention of the laser, reported in the October 1993 issue of *Physics Today* in an article by Nicolaas Bloembergen, who first reminds us of the ubiquity of the laser:[19]

19. Lew Counts drew my attention to an article entitled "The Shock of the Not Quite New" in *The Economist* of June 18th, 1994, in which it is noted that "lawyers at Bell Labs were initially unwilling to even apply for a patent of their invention, believing it had no possible relevance to the telephone industry." This brief article is well worth reading. It includes several other illustrated examples of innovations which went unrecognized until much later, including the steam engine, the telephone, radio, the computer, and the transistor.

"The widespread commercial applications of lasers include their use in fiber optic communication systems, surgery and medicine, printing, bar-code readers, recording and playback of compact discs, surveying and alignment instruments, and many techniques for processing materials. Laser processing runs the gamut from sculpting corneas by means of excimer laser pulses, to the heat treatment, drilling, cutting and welding of heavy metal in the automotive and shipbuilding industries by CO_2 lasers with continuous-wave outputs exceeding 10kW . . . Lasers have revolutionized spectroscopy, and they have given birth to the new field of nonlinear optics. They are used extensively in many scientific disciplines, including chemistry, biology, astrophysics, geophysics and environmental sciences . . ."

Of course, it would be foolish to suggest that all of these "devices that satisfy real needs" were foreseen by the inventors. But, just what did motivate them? Bloembergen goes on:

"[T]he physicists who did the early work were . . . intrigued by *basic questions* of the interaction of molecules and magnetic spins with microwave and millimeter-wave radiation. Could atoms or molecules be used to generate such radiation, they asked themselves, and would this lead to better spectroscopic resolution?" [Italics mine]

The motivation in this case seems to have arisen from the desire to find a way to greatly improve an existing technique (spectroscopy) and thus open up new possibilities (higher resolution). That sounds like incrementalism. On the other hand, although we cannot be sure, it is doubtful that the laser was the result of a survey of other physicists as to what *they* perceived would be useful. Rather it appears to have been a spontaneous invention by physicists *who knew what would be useful to other physicists out of their own experience*.

Cannot we do the same sort of thing? Are not we aware of advances that, even though not yet expressed by our users, are nevertheless known to be valuable? Perhaps the development of new integrated circuits ahead of market demand cannot be compared to such a monumental leap forward as the invention of the laser. Still, there is no reason why the same *spirit of leadership* cannot be present even in this humble endeavor. Furthermore, the essential idea of *innovating out of a broad knowledge of the possibilities and utilities of one's technologies* applies equally in both cases.

My second example is of another 'big' idea, the invention of nuclear magnetic imaging (NMI), whose full potential is only just beginning to be realized: indeed, it is thought by some that NMI will soon surpass X rays in medical diagnosis. NMI came out of nuclear magnetic resonance (NMR) techniques which were originally developed to investigate nuclear properties. In *Science* (10 December 1993), George Pake is quoted as having this to say about the sources of the ideas:

> "Magnetic resonance imaging could arise only out of the *nondirected* research, not focused on ultimate applications, that gave rise to what we know today as NMR. The key was the series of *basic quests* to understand the magnetic moments of nuclear spins; to understand how these nuclear magnets interact in liquids, crystals and molecules and to elucidate the structure of molecules of chemical interest. Out of these *basic quests* came the knowledge that enabled a vision of an imaging technique. Without the *basic research*, magnetic resonance imaging was unimaginable." [Italics mine]

I'm not suggesting that our primary mission as individual integrated-circuit designers, or as team members, or as this or that microelectronics corporation, is to conduct basic research. But even in our industry, we cannot allow these 'basic quests' to be ignored. This requires that we constantly reflect on the utility of new circuit functions, or consider new topological realizations, or pursue advanced silicon processes, or time-saving testing techniques, *before* their need has been articulated by our customers, and to relentlessly search for novel ways of using our technologies to produce "devices that satisfy real needs." Sure, the pressures to meet even *known* market demands are unrelenting, and seem to consume all available resources. Nonetheless, some of our time *must* be spent in 'nondirected research' if we are to continuously strive toward leadership.

Many Voices

Nowadays, as already noted, we are more than ever being urged to pay close attention to the Voice of the Customer. And, as already noted, there are frequently situations in which this makes eminently good sense. Faced with the need to respond to an emerging market requirement about which we may know little, it is valuable to solicit would-be customers about their specific needs. Of course, if we had been practicing good gatekeeping skills, accumulating a large body of *relevant knowledge* about our industry, and keeping abreast of new standards by representation on relevant committees, the criticality of the customer interview would be substantially reduced. Furthermore, we could address our customers as *equal partners*, with advice to offer proactively, and solutions readily at hand.

By contrast, the textbook VOC technique requires a neutral interview procedure, using two representatives, one of whom poses a series of previously formulated questions (invariant from customer to customer) while the other takes notes as the customer responds. I can think of no more infertile approach to understanding the true needs of the customer, and hope that *actual* VOC practice differs substantially from this inflexible characterization, which would represent the antithesis of leadership.

Responding to the market makes sense in certain cases. Clearly, it would be foolishly presumptuous, and very risky, to imagine that we can lead out of our own superior knowledge in every situation. But risks remain, even with the most enlightened and fastidious market research. One obvious danger is that, if we depend excessively on the customers' inputs to fuel our innovation, we have no advantage over our competitors, who can just as easily work through the same VOC procedures and presumably arrive at an equally potent assessment of a particular opportunity. Further, the 'blank page' approach could lead our customers to believe that we know little or nothing about their requirements, and that we are therefore unlikely to be in any position to offer novel or cost-effective solutions.

The value of the VOC process is presumed to lie in its efficacy in extracting key nuggets of knowledge from the customer. This may be illusory; customers may be quite unable to imagine a better way of solving their system problems, and may doggedly present what are believed to be needs (stronger gas mantles) while failing completely to appreciate that there may be several better alternatives.

Indeed, if the VOC process is constrained to a question-and-answer format, we may actually be prevented from volunteering our views about novel approaches, like Edison with his vision of Electric City, much less show how excited we are about these. Sometimes, customers may decide to withhold critical information from us, for various reasons. For example, they may have become tired of spending their time with an endless stream of VOCers signing in at their lobbies; it may be that the individuals being interviewed had been told by their supervisor not to reveal the intimate details of some project; they may have already made up their minds that National Maxilinear of Texas, Inc. is going to be the vendor, because of all the good ideas they presented, and the leadership image they projected, at their last on-site seminar; and so on.

Thus, the formal VOC process is inevitably of limited value. It is merely a way of *responding* to the marketplace, and as such is bound to be *lagging* the true needs of the market. Though important, it clearly is not the primary path of leadership, which requires the constant anticipation of future needs. I am not, of course, advocating the abandonment of customer interviews, merely noting that they are only one of *numerous* gatekeeping activities with which *all* key contributors in an innovation-based company—not just those formally designated as 'strategists'—must be involved.

Musings from System Theory

Since this is being written for the enjoyment of those in the microelectronics community, we might perhaps invoke some familiar ideas from control system theory, and liken the VOC process to a feedback system. The basic objective of negative feedback is to minimize the error between some desired set point (in this case, the customer's specifications) and the current state of output (the products we have in the catalog). Signals at various points along the path (products under development, new concepts in our portfolio, and the like), and the nature of the path (business and technical procedures) also determine the state of the control system, which, to be effective, requires *continually* sensing the customer's most recent needs.

The output of this system also has some noise (the uncertainty of local, national, and global economies, lack of knowledge about competitors' product plans, resource collisions, and so on), requiring that decisions about optimal actions be based on incomplete or corrupted data. In fact, this 'market-responding' system has a *great deal* of noise in it, which translates to a significant dependence on judgment in dealing with its indications.

The seductive promise of a feedback system is that one eventually ends up with essentially no error between the 'set point' and the state of the system (that is, we meet the customer's clearly articulated needs completely). However, as is well known, the inertia inherent in any control system, mainly due to the presence of delay elements in the loop, can lead to long settling times or even no stable solution at all. Furthermore, feedback systems are less successful in coping with inputs (market demands) that are constantly changing, due to this very inertia. Sometimes, when a sudden large change is needed, they exhibit slew-rate limitations (that is, there's a long ramp-up time to get to the solution, as when a new package style may 'suddenly' be needed).

I'd like to suggest that the leadership approach is more like an open-loop system. Such systems can be made extremely fast and effective, at some expense in final accuracy, which is bounded by the quality of the input data (now based on a trust of one's key technologists, and their broad, rather than specific, market knowledge) and by the accuracy of the implementing system (knowledge about how to optimally achieve the final state, that is, *practical engineering knowledge*). Noise is still there, but being based on *long-term* data (fundamental physical limitations of devices and technologies, durable principles of design, long familiarity with a wide variety of customer needs, well-established standards which will impact a large number of customers to result in similar demands, and so on) *the noise is heavily filtered* before it enters the system.

Thus, with a stronger emphasis on leadership, the reliance on a low-bandwidth, possibly oscillatory closed-loop system must be replaced by a dependence on a fast, direct response based on a comprehensive, sure-footed knowledge of the market in rather general terms, and the technologies and the design skills which can quickly be deployed in an anticipatory manner.

Open-loop (predictive, feedforward) systems are well known for their inherent stability and for being able to track rapidly changing inputs; in our analogy, this means that we are ready with that special package before the product is nearing release—because *its need was anticipated*, knowing of the current trends in manufacturing techniques among one's customers—and that one is ready with the *next* product at about the same time that the latest part is being released.

Incidentally, this raises an interesting strategic challenge: How soon should a company introduce follow-up products, in an 'open-loop' fashion (before the demand is obvious) so as to stay on the competitive curve, knowing that these will inevitably cause some erosion in the sales of earlier products? Historically, many IC companies haven't been particularly adept in addressing this question. Clear opportunities for follow-up action are often neglected, because of the concern that some product "released only last quarter" might be obsoleted too soon. To be competitive, one doesn't have much choice: leadership requires making those decisions without waiting for the clamor from the customer. They should be based on a sound understanding of trends and in anticipation of market needs, rather than waiting for that coveted million-piece order to be delivered to the doorstep.

But this is not really an either-or situation. One needs a judicious balance of both approaches (leading and following, anticipating and responding) to be completely effective. However, current philosophies and policies in the microelectronics industry, designed to improve the success rate of new products and minimize investment risk, point *away* from the traditional emphasis on leadership-based innovation and the freedoms granted the product champion that proved so successful in earlier times. Are the practices of those times still relevant? That's not clear. Nevertheless, it seems to me that it is preferable to do business based on long-term *internal* strengths than to depend too much on going "out there" to get the critical information needed to make reliable business decisions.

Leadership is required to be successful in all product categories. For standard products, the challenge is to constantly be on the watch for competitive threats, and have an arsenal of next-generation solutions always on hand. These products need a high level of predictive innovation, motivated by a keen awareness of what the customer will probably need two to five years from now, as well as what emerging technologies will become available in one's own factory, in competitors' factories, and in advanced research houses. This requires judgment about trends, and a good sense of future product value and utility. In the control theory analogy, the development of standard products is likely to benefit from the 'feedforward' approach. It will be the Voices-of-Many-Customers that are here important, as well as the Voices-of-Many-Competitors, as indirectly articulated in their ads, their data sheets, and their application notes.

Special-purpose ICs, on the other hand, are clearly more likely to benefit by listening to the Voice of sometimes just One Key Customer, maybe two or three, as well as the Voices of Committees (writing standards, recommending certain practices), and the Voices of Consultants (people hired to advise a company about some new and unfamiliar field or specialized domain of application) and finally, because of the probably low margins of most high-volume products (almost by definition), one needs to listen to the Voice of Caution. The challenge here is first, to grasp some less familiar new function, or set of functions, or a whole new system; second, to achieve a higher level of system integration (manage complexity); third, to achieve a very low solution cost (since one is competing with existing well-known costs and/or other bidders); fourth, to get to product release fast on a very visible schedule; fifth, to ramp up quickly to volumes of many thousands of parts per month. In the development of special-purpose ICs, the customer is, of course, the primary reference, the 'set-point' in the innovation feedback system.

Innovation and TQM

Product quality has always been important, but it is especially critical in modern microelectronics, as competitive pressures mount and market expectations for commercial-grade parts now often exceed those required only by the severest military and space applications a few years ago, but at far lower cost. During the past few years, the airport bookstores have been flooded with overnight best sellers crowing about the importance of 'excellence' in modern corporations, and how to foster a culture in which excellence is second nature. Sounds like a good idea. But excellence alone is not enough to ensure success:

> "Excellence . . . will give [companies] a competitive edge only until the end of the decade. After that, it becomes a necessary price of entry. If you do not have the components of excellence . . . then you don't even get to play the game."

says Joel Arthur Barker.[20] Quality for quality's sake doesn't make much sense. It wouldn't help to have perfect pellicles, 100% yields, zero delivered ppm's and infinite MTBFs unless the products that these glowing attributes apply to have relevance in the marketplace, and are introduced in a timely fashion. They are, to paraphrase Barker's comment, "merely essential" requirements of the business.

The need for a strong focus on quality is self-evident, widely appreciated, and has received a great deal of attention in recent times. This emphasis, commonly referred to as Total Quality Management, or TQM, is

---

20. Joel Arthur Barker, *Paradigms: The Business of Discovering the Future* (1994). It appears that this book was previously published in 1992 under the title *Future Edge*. I guess by that time anything with the word "Future" in its title was becoming passé—so perhaps it enjoyed only lackluster sales; by contrast, "Paradigms" became a very marketable word in 1994.

clearly essential and must be relentlessly pursued. One of the many sub-goals of TQM (all of which have 3LAs and 4LAs that are very effective in numbing the mind to the importance of their underlying concepts) is Design for Manufacturability, or DFM.

However, some seasoned designers do not respond favorably to the *formalism* of TQM. This is probably because they feel slightly insulted that anybody would assume they were prone to overlook the obvious importance of such things as DFM. They are also bemused by the apparent 'discovery of quality' as a new idea. They may feel that the notion that one can legislate quality by the institution of formal procedures, such as checklists of potential mistakes and omissions, is somewhat naive. Many of the rituals, observed with near-religious fervor, and recommended practices seem overly regimented. Thus we read, in a much respected manual of instruction[21]

"3. Ten to twelve feet is the distance at meetings and seminars. In a meeting or seminar situation, try having the speaker first stand about 15 feet from listeners and then stand 30 feet from listeners. Moving farther away from listeners noticeably changes the speaker's relationship to the audience. During a meeting the instructor should be about 10 to 12 feet from most of the participants. After the formal session, the instructor can move to the 4 foot distance for an informal discussion and refreshments."

Are you listening, Mr. Edison? If your ghost is ever invited to make an after-dinner presentation at a modern company, you'd better get that little matter straight. It's this sort of 'institutionalizing the obvious,' and the evangelical emphasis of method over content, of process over product, that many of us find irritating and counterproductive in contemporary TQM methodologies.

There's also the old-fashioned matter of pride of workmanship at stake. Skilled designers believe they have an innate sense of what is manufacturable and what is not, and they exercise constant vigilance over the whole process of finding an optimal solution *with manufacturability very much in mind*. For such persons, it is quite futile to attempt to mechanize the design process, if this means applying a succession of *bounds* on what methods can and cannot be used. Strong innovative concepts and products cannot thrive in a limiting atmosphere.

Design quality is never the result of completing checklists. It is even conceivable that by instituting a strong formal mechanism for checking the design one could impair this sense of vigilance, replacing it with the absurd expectation that mistakes will assuredly be trapped by checking

---

21. Shoji Shiba, Alan Graham, and David Walden, *A New American TQM* (Productivity Press, 1993), 298.

procedures. Nor does quality increase when the number of signatures increases. There is no disagreement with the idea that *design checking* is important. It can catch errors which might easily be overlooked and allow designers to benefit from hard-won experience. But this function needs to be integrated into one's workplace, and be active, in a background mode, continuously throughout the development.

In the long run, this quality-enhancing process will benefit by being automated, using our workstations and company-specific 'experience databases.' For now, this will be limited to what can be done with present-day computers. It will depend on giving all who need it essentially instant access to massive amounts of knowledge about our business; it will depend on building more helpful monitoring agents into our tools, that catch anomalies; it will often depend on providing quite simple pieces of code to reduce a keyboard-intensive task to a single keystroke; it will depend on the pioneering use of teleconnections of all sorts to link together geographically disparate groups.

In a future world, the quiet time that our machines have when (if) we go home at night will be used to review our day's work: in the morning, there'll be a *private* report of our oversights, indiscretions, and omissions, for our personal benefit. To err is humiliating, and particularly so in public; but to have one's errors pointed out in private can be enriching. I sincerely believe that such aid will eventually be available, but of course it is far from practicable today. Nevertheless, there are many 'intelligent' ways in which automated assistance *can* be built into design tools, such as circuit simulators. Some are very simple 'warning lights' that would advise of improper operating conditions in a circuit; others will require substantial advances in artificial intelligence before they can be realized, such as agents that can detect the possibility of a latch-up, or a high-current fault state, in some topology.

Is Design a Science or an Art?

Should one emphasize the science of design over the art of design, or vice versa? This is of considerable interest in academia, where the challenge is not usually to pursue excellence by participating in the actual design of innovative, market-ready products, but rather, by choosing the best paradigm to instill in the minds of students who wish to become good designers in industry.

The distinction between science and art is quite simple. Science is concerned with observing 'somebody else's black box' and about drawing conclusions as to how this black box (for example, the physical universe) works, in all of its various inner compartments. Science is based on experiment, observation, and analysis, from which basic material scientists then suggest hypotheses about the underlying laws which might plausibly lead to the observed behavior. These hunches can be tested out, by dropping sacks of stones and feathers off the Tower of Pisa, or by hurling unsuspecting protons around electromagnetic race tracks. The experimental results acquired in this way require further analysis. If all goes well, the humble hypothesis gets promoted to the status of Theory.

While hypothesis-generation and the creation of grand theories can be said to be a constructive, even artful, endeavor, it often amounts to little more than trying to find ways to figure out how 'somebody else's pieces' fit together. Science is (or should be) primarily a *rational, analytic activity*. In electronics, we might speak of reverse engineering (referring to the process of finding out how your competitor was able to achieve performance that you thought was impossible, by tracing out the schematic of his circuit) as a 'science.' However classified, though, *that* practice represents the antithesis of innovation, by general agreement.

Art, on the other hand, is about seeing the world in an ever-fresh way. The artist often scorns previous conventions as worthless anachronisms. The challenge to the artistic temperament is to *create* a new reality, even while building on old foundations. Thus, the painter sees the blank canvas as an opportunity to portray his or her personal vision of our world (or some other non-world); the composer sees the keyboard beckoning to be set afire with an exciting new musical statement. Certainly, there is practical skill needed in handling art media (a knowledge of how to mix paints, for example, or of the principles of harmony and counterpoint), but the artist's primary locus is a *creative, striving, synthetic activity*. In the artist's life pulses the ever-present belief that the old conventions can be pushed far beyond their known limits, or even be overthrown completely. Analysis of the kind pursued by technologists is foreign to the mind of the artist.

The painstaking process of innovating sophisticated, competitive IC products embraces a considerable amount of 'art.' This is not a popular idea, for it evokes such images as 'ego-trip,' 'open-loop behavior,' 'loose cannon,' 'disregard for community norms,' 'abandoning of sound practice,' and other Bad Things. Which is a pity, because designing an integrated circuit is *very much* like painting in miniature, or writing a piano sonata: it's the creation of a novel entity, the distillation of our best efforts into something small in scale but big in importance; it is craftsmanship at the limits of perfection, and at its best, transcending these limits into new realms of expression. When this impulse is faithfully acted upon, quality of design will be essentially automatic. The science has been sublimated; it's still there, like the knowledge of paints or harmony, but it permeates the whole creative process without needing to be raised to the top levels of consciousness.[22]

True, engineers are not explicitly paid to be artists, and admittedly we'd be in deep trouble if we designed our ICs just so that the layout

---

22. We can be thankful that Rembrandt and Beethoven or Shakespeare did not have to sign off quality checklists before their works could be released to the world.

looked pretty. But that is not *at all* what I have in mind. Perhaps a better word, for now, might be 'artfulness,' that is, an approach to design which cannot readily be captured in a formal set of procedures. An artful design is one that calls on a wide range of deeply felt ideas about the *intrinsic correctness* of certain techniques, deviating from the ordinary just far enough to open new doors, to push the envelope gently but firmly toward ever more refined solutions, in lively anticipation of tomorrow's demands. This view about the relevance of art in design is shared by Joel Arthur Barker, who writes[23]

> "Some anticipation can be scientific, but the most important aspect of anticipation is *artistic*. And, just like the artist, *practice and persistence* will dramatically improve your abilities. Your improved ability will, in turn, increase your ability in dealing with the new worlds coming." [Italics mine]

Perhaps one can criticize this 'artistic' view of the design challenge as having an appeal only to a certain kind of mind, although it seems that a love of engineering and a love of the visual and musical arts often go hand in hand. I happen to believe it is *central* to the quality theme, and that it is overlooked because we work in a business—indeed, live in an age—where it is presumed that everything can be measured, codified, and reduced to a simple algorithm, and that profound insights can be mapped on to a three-inch-square Post-It™ and comfortably organized within an hour or so into a coherent conclusion and set of action items. This is a deeply misinformed philosophy; it's part of the 'instant' culture that sadly we have become.

Certainly, there are times when team members need to get together and look for the 'most important themes,' to help us simplify, as much as possible, the various challenges that beset us. The problem, it seems to me, is that the process takes over; the need to use the 'correct method,' under the guidance of a 'facilitator,' who alone knows the right color of marker pen to use, gets slightly ludicrous. But who dares speak out in such a sacred circle? I personally believe that corporations which put a high value on such rituals will one day look back and wonder how they could have been so silly, though I realize this is not a politically correct thing to say.

With all of the 'quality improvement methods' now being pursued like so many quests for the Holy Grail, there is little likelihood that the science will be overlooked; it is far more likely that we will dangerously underestimate the value of the art of design. Those various 'quality algorithms' should be regarded as only guidelines. They cover some rather obvious principles that *always* need to be observed. But they also overlook some very subtle, equally crucial, issues, which are often specific to

---

23. Ibid.

a particular product or design activity, and are usually *extremely difficult* to articulate in algorithmic form. The innovation of competitive high-quality microelectronic components is both a science and an art. Neither is more important than the other.

Innovation in microelectronics is not, of course, limited to product design. There is need for innovative new approaches across a broad front, particularly in the development of new IC fabrication processes, involving many team members. At Analog Devices, the utilization of bonded wafers as a means to manufacture a dielectrically isolated substrate is a good example from recent history. This was a step taken *independently* by the process development team, and had no VOC basis, though the technology it now supports certainly had. They even had to make their own bonded wafers in the early days, and I'm sure it was out of a conviction that here was a brand-new approach that promised to allow a step-function advance to be made in IC technology.

The eventual result of this *anticipatory* research was an outstanding technology (XFCB) which unquestionably enjoys a world-class leadership position. It includes an innovative capacitor technology and retains the thin-film resistors that have been a distinctly innovative aspect of Analog's approach to IC fabrication for more than twenty years. The perfection of these ultra-thin resistors was another hard-won battle, undertaken because of the dogged conviction on the part of a handful of believers that the benefits were well worth fighting for.

Sometimes, innovation involves the bringing together of many loosely-related processes into a more potent whole, such as the laser-trimming of thin-film resistors at the wafer level.[24] This required an innovative synergy, combining significantly different design approaches, altered layout techniques (the use of carefully worked-out resistor geometries), and novel test methods (involving the use of clever on-line measurement techniques to decide what to trim, and by how much).

At each of these levels, there was also the need for independent innovation: thus, the precise formulation of variants of the resistor composition needed to achieve a very low temperature-coefficient of resistance (TCR); the control of the laser power to minimize post-drift alteration of this TCR, and thus maintain very accurate matching of trimmed networks over temperature; understanding the importance of oxide-thickness control to prevent phase-cancellation of the laser energy; development of new mathematical methods to explore potential distributions in arbitrarily-bounded regions; the realization of the potency of synchronous demodulation as a better way to trim analog multipliers; and so on. These, and many other advances by numerous contributors, were needed to bring the science of laser-wafer trimming to a high art.

---

24. Dan Sheingold reminded me of this, in reviewing a draft of this manuscript, and suggested LWT as an example of what might be called collaborative innovation.

Knowledge-Driven Innovation

I would like to suggest some ways in which we might raise the rate and quality of innovative output in a real-world IC development context. Obviously, good management and mentoring on the part of skilled seniors is important, but to a significant extent, *the management of innovation is largely about the management of knowledge*. And electronics, as already noted, has become an indispensable servant of knowledge. Colin Cherry writes[25]

> "Man's development and the growth of civilizations have depended, in the main, on progress in a few activities—the discovery of fire, domestication of animals, the division of labor; but above all, in the evolution of means to communicate, and to record, his knowledge, and especially in the development of phonetic writing."

Just as the invention of writing radically altered all facets of human endeavor, today's computers can help us in numerous knowledge-related contexts to achieve things which only a few years ago were quite impossible. This is hardly news. The question for us here is, how can we make more effective use of computers to put knowledge at the disposal of the innovator?

We noted that the creative spark may well be a random event, mere cranial noise, but it is only when this is coupled into a strong body of experience and encouraged by a lively interest in anticipating—and actually realizing—the next step, that we have the essential toolkit of innovation. Unfortunately, many of us are quite forgetful, and even with the best record-keeping habits cannot quickly recall all that we may have earlier learned about some matter.

It is often said that today's computers still have a long way to go before they can match the human mind. That's obviously true in some important respects; lacking afferent appendages, it's hardly surprising that they are not very streetwise. And they have been designed by their master architects to be downright deterministic. But they are possessed of prodigious, and infinitely accurate, memories, unlike our own, which are invariably fuzzy, and depend a great deal on reconstructive fill-in. They are also very quick, giving us back what is in RAM within tens of nanoseconds, and knowledge fragments from several gigabytes of disc within milliseconds. Obviously, in accuracy of memory recall, and possibly even in actual memory size, computers *really have* become superior, and I don't think there's much point in trying to deny that particular advantage.

Computers are also very good at relieving us of the burden of computation. There is no virtue in working out tables of logarithms (as Charles

---

25. Colin Cherry, *On Human Communication* (New York: Wiley, 1957), 31.

Babbage noted, and decided to do something about, with the invention of the 'difference engine'[26]) and there is no virtue in using those tables, either; we can, and should, leave such trivia to our silicon companions. Not only are they flawless in calculation, they are very, very quick. These calculations often go far beyond primitive operations. When running SPICE, for example, we are invoking many, many man-years of experience with the behavior of semiconductor devices. Who reading this has memorized all of the equations describing current transport in a bipolar transistor, or would wish to manually develop the numbers for insertion into these equations? Here again, I do not think it silly to assert that computers are far better than we. Sure, they aren't painting still-lifes like Van Gogh, or writing tender sonnets, but they can run circles around all of us when it comes to sums. Round two to computers.

They have another advantage over us. They will work night and day on our behalf, and never complain nor tire. While we sleep, the networks chat; updates of the latest software revisions and databases silently flow and are put into all the right places, ready for us to do our part the next day. We surely need to be honest in acknowledging that in this way, too, they definitely have the edge; we have to black out for a considerable fraction of each and every day. We need to take full advantage of these valuable knowledge-retaining, knowledge-distributing, and knowledge-based-calculating attributes of our indefatigable silicon companions.

Modern innovators have a critical dependence on operational knowledge across a broad front. This includes knowledge of the microelectronics business in general terms, knowledge of one's specific customers (who they are, where they are, and what they need), knowledge of one's IC process technologies, of semiconductor device fundamentals, of circuit and system principles, of one's overall manufacturing capabilities and limitations, and on and on. It is widely stated these days that knowledge is a modern company's most valuable asset. That much ought to be obvious. But while a lot has already been done to automate manufacturing processes using large databases, progress in making design knowledge widely available to engineering groups has been relatively slow in becoming an everyday reality.

Most IC designers will readily be able to recall numerous instances of having to spend hours chasing some trivial piece of information: What is the field-oxide thickness on the process being used for a certain product? What is the standard deviation of certain resistor widths? Where is there a comprehensive list of all internal memos on band-gap references? Where is there a scale-drawing of a certain IC lead-frame? Each of these could be reduced to a few keystrokes, given the right tools. Instead, the quest

---

26. See, for example, Babbage's memoirs *Passages from the Life of a Philosopher* published by Rutgers University Press, with an introduction by Martin Campbell-Kelly (1994) for the background to the invention of the "difference engine."

for these rudimentary fragments of essential knowledge can (and often does) erode a large fraction of each day.

At Analog Devices, we have developed a database program which represents an excellent start toward providing answers to these sorts of questions. In the long run, our effectiveness in reducing development time, in lowering costs, in enhancing quality and much else, is going to increasingly depend on *much more massive* and *interactive* databases— better viewed, perhaps, as knowledge networks. While primarily having a technology emphasis, such networks would allow querying in a wide variety of ways, about the business in general, and they would also be *deeply integrated* into our development tools.

They would be available at all levels and throughout the entire development cycle, starting with access to relevant public standards, the market research process, product definition phase and prior art searches, through actual component design and checking, layout and verification, wafer fab, early evaluation, test development, packaging, data sheet preparation, applications support, and beyond, to customer feedback, and so on. These electronic repositories would provide information on many other vital matters, such as resource scheduling, all in one place, available for searching, browsing, consulting, and interacting with, anywhere, anytime, as instantly as keystrokes.

The data itself would be in many forms (text, hypertext, graphics, schematics, drawings, schedules, sound bites, and, as multimedia capabilities expand, video clips). It would represent the amassed experience of hundreds of contributors. The operational shell should support much more than a mere searching function—it would be interactive and anticipatory, pointing to other sources of relevant or coupled information; that is, it will work with *relational databases*. It should be possible for anybody with the necessary level of authority to make additions to the databases, and it should allow masking of the field of inquiry; that is, the interaction process would also be amenable to *personalization*.

Because of their immense commercial value, many parts of such databases would require protection of access. Many individuals would be given access to only certain databases; as a general rule, the new hire would be given minimum access to critical information, while senior employees would have access to a very wide range of knowledge about the whole business. The whole question of security is fraught with contradictions and dilemmas: knowledge which is so potent that one's commercial success depends on it would obviously be very dangerous in the wrong hands. However, that cannot be raised as a fundamental reason for not providing access to 'world-class' knowledge. In all likelihood, developers of such databases will need themselves to exhibit considerable innovation in developing ways to temper this two-edged sword. But I cannot imagine how one can be competitive in the long run without serious attention to such a knowledge network.

It would take much imaginative planning and immense effort to turn this Promethean undertaking, easily stated here, into reality. Clearly, this is about the development of a resource that is much more than a way of finding valuable bits of information without significant delay. I see it as being the basis for propagating ideas throughout a corporation, as a mentoring vehicle, as a way to keep track of one's project schedules (and ensure that all their interdependencies are properly calculated) and much else. I would expect it to take advantage of the most recent developments in the management of large depositories of knowledge, and to make use of the latest multimedia hardware.

This is something that decidedly cannot be achieved by one or two new hires with degrees in computer science, or even a highly motivated and well-qualified software innovator. It will require the full-time efforts of a large team, headed up by a respected leader in this field reporting into a high level of the company. Is it too far-fetched to look forward to the time when companies have a VP of Knowledge? This person would not necessarily come from the world of engineering or computer science. Because the objectives set before this person would be so broad and so important, they could not be left to generalisms and rhetoric; they would need very careful articulation as precise deliverables.

More than any other initiative, I see this as being one that is most likely to bring about real change and be most effective in coping with the vicissitudes of the modern microelectronics world. And I would go so far as to assert that is it precisely because the task is *so monumentally difficult* that one may be inclined to tinker with the latest management methodology instead, in the hope (funny, I first typed that as h-y-p-e) that there's still another one or two percentage points yet to be squeezed out of the guys and gals on the production floor through the implementation of another new procedure with another mystical name. Perhaps I just don't get it.

Enhancing Innovation

Is there anything else that can be done to encourage, elevate and propagate the innovative spirit? How might our high-quality innovative output be enhanced? I think there are many ways. First, a little early success as we start out on our career can make a big difference. I recall how valuable it was to me to be heartily praised for my minor (and often deviant!) accomplishments as a new boy at Tektronix. It immensely strengthened my resolve to do something the next day that was *truly* deserving of praise! And it was so different from the bureaucratic, authoritarian, *rule-based* structure which I'd worked under as a junior in England.

Those of us with managerial and mentoring responsibilities need to do all we can to help new hires to see tangible proof of their value to the company, as quickly as practicable. From the very start, we need to provide and sustain *an elevated sense of the possible*. This trust-based cultivation of a sense of worth, character, responsibility, and potency is of prime importance, not only in raising expectations, but in actually fulfilling them. Analog Devices has traditionally succeeded very well in this respect.

Second, for all of us charged with innovation, getting out into the field and talking one-on-one with customers, not just as voices on the phone, but as people in their own working environment, is crucially important. Our customers are not *always* right, but when they are we can't afford to miss their message. However, it will by now be abundantly clear that I believe this is too narrow a description of the challenge, a view that is shared by many of my fellow technologists. In addition to listening to our customers, we need to pay attention to numerous other voices, including the all-important internal Voice of Conviction about which projects make sense and which are likely to be dead ends.

The provision of a supportive corporate infrastructure is also of immense value; if we feel we are *trusted* to make good decisions, *empowered* to achieve great results, and then provided with *powerful tools*, we almost certainly will succeed. A palpable interest from the top is of inestimable value. Working at Tektronix in the mid-sixties, I was impressed by the fact that its then-president, Howard Vollum, and many of the VPs, would frequently tour the engineering areas, usually dressed down in jeans and sneakers, and talk with us designers about our latest ideas at some length. They would push buttons, twiddle knobs, and administer words of praise, advice, and encouragement. That kind of interest, visibility, and personal involvement on the part of senior managers is often lacking in modern corporations, much to their loss.

The element of risk is an essential ingredient of innovation. Once we allow ourselves to believe that there are textbook ways of achieving greatness, we are doomed. Strong-mindedness, conviction, and commitment can compensate for a lot of missing data and counterbalance a certain amount of misjudgment, an idea echoed by these words by Analog Device's Ray Stata:

"In the case of Nova Devices [now ADI] there couldn't be a better example of the necessity of a lot of will power and commitment because it was a very, very rocky experience. In these companies which are basically high risk in nature, you really have to have somebody who decides on a course of action—I don't know whether fanatical is the word—but with tremendous conviction in terms of what they want to do and why it's necessary to be done. . . . All the reasons why it cannot be done are somehow submerged, even those with validity. There has to be a capacity to take great risks and not all that much concern about the fact that you might not make it."

—(From an interview conducted by Goodloe Suttler, at the Amos Tuck School of Business, 1980)

Computers: Tools or Companions?

I have already exposed my views about the superiority of computers in certain activities, but have a couple of other things to add about the ergonomics of innovation. The designer's most important tool is the high-speed workstation. Time-to-market considerations, increased circuit complexity, accuracy of simulation, and design for manufacturability demand that our machines be state-of-the-art. A study of work habits would almost certainly reveal that a circuit designer[27] is seriously bounded by machine speed, and spends a large part of the day simply waiting for results.

This seems like a confession of poor work habits. It may be asked: Why don't you do something else during that time? The answer is simple. First, many simulations are of fairly small cells undergoing intensive optimization: there is a lot going on in one's mind as each simulation is launched; small changes are being explored, the consequences compared; and while that is happening, the next experiment is already being assembled in the shunting yard of the mind. The process is a fluttering dynamic, demanding instant resolution. We want to be at all times mind-limited, not machine-limited.

Typically, what happens is this. A simulation is launched, and the result is expected to be available in perhaps ten seconds, perhaps twenty seconds, perhaps half a minute. None of these intervals is long enough to start another project of any magnitude. So instead of being completely idle, we may on occasion find ourselves pecking away at some text file in another window on our CRT. But the design process requires strong focus and full concentration to achieve our rapidly developing objectives. It is difficult to deflect one's attention from a flood of conscious thought about these goals toward some secondary cerebral occupation. These machine delays evoke a frustration not unlike trying to enjoy an exciting adventure movie on a VCR with the pause button depressed for much of the time by a mischievous prankster.

We have a long way to go before we can be completely happy with the performance of workstations in a circuit development context. We have seen significant improvements in such things as memory space: the most advanced workstations (such as those from Silicon Graphics Inc.) provide up to 512 megabytes of RAM, and several gigabytes of hard disk. Nowadays, raising CPU speed, and the use of superscalar instruction cycles and multiple parallel processors, represent the new frontier. Hopefully,

---

27. Other computer users, such as layout designers and test engineers, also need fast machines, but it is the computationally intensive aspect of circuit simulation that most seriously delays circuit development.

IC designers will only be limited by what the computer suppliers can provide, and never by poor judgment on the part of managers as to how much one can "afford" to spend on fast machines.

We might reflect that our competitors are faced with exactly the same limitations as we (unless their computer budget is significantly larger) and thus the challenge facing each of us is to find ways of improving our efficient use of the machines we already have. Part of the solution may be in revising our work habits, although the problems of machine-gated creativity, just described, are real. Another piece of the solution, though, is to continue to emphasize the value of proprietary software.

When one considers the critical role played by computers and software in today's competitive arenas, and the importance of operational knowledge, there can be little doubt that the most important way in which management can help to advance one's innovative potency is through the establishment of a much larger CAD activity. I do not think this is the time to be winding down or holding steady, relying exclusively on third-party vendors of 'turn-key' (ha!) software. IC companies need to be especially careful about harboring the naive belief that large software houses are exclusively capable of providing the tools needed for making the future. One may on occasion choose to buy some standalone software, but it is axiomatic that, being forced to use generally the same software as everyone else, and to an increasing extent, obliged to use the same technologies as everyone else (such as foundry-based sub-micron CMOS) one's competitive advantage will be limited to what can be achieved with marketing prowess and design skills alone.

Thus, in my view, the future success of any company that aspires to a high rate of innovation will significantly depend on a *very strong in-house CAD activity*. A major and urgent objective of that CAD Group would be the implementation of an interactive knowledge network embodying massive amounts of essential information, organized in such a way as to be not only readily accessible, but also in some way to offer help proactively. It will be the incredible potential of networked computers to tirelessly inform and illuminate our lives as engineers, as well as their continued use as calculating tools, that will bring about the largest improvements in innovative productivity. A more effective union of thinking machines and cerebrally-sparkling human minds promises to radically alter everything we do.

But we should not imagine that the demands on human energy and the need for creative thrust and sparkle will be lessened. Norbert Wiener, in *God and Golem Inc.,* has this to say:

"The future offers very little hope for those who expect that our new mechanical slaves will offer us a world in which we may rest from thinking. Help us they may, but at the cost of supreme demand upon our honesty and intelligence. The world of the future will be an ever more demanding struggle against the limitations of our intelligence, not a comfortable hammock in which we can lie down to be waited upon by our robot slaves."

Nevertheless, the computers we will be using as we pass through the portals into the coming millennium, some 5,000 years since the invention of writing, will, I am convinced, be more like silicon companions than mere tools, even less like "robot slaves." Before that can happen, we will need to radically revise our ideas about what our machines *ought* to be allowed to do, and ideas about how much *free will* we wish to impart to them. This is destined to be an area of tremendous innovation in its own right. Computer experts may disagree. Many seem to wish machines to be forever deterministic. They would argue that if, for example, one enters a command with a slightly deviant syntax, or points to a non-existent directory, or allows a spelling error to creep into a file name, it is not up to the machine to look for a plausible meaning and offer it back to the human for consideration. That might lead to anarchy.

I strongly disagree with that view. Please!—*Let* the computer make these suggestions, and help me, its fumbling, memory-lapsing human user. Many of these 'little things' can be, and are, easily performed on present-day machines. Thus, the UNIX command *set filec* will usefully expand a truncated file name into its completed form.[28] But on other occasions, even using the most recent workstations, we get very nearly the same old dull reactions to our aberrant requests as we did back in the old DOS days. A handful of heuristics is invariably helpful. That's often the human's most important way forward; why shouldn't machines be given the same advantage?

Some believe that there is little point in attempting to make machines "like us." Erich Harth writes[29]

"It is still intriguing to ask the question 'What if?' *What if* our engineers succeed in constructing a truly thinking computer? And what if, to complete the illusion, we could clothe it in an audio-animatronic body, making a perfect android, a human recreated in silicon hyperreality? Would it have been worth the effort? Certainly there is value in the exercise, the challenge to our ingenuity. But the final product would be as useless as Vaucanson's duck. The ultimate kitsch! There are easier ways of making people, and anyway, there are too many of us already."

The image of "a perfect android" is not what I have in mind; such an entity might indeed be of as much value as a distinctly dull-minded junior assistant. This description completely fails to take into account what a "silicon *hyper* reality" *might* do. Freed of our own frail forgetfulness, and our emotional variability, endowed with a bevy of Bessel functions in the

28. If one believes that creativity is merely what happens "when normally disparate frames of reference suddenly merge," as Koestler believes, then could one say that in some tiny way the machine is doing a creative act in making this decision on our behalf?

29. Erich Harth, *The Creative Loop; How the Brain Makes a Mind* (Reading, MA: Addison-Wesley Publishing Company, 1993), 171–172.

bowels and Fourier integrals at the fingertips, knowledgeable of all of the best of Widlar's and Brokaw's circuit tricks, our *imperfect*, but *highly specialized*, android, *The KnowledgeMaster Mk. I*, could be a tremendous asset. He need not move; but remote sight would be useful (in scanning those old papers of Widlar that we leave on the desk), and hearing may be essential, not only in freeing up our fingers, but in eavesdropping on the engineering community (à la HAL, in *2001: A Space Odyssey*, which, incidentally, was another vision from the neurally noisy mind of Arthur C. Clarke).

A brief consideration of earlier projections of what computers might "one day" do leads us to be struck by how limited these visions often were. We've all heard about the early IBM assessment of the U.S. market for computers being about seven machines. Isaac Asimov, another noted visionary, imagined a time when robots might check our texts but didn't seem to anticipate how utterly commonplace and powerful the modern word processor, and in particular, the ubiquitous spelling-checker, would become. In his science-fiction story[30] *Galley Slave* he portrays an android named Easy who specialized in this task, and has the storyteller marvel at how

"With a slow and steady manipulation of metal fingers, Easy turned page after page of the book, glancing at the left page, then the right . . . and so on for minute after minute . . . The robot said, 'This is a most accurate book, and there is little to which I can point. On line 22 of page 27, the word "positive" is spelled p-o-i-s-t-i-v-e. The comma in line 6 of page 32 is superfluous, whereas one should have been used on line 13 of page 54. . . .'"

I wonder how many young users of the program I'm using to write this essay—Microsoft™ *Word*—know that, less than forty years ago, its capabilities were solely the province of sci-fi? Probably very few people living back then would have believed that robots who could correct our spelling and even our grammar would become commonplace so soon. A page or two later in Asimov's story we hear the robot's promoter say, over objections about allowing such powerful machines to enter into our daily affairs:

"The uses would be infinite, Professor. Robotic labor has so far been used only to relieve physical drudgery [in the futuristic setting of this story-BG]. Isn't there such a thing as mental drudgery? [You'd better believe it-BG]. When a professor capable of the most creative thought is forced to spend two weeks painfully checking the spelling of lines of print and I offer you a machine that can do it in thirty minutes, is that picayune?"

---

30. *Galaxy* (December 1957).

Thirty minutes! We are already irritated if it takes more than a few seconds to perform a no-errors spelling check on something of about the length of this essay. Users of Microsoft™ *Word 6.0* can now have spelling errors trapped on the fly, with their author's most probable intentions proffered for consideration (another one of those examples of an emergent capability for Koestler's creative conjugations of frames of reference, perhaps?). Modern word processors can also do a tolerably good job of correcting bad grammar. Note in passing how much we depend on being able to *personalize* the dictionaries and rules behind these checkers; my little *PowerBook 180*, on whom I daily cast various spells, has already become a serviceable, though still rather dull, companion.

Are we being equally shortsighted in seeing how tomorrow's connection machines will be capable of serving our needs as innovators? In visualizing the many further ways they could perform more than mere 'spelling checks' on circuit schematics (that is, going beyond catching just gross errors—roughly equivalent to grammar checking)? Even without an independent spirit, there is much they could, and I think, will help us with. Eventually freed from the frustrations of not being able to find the information we need to do our job, aided by more liberally minded machines, and allowed to operate in a strongly anticipatory mode, designers in all fields could make great strides toward more rapid, more accurate, and more effective development of new products. Our visionary use of the leverage afforded by prodigious auxiliary minds could make an immense difference.

Ultimately, we may even decide that it's not so stupid to build into these machines, very cautiously at first, *some* sections which are 'afflicted' by noise. We will have to get used to the idea that these bits may not behave in the same way every day, that they may even cause our silicon companion to have moods. It is this propensity for unpredictability and irrationality that makes people interesting. Like latter-day Edisons, we are, insofar as machine intelligence is concerned, just on the threshold of a whole new world of opportunity, a future (not *so* far off, either) where we will, for the first time in human history, need to be sensitive to the emerging question of machine rights. . . . There are no ready-made solutions, ripe for exploitation, in this domain; we will need to decide what kind of assistance we, as innovators, want our knowledge-gatherers and collators to give us, and just how much of the excitement of engineering we want to share with them.

A better vision of this future is found in a new book[31] by David Gelernter, who writes

"But why would anyone *want* to build a realistic fake mind? Is this really a good idea? Or is it pointless—or even dangerous?

"That's an important question, but in one sense also irrelevant. The urge to build fake minds stands at the nexus of two of the most

31. David Gelernter, *The Muse in The Machine: Computerizing the Poetry of Human Thought* (New York: The Free Press, 1994), 48.

powerful tendencies in the histories of civilization. These two are so powerful that it's pointless even to contemplate *not* pursuing this kind of research. It will be pursued, to the end.

"*People have always had the urge to build machines*. And *people have always had the urge to create people*, by any means at their disposal—for example, by art. . . . The drive to make a machine-person is . . . the grand culminating *tour de force* of the history of technology and the history of art, simultaneously. Will we attempt this feat? It is predestined that we will." [original italics]

What we do with these fake minds is up to us (at least, that's what we think today . . .). In less dramatic ways, we already see it happening, and there is no doubt in my own watery mind that since machines came on the scene, I've been a much more effective innovator. No single microelectronics corporation can undertake vast journeys of exploration and discovery into the world of artificial intelligence. For now, we just have to recognize that we can become more effective only by putting design and marketing knowledge into the hands and minds of every person in our design teams.

Our innovating descendants will probably still be teaching the value of VOC techniques well into the next century. But to them, this dusty acronym will have long ago become a reference to the wisdom of listening to the Voice of the Computer (the old-fashioned name we would use today), or rather, reflecting the diminution of its erstwhile merely-calculating function, and the by-then commonplace acceptance of the total symbiosis with, and essential dependence on, these sentient adjuncts to human minds, *The Voice of the Companion*. Long live VOC!

19. The Art and Science of Linear IC Design

I have been asked several times by other integrated circuit (IC) design engineers, "How do you come up with ideas?" And my answer was usually something flip, like "Beats me, it just happens." Later, I began to think more seriously about the actual process that I went through to come up with new ideas for designs. My motive for figuring out the process was mostly curiosity, but I also wanted to document from new design ideas and the satisfaction of seeing successful products going out the door.

What I decided after a little pondering was that good IC design depends on a healthy disrespect for what has been, and lots of curiosity for what might be. By this I mean that one must assume that we have seen only a tiny part of the secrets in silicon, and therefore there are endless discoveries to be made. We must keep ourselves from thinking in terms of perceived limitations, and instead strike off on new paths, even if they don't seem to be going anywhere. On the other hand, engineering is based on fundamental laws that stubbornly refuse to let bad designs work well. I am continually amazed by engineers who hang on to a concept even when it clearly requires the laws of physics to bend. The human brain has a wonderful ability to combine what is into what might be, and a good engineer must let this process charge along, then apply reality checks so that mistakes, dead ends, and dumb ideas get cast aside.

When I tested this philosophy on other engineers, it soon occurred to me that from their viewpoint it seemed more like rhetoric than revelation. What was needed was details—the engineer's stock in trade. To that end I tried to create a list of specific techniques that can be used in analog IC circuit design. This probably leaves me wide open to the criticism of egotism, but it's been my observation that many of the best engineers have monstrous egos, so possibly it somehow aids in the design process. I hope the following ideas are helpful. If they're not, at least I finally made Jim Williams happy by coming through on my promise to do a chapter for this book.

The first section is on inspiration, so it is kind of vague and slippery, much like the process itself. The next section is more down to earth, and obviously exposes a litany of mistakes I made along the way. We learn and remember from our own mistakes, so maybe force feeding them isn't too helpful, but that's the way it came out. Good luck.

Inspiration: Where Does It Come From?

Free Floating Mind

Many of the best IC designers agree that some of their great design ideas occur outside of the workplace. I know it is true for me, and in my case it is usually someplace like the car, shower, or bed. These are places where only minimal demands are being made on your mind, and interruptions are few, unless you get lucky. (I commute on autopilot. I think there is a special part of the brain allocated just for getting back and forth to work. It can accomplish that task with only 128 bits of memory.) You can let your mind float free and attack problems with no particular haste or procedure, because you own the time. It doesn't matter that ninety-nine times out of one hundred nothing comes of it. The key is to have fun and let your mind hop around the problem, rather than bore into it. Don't think about details. Concentrate on broader aspects like assumptions, limitations, and combinations. Really good ideas often just pop into your head. They can't do that if you're in the middle of some rigorous analysis.

Trials at Random

Colleagues think I'm really weird for this one, but it does work sometimes when you have spare time and pencil and paper. I connect things up at random and then study them to see what it might possibly be good for. It's mostly garbage, but every so often something good shows up. I discovered an infinite gain stage, a method for picoamp biasing of bipolar transistors, and several new switching regulator topologies this way. Unlike the free floating mind mentioned earlier, here you concentrate totally on the details of what you've done to see if there's anything useful in it.

One good thing about this simple-minded technique is that it teaches you to analyze circuits very quickly. Speed is essential to maximize your chance of finding something useful. The other good thing about it is that when you do come up with something useful, or at least interesting, you can drive people crazy with the explanation of how you thought of it.

Backing In from the End

A natural tendency for design engineers is to start at the beginning of a design and proceed linearly through the circuit until they generate the desired output. There are some situations where this procedure just doesn't work well. It can work where there are many possible ways of accomplishing the desired goal. It's kind of like a maze where there are many eventual exits. You can just plow into the maze, iterate around for a while, and voilà, there you are at one of the exits.

There are other situations where this beginning-to-end technique doesn't work because the required result can only be obtained in one of a few possible ways. Iteration leads you down so many wrong paths that nothing gets accomplished. In these cases, you have to back into the design from the end.

The "end" is not necessarily the desired circuit output. It is the restrictions that have been placed on the design. If the circuit must have some particular characteristic, whether it be at the input, in the guts, or at the output, sketch down the particular device connections which must be used to accomplish the individual goals. Don't worry if the resulting connections are "bad" design practice. The idea here is that there is only one or at most a few ways that you can get to where you need to be. After you have all the pieces that solve particular parts of the problem, see if it is possible to hook them together in any rational fashion. If not, alter pieces one at a time and try again. This is a parallel design approach instead of the more conventional serial method. It can generate some really weird circuits, but if they work, you're a hero.

Testing Conventional Wisdom

Bob Widlar taught me to consistently and thoroughly mistrust anything I hadn't proved through personal experience. Bob wasn't always right about things, but partly by refusing to believe that anyone else knew much about anything, he made great advances in the state of the art. Conventional wisdom in the late '60s said you couldn't make a high current monolithic regulator. The power transistor on the same die with all the control circuitry would ruin performance because of thermal interactions. He did it anyway, and the three terminal regulator was born. The funny part of this story is that Widlar said at about the same time that no IC op amp would ever be built with a useful gain greater than 50,000 because of thermal interaction limitations. Not long after that, op amps appeared with gains greater than 500,000. Some designer obviously didn't believe Bob's rhetoric, but believed in his philosophy.

Conventional wisdom is something that constantly intrudes on our ability to make advances. Engineers are always using "rules of thumb"[1] and too often we confuse useful guidelines with absolute truth. By constantly questioning conventional wisdom I irritate the hell out of people, but sometimes it pays off when a new product is born that otherwise wouldn't have happened. This doesn't mean that you should bash around trying to get away with designs that are nearly impossible to produce with good yield. It means that you should ask people to detail and support the limitations they place on you, and then do your damnedest to find a hole in their argument. Try to remember your childhood years, when the most-used expression was "But why not?" Remain intellectually honest and maintain good humor while doing this and you should escape with your life and some great new products.

---

1. In the not so distant past, men were allowed to use a stick no larger in diameter than their thumb to beat their wives. This useful guideline fell out of general use when the Supreme Court decided that wives could not use anything larger than a .38 to defend themselves.

Find Solutions by Stating the Problem in Its Irreducible Terms

This technique has been helpful on several occasions. The idea is to clarify the possible solutions to a problem by stating the problem in its most basic terms. The LM35 centigrade temperature sensor, developed while I was at National Semiconductor, came about in this way. At that time, monolithic sensors were based on designs that required level shifting to read directly in degrees centigrade. I wanted to create a monolithic sensor that would read directly in centigrade. More importantly, it needed to be calibrated for both zero and span at the wafer level with only a single room temperature test. This flew in the face of conventional wisdom, which held that zero and span accuracy could only be obtained with a two-temperature measurement.

I found the solution by expressing the desired output in its simplest terms. A PTAT (Proportional to Absolute Temperature) sensor generates an inherently accurate span but requires an offset. A bandgap reference generates a precise zero TC offset when it is trimmed to its bandgap voltage, which is the sum of a PTAT voltage and a diode voltage. A centigrade sensor therefore is the difference between a first PTAT voltage and a reference consisting of a second PTAT voltage added to a diode voltage. Subtracting two PTAT voltages is simply equal to creating a smaller PTAT voltage in the first place. Also, it was obvious that creating a centigrade signal by using span-accurate PTAT combined with zero TC bandgap would create a sensor which still had accurate span. By thinking of the problem in these terms, it suddenly occurred to me that a centigrade thermometer might share symmetry with a bandgap reference. Instead of the sum of two opposite-slope terms giving zero TC at a magic (bandgap) voltage, it might be that the difference of two opposite-slope terms would generate a fixed slope, dependent only on the difference voltage. This means that a simple calibration of difference voltage at any temperature automatically defines slope. Sure enough, the same equations that predict bandgap references show this to be true. The LM35 is based on this principle, and produces very high accuracy with a simple wafer level trim of offset.

Philosophical Stuff

Things That Are Too Good to Be True

Many times I have been involved in a situation where things seemed better than they ought to be. Eventually a higher truth was revealed, and along with the embarrassment, there was much scrambling to limit the damage. This taught me to question all great unexpected results, sometimes to the point where my colleagues hesitate to reveal good fortune if I am in earshot. The point here is that the human ego will always try to smother nagging little inconsistencies if a wonderful result is at stake. This has shown up in recent high-profile scandals involving such diverse fields as medicine, physics, and even mathematics.

When the situation arises, I try to make a judgment about the worst case downside of embracing results that seem just a little too good. If the potential downside is sufficiently bad, I refuse to believe in good fortune until every last little inconsistency has been resolved. Unfortunately, this sometimes requires me to say to other engineers, "I don't believe what you're telling me," and they are seldom happy with my "too good to be true" explanation.

A good example of the danger in embracing wonderful results appeared in a recent series of editorials by Robert Pease in *Electronic Design*. He took on the hallowed work of Taguchi, who seeks to limit production variations by utilizing Statistical Process Control. Taguchi believes that most production variation problems can be solved by doing sensitivity analysis and then arranging things so that the sensitivities are minimized. He used an example in his book of a voltage regulator whose output was somewhat sensitive to certain resistors and the current gain of transistors. After some fiddling with the design, Taguchi was able to show that it was no longer sensitive to these things, and therefore was a "robust" design. Unfortunately, Mr. Taguchi didn't bother to check his amazing results. Pease showed that the output was insensitive simply because the circuit no longer worked at all!

If this was just an academic discussion, then one could indulge in whatever level of delusion one liked, but the IC design business is extremely competitive, both professionally and economically. A small mistake can cost millions of dollars in sales, not to mention your job. I remember an incident many years ago when a new micropower op amp was introduced which had unbelievably low supply current. I questioned how the current could be so low, especially since the start-up resistor alone should have drawn nearly that much current. I studied the schematic, and sure enough, there was no start-up resistor! The circuit needed only a tiny trickle of current to start because it had closed loop current source biasing that needed no additional current after starting. This tiny current was apparently supplied by stray junction capacitance and the slew rate of the supplies during turn on. This seemed too good to be true and the data sheet made no mention of starting, so we purchased some of the amplifiers and gave them the acid test; slow ramping input supplies at the lowest rated junction temperature. Sure enough, the amplifiers failed to start. I heard later that irate customers were returning production units and demanding to know why there was "no output." It takes only a few of these incidents to give a company a bad reputation.

Unfortunately, some engineers become so fearful of making a mistake that they waste large amounts of time checking and cross-checking details that would have little or no impact on the overall performance of a circuit. The key here is to emulate the poker player who knows when to hold 'em and when to fold 'em. Ask yourself what the result would be if the suspect result turned out to be bogus. If the answer is "no big deal," then move on to other, more important things. If the answer is "bad news," then dig in until all things are explained or time runs out. And don't be shy about discussing the discrepancy with other engineers. As a class, they love a good technical mystery, and will respect you for recognizing the inconsistency.

Checking Nature's Limits

Many of the important advances in linear ICs came about because someone decided to explore just exactly what nature's limits are. These ideas were developed because someone asked himself, "How well could this function be done without violating the basic physical limits of silicon?" Studying the limits themselves often suggests ways of designing a circuit whose performance approaches those theoretical limits. There's an old saying that is true for linear IC design—once you know something can be done, it somehow becomes a lot easier to actually do it. Until you know the real limits of what can be done, you can also make the error of telling your boss that something is impossible. Then you see your competition come out with it soon after. A classic example of this is the electrostatic discharge (ESD) protection structures used to harden IC pins against ESD damage. A few years ago no one thought that you could provide on-chip protection much above 2,000V, but no one really knew what the limits were. Our competition suddenly came out with 5,000V protection, but got smug. We scrambled to catch up and discovered a way to get 15,000V protection. We still don't know what the limits are, but we're sure thinking about it a lot more than we used to.

When I worked in the Advanced Linear group at National Semiconductor, we had a philosophy about new design ideas; if it wasn't a hell of a lot better than what was already out there, find something better to do. This encouraged us to think in terms of the natural limits of things. It wasn't always clear that the world wanted or needed something that was much better than was already available, but it turned out that in most cases if we built it, they bought it. It is my observation that customers buy circuits that far exceed their actual needs because then they don't have to waste time calculating the exact error caused by the part. They can assume that it is, at least for their purposes, a perfect component. Customers will pay to eliminate worry simply because there are so many things to worry about in today's complex products.

What to Do When Nothing Makes Any Sense

Everyone has been in group situations where no one can agree on the truth of the matter under discussion. This often happens because no test exists which can prove things one way or another. In some cases when I suggest a test that might prove who's right and who's not, the response is total apathy. Evidently, human nature sometimes loves a good argument more than truth, and I suppose that if life, liberty, and cable TV are not at stake, one can let these arguments go on forever. Engineering is not nearly so forgiving. We find ourselves in situations where the cause of some undesirable phenomenon must be discovered and corrected—quickly. The problem gets complicated when nothing makes any sense.

An engineer's nightmare consists of data that proves that none of the possible causes of the problem could actually be the real cause. My favorite phrase after an engineer tells me that all possibilities have been exhausted is "Hey, that's great, you just proved we don't have a problem!"

Of course life is not that simple, and the challenge is to identify a new series of tests which will clearly show what is going on. The great thing about this mental process is that it sometimes leads to a solution even before the tests are run. Defining the tests forces you to break down the problem into pieces and look at each piece more carefully. This can reveal subtleties previously hidden and suggest immediate solutions.

The first step is to challenge all the assumptions. Ask all of the people involved to state their assumptions in detail and then make it a game to blow a hole in them. A good engineer is more interested in solving problems than protecting ego, so give and take should be welcomed.

The classic mistake in problem solving is mixing up cause and effect. I have been in many meetings where half the crowd thought some phenomenon was a cause and the other half considered it an effect, but no one actually expressed things in these terms, so there was much pointless arguing and wasted time.

Order of the testing is critical when time is short. Tests with the highest probability of success should get priority, but you should also consider the worst-case scenario and start lengthy tests early even if they are long shots. Nothing is more career-threatening than explaining to your boss well down the road that your pet picks came up empty, and that you will now have to start long term tests.

The final step is to pre-assign all possible outcomes to each of the tests. This sometimes reveals that the test won't prove a damn thing, or that additional tests will be needed to clarify the results. My rough estimate is that 30–40% of all tests done to locate production problems are worthless, and this could have been determined ahead of time. If we were in the pencil making business, it wouldn't be a big deal, but the IC business runs in the fast lane on a tight schedule. I have seen fab lines throw mountains of silicon at a bad yield problem simply because they have no choice—the customer must get silicon. All lines have problems, but what separates the winners from the losers is how fast those problems get fixed.

Gordian Knots

There are certain kinds of problems with circuits that defy all attempts at clever or sophisticated analysis. Cause and effect are all jumbled, complex interactions are not understood, and no tests come to mind that would isolate the problem. These electronic Gordian knots must be attacked not with a sword, but with the same technique used to untangle a jumbled mess of string. Find an end, and follow it inch by inch, cleaning up as you go until all the string is in a neat little ball. I find that very few people have the patience or concentration to untangle string, but for some reason, I get a kick out of it. The electronic equivalent consists of taking each part of the circuit and forcing it to work correctly by adding bypass capacitors, forcing node voltages or branch currents, overriding functions, etc. When you have the circuit hogtied to the point where it is finally operating in some sane fashion, it is usually much easier to isolate cause and effect. Then you can start removing the Band-Aids one at a time. If removing one causes the circuit to go crazy again, replace it and try another. Try to remove as many of the unnecessary Band-Aids as possible, checking each one to make sure you understand why it is not needed. Hopefully, you will be left with only a few fixes and they will paint a clear picture of what is wrong. If not, take your children fishing and practice on backlashes.

Don't Do Circuits That Try to Be Everything to Everybody

I have seen many linear IC products introduced which are touted as a universal solution to customer needs. These products have so many hooks, bells, and whistles that it takes a 20-page data sheet just to define the part. The products often fail in the marketplace because: (1) They are not cost effective unless most of their features are used. (2) Engineers hate to waste circuitry. (3) Customer needs change so rapidly that complex products become obsolete quickly. (4) Engineers subconsciously tend to allow a certain amount of time for learning about a new product. If they perceive that it will take much longer than this to be able to design with a new circuit, they may never get around to trying it.

The most successful linear IC products are those which do a job simply and well. The products themselves may be internally complex, such as an RMS converter, but externally they are simple to use and understand. Flexibility should not be provided to the user by adding on a pile of seldom-used optional features. Instead, the chips should be designed to operate well over a wide range of temperature, supply voltage, fault conditions, etc. A well-written data sheet with numerous suggestions for adapting the chip to specific applications will allow users to see the usefulness of the part and to make their own modifications that give them ownership in the final application.

Use Pieces More Than Once

For reasons I have never figured out, I love to make pieces of a circuit do more than one function. And like love, this can be both dangerous and exciting. Actually, before ICs it was standard procedure to make tubes or transistors do multiple duty, either because they were expensive, or because of space limitations. Engineers became heroes by saving one transistor in high-volume consumer products. Nine-transistor radios performed nearly as well as modern IC designs that use hundreds of transistors. Transistors on a monolithic chip are literally a penny a dozen, and they are tossed into designs by the handful. Even discrete transistors are so cheap and small that they are considered essentially free.

So why should designers discipline themselves in the archaic art of not wasting transistors? The answer is that like any other skill, it takes practice to get good at it, and there are still plenty of situations where minimalist design comes in very handy. One example is when a change must be made to an existing design to add an additional function or performance improvement, or to fix a design flaw. To avoid expensive re-layout of a large portion of the IC, it may be necessary to use only the components already in the design. A practicing minimalist can stare at the components in the immediate area, figure out how to eliminate some of them, and then utilize the leftovers to solve the original problem. He's a hero, just like in the old days.

Micropower designs are another example where double duty comes in handy. Every microampere of supply current must do as much work as possible. A transistor whose collector current biases one part of the circuit can often use its emitter current to bias another part. The bias current for one stage of an amplifier can sometimes be used for a second stage by cascoding the stages. There are certain classes of bandgap reference design where the reference can also do double duty as an error amplifier. These and many other examples allow the designer to beat the competition by getting higher performance at lower current levels.

Often, I don't see many of the minimizing possibilities until a circuit is well along in design, but that is the best time to look for them. All the pieces are in front of you and it is much easier to see that two pieces can be morphed[2] into one. If you do this too early, you tend to waste time bogged down in details. At the very end of the design such changes are risky because you might forget or neglect to repeat some earlier analysis that would find a flaw in the design change. Keep in mind also that future flexibility in the design may be compromised if too much fat is removed originally.

I Never Met a Burn-in Circuit I Liked

One of my pet peeves concerns testing reliability with burn-in. This is standard procedure for all IC designs and the typical regimen during product development is a 125°C burn-in on 150 pieces for 1000 hours at maximum supply voltage. Burn-in is supposed to detect whether or not the IC has any design, fabrication, or assembly flaws that could lead to early field failures. In a few cases, the testing does just that, and some built in problem is discovered and corrected. Unfortunately, with highly reliable modern linear IC processing, most burn-in failures turn out to be bogus. The following list illustrates some of the ways I have seen perfectly good parts "fail" a burn-in when they should not have.

1. IC plugged into the socket wrong.
2. Burn-in board plugged into the wrong power supply slot in the oven.
3. Power supply has output overshoot during turn-on.
4. Power supply sensitive to AC line disturbances.

---

2. From image processing computer programs that combine images.

5. Power supplies sequence incorrectly.
6. IC is inserted in test socket incorrectly after burn-in and gets destroyed.
7. IC fails to make good contact to all burn-in socket pins, causing overstress.
8. Burn-in circuit allows so much power dissipation that IC junction temperature is outrageously high.
9. Burn-in circuit applies incorrect biasing to one or more pins.
10. IC oscillates in burn-in circuit. (With hundreds of parts oscillating on one board, power supply voltages can swing well beyond their DC values.)
11. Some parameter was marginal and a slight change during burn-in caused the IC to change from "good" to "bad."
12. IC was damaged by ESD before or after burn-in.

These twelve possibilities could probably be expanded with a poll, but they serve to illustrate a serious problem with burn-in; namely, most of the failures have nothing to do with reliability issues. Even one burn-in failure is considered serious enough to warrant a complete investigation or a new burn-in, so bogus failures represent a considerable waste of time and money. Delay in time-to-market can multiply these direct costs many times over.

An IC designer has control over items 7 through 11, and these represent a large portion of the bogus failures. Considerable thought should be given to the design of the burn-in circuit so that it does not overstress the part in any way, even if one or more IC pins do not make contact to the burn-in socket. Remember that you are dealing with thousands of socket pins which see thousands of hours at 125°C. Some of them will fail open through corrosion, oxidation, or abuse. The chance that an open pin will be identified as the cause of a burn-in failure is very slim indeed, so you must protect the IC from this fate with good design techniques.

The fully stuffed board should be transient tested if there is any question about oscillations. ICs which dissipate any significant power should be analyzed very carefully for excess junction temperature rise. This is complicated by the complex thermal environment of a maze of sockets coupled to a common board with poorly defined air movement. I often just forget calculations and simply solder a thermocouple to one of the IC leads. Testing is done with a fully stuffed board in the burn-in oven sandwiched in between other boards to minimize air flow. Finally, use good judgment to define fail limits so that small, expected changes through burn-in do not trigger failures. Many linear ICs today are trimmed at wafer test to very tight specifications, and this may necessitate a more liberal definition of what is "good" and "bad" after burn-in.

Asking Computers the Right Questions

Computers are without a doubt the greatest tool available to the IC designer. They can reduce design time, improve chances of silicon working with minimal changes, and provide a reliable means of documentation. Computers don't create, but by analyzing quickly, they can allow a designer to try more new ideas before settling on a final solution. A good working relationship with a computer is critical to many designs where classical breadboards are out of the question because of issues such as stray capacitance, extreme complexity, or lack of appropriate kit parts.

A nagging problem with computers is that they only do what they're told to do, and in general, they only do one thing at a time. This is reassuring from a confidence viewpoint but it leads to a fatal shortcoming: the computer knows that something is wrong with a design, but steadfastly refuses to tell you about the problem until you ask it nicely. A particular set of conditions causes the circuit to react badly, but those conditions are never analyzed by the computer. With breadboards, it is much easier to spot problems because it is easy to vary conditions even on a very complex circuit. You can adjust input signal conditions, power supply voltage, loads, and logic states over a wide range of permutations and combinations in a relatively short time, without having to figure out which combinations are worst case. The results can be observed in real time on meters and oscilloscopes. Temperature variation takes longer, but is still quite manageable. This ability to quickly push the circuit to "all the corners" is invaluable when checking out a design.

Computer analysis is typically very slow compared to a live breadboard, especially on transient response. This can lead to a second hazard. The designer knows what analysis he should do, but when confronted with extremely long run times, he saves time by attempting to second-guess which conditions are worst case. One of the corollaries to Murphy's Law states that fatal flaws appear in a design only after the analysis that would have detected them is deemed unnecessary.

How do you select the proper questions to ask the computer to ensure that potential design flaws are detected? This decision is critical to a successful design and yet many engineers seem very blasé about the whole procedure and do only token amounts of analysis. They become the victims of the lurking flaw and have to cover their butts when the boss asks if the silicon problem shows up on simulations. Others waste enormous amounts of time doing analysis that generates huge reams of redundant data. They get fired when the design is hopelessly behind schedule. The following list of suggestions are my version of a compromise, and limit nasty surprises to those the simulator doesn't predict anyway.

Do a Thorough Analysis of Small Pieces Separately. "Small" is defined in terms of computer run time, preferably something less than a few minutes. This allows you to do many tests in a short period of time and forces you to concentrate on one section of the design, avoiding information overload. Things go so quickly when the number of devices is low that you tend to do a much more thorough job with little wasted time.

The lowly biasing loop is a good example of why analyzing small pieces is helpful. In modern linear IC design, the biasing loops often use active feedback to control currents accurately over wide supply variations, or to tolerate variable loading. I have seen many cases where the bias loop had very poor loop stability and this did not show up on full-circuit transient or small signal analysis. In other cases the peaking in the bias loop did show up as an aberration in circuit performance, but was not discovered as the cause until hours or days of time were wasted. A simple transient test of the bias loop by itself would have saved time and teeth enamel.

Beware of Bode Analysis. Many designers use Bode analysis to determine loop stability. This technique has the advantage of defining response over the full range of frequencies and it gives a good intuitive feel for where phase and gain problems originate. The problem is that with some loops, it is nearly impossible to find a place to "break" the loop for signal injection. The sophisticated way to inject the test signal is to do it in a way that maintains correct small-signal conditions even when large changes are made to components or DC conditions. This allows rapid analysis of various conditions without worrying about some "railed" loop condition. There are many possible ways to inject the signal that accomplish this, but correct Bode analysis requires that the impedance on one side of the signal be much larger than the other overall frequencies of interest. This is often not the case, and a Bode plot that seems to be giving reasonable answers is actually a big lie. It turns out that the impedance requirements typically fall apart near unity gain, just where they do the most harm. (Murphy is in control here.) If you have any doubts about the impedance levels, you can replace the voltage source with two low-value resistors in series. Inject a current test signal to the center node and ask for the ratio of the two resistor currents over all frequencies. If the ratio is less than 10:1 at any frequency, the analysis is flawed. (Actually, it turns out that there is a way to do an accurate Bode analysis with arbitrary impedance levels. This is detailed in *Microsim PSpice Application Design Manual*, but it is a fairly tedious procedure.) Another sanity check is to do a small-signal transient test of the loop and compare results with the Bode test. (See section on transient testing.)

A second problem in Bode testing is multiple feedback paths. As linear circuits get more sophisticated, it is not unusual to find that there is more than one simple loop for the feedback signal to travel. A typical example is a bandgap reference where most of the circuitry uses the regulated output as a supply voltage. Signals from the output can feed back to intermediate nodes in the gain path via load terminations and bias loops. This can cause some really strange effects, like common emitter stages that have zero phase shift at low frequencies instead of the expected −180. It seems impossible until you realize that the current source load is changing enough to cause the collector current to increase even though the base emitter voltage is decreasing. The result is that the net impedance at the collector node is negative, and this causes the phase to flip at low frequencies. The overall loop still works correctly with flipped phase because of overall feedback through the normal feedback path. Phase returns to normal (–270) at higher frequencies because capacitance dominates impedance. A second problem occurs at high frequencies where capacitive feedthrough in the extra loops can cause main-loop oscillations. A standard Bode plot may not show a problem, whereas a transient test usually does. It works both ways, of course. I have seen circuits where the Bode plot predicts oscillations, but the circuit is actually quite stable because of a secondary high-frequency feedback path.

Transient Testing Can Also Fool You. I used to think that transient testing was a foolproof way to judge loop stability. It didn't require any interpretation—either the response looked clean or it didn't. Now I know of several ways to get fooled. The first is to inject the test signal at the wrong point or to use voltage when you should use current. There are some points in a feedback loop that smother the test signal with a low-pass network that allows only the lower frequencies in the test pulse to get into the main part of the loop. The result is a very benign-looking output response that does not show dangerous high-frequency ringing problems. My experience shows that this problem almost never occurs if you inject a current into a low-impedance node in the loop. Typically, this would be the output, but a more general guideline is that it be a node that the loop is trying to hold to a constant voltage. In a switching regulator, for instance, do not inject the signal into the post-filter output if that filter is outside the main feedback loop.

A second way to get fooled is to use the wrong test frequency. A loop that rings at 50KHz will not look ringy when excited at 100KHz. This may seem obvious, but many loops have more than one frequency where phase margin is poor. If you concentrate only on the high-frequency portion, you might miss that little slow-settling tail that bites you later. Likewise, if the test frequency is too low, you might miss a very high-frequency buzz that washes out in the screen resolution. A frequent cause of these buzzies is a minor internal loop which has a bandwidth much higher than the main loop. Zoom in on edges if there is the slightest hint of raggedness.

Use Temperature to Test Robustness. Sometimes one has to do exhaustive analysis of a circuit to prove out the design. You might have to vary supply voltages, component values, device parameters, load conditions, logic and signal levels, operating frequencies, and on and on. This is very time consuming, in some cases much more so than if one had a real breadboard to test in the lab. When a change is made to the design, one has to carefully consider how much of the previous testing will have to be repeated. But engineers are human, and when they get lazy or rushed, design flaws are missed simply because the designer decided not to repeat a previous test after a "tiny" change was made to the design.

I believe that one way to help ensure a "robust" design is to have the computer analyze the circuit at temperatures well beyond the expected operating range. The reason this works so well is that temperature has an effect on nearly everything in the circuit if the components are modeled correctly for temperature dependence. This has the desired effect of varying more than one thing at a time and greatly reduces analysis time, especially if you just want to verify that nothing got screwed up by a tiny little change. I force the circuit to as many simultaneous worst-case conditions as I can, then vary temperature from –80°C to +200°C to see where things fall apart. This usually points out any design weaknesses which may be occurring dangerously close to the desired operating temperature. A good rule of thumb is that the circuit should be a healthy 25°C below its minimum expected temperature and 50°C above the maximum expected temperature. Circuits which are checked in this manner also tend to be very tolerant of those nasty little fab variations that haunt all linear designers.

Look at Transistor Base Currents to Detect Incipient Saturation. Bipolar transistor saturation has become more of a problem with modern analog circuits that have to work at very low supply voltages. Even in older designs, the collector-to-emitter voltage of an amplifying transistor was often the base-to-emitter voltage of a second transistor. This is problematic because the collector-to-emitter voltage required to avoid saturation is proportional to absolute temperature (+0.33%/C), and the voltage actually forced on it by a base emitter voltage decreases with temperature. At some high temperature these two requirements clash and the result is at least partial saturation of the first transistor. For example, if 250mV is required to keep a specific transistor out of saturation at 25°C, it will take 354mV at 150°C. A Vbe of 600mV at 25°C will decrease to 350mV at 150°C. Therefore, at temperatures above 150°C, saturation will occur.

Regardless of the exact cause of saturation, the simplest and most sensitive way to look for the problem is to plot base currents versus temperature. A sudden increase in base current at some temperature is a good indication of saturation. This is especially critical in precision applications, such as bandgap references, operational amplifiers, and comparators. One word of warning: computer models can do a poor job of predicting saturation problems when certain model parameters are adjusted to make other things come out right. Have the computer plot Ic versus Vce with constant base current and compare this plot with curve tracer readings. Discrepancies will have to be accounted for, or model changes made.

Force Input and Output Signals Beyond Their Expected Range. There are all kinds of nasty surprises that can pop up when signals go beyond their expected range. The best example is phase reversal in a single supply input stage. A simple PNP differential input stage with a grounded emitter NPN as the second stage will exhibit phase reversal when one of the PNPs has zero volts on its base. If the result of phase reversal is that the PNP base remains at zero, a nonrecoverable latch occurs. I have seen this problem get to final silicon many times because zero volts was not a "normal" operating condition, and the designer failed to consider start-up or fault situations.

A second example is regulator output polarity reversal. One normally would not expect the output of a voltage regulator to see reverse voltage, but this occurs quite often in cases where both positive and negative regulators are used in a system. If power is delivered to one regulator before the other, and loads are connected across the regulator outputs, the powered regulator will force the unpowered regulator output to a reverse voltage via the common load. System designers routinely protect against this condition by connecting diodes from each regulator output to ground to limit reverse voltage to one diode drop. Imagine their consternation to find out that this doesn't work with some IC regulators because these regulators refuse to start when power is applied with the output reverse biased by one diode drop. During simulations, I always force the output of regulators to 1.5V reverse voltage and check for proper start-up and full output drive current. After layout, I check saturated transistors in this state to make sure they don't inject to some nearby structure that would cause problems, a situation that won't show up on simulations!

Living in Fear of LVceo

Many linear designers make the mistake of assuming that circuits will not work properly if the voltage across bipolar transistors exceeds LVceo (latching voltage, collector-to-emitter, with the base open). In discrete design, one can simply specify transistors with high breakdown voltages, but with a given IC process, the only way to increase LVceo is to reduce gain (hFE). More times than I care to remember I have seen fab lines struggling to keep hFE in a very narrow range because the circuit designer demanded an unreasonable combination of hFE and LVceo. The truth of the matter is that transistors are quite happy to operate well beyond LVceo if there is provision to handle reverse base current. The graph in Figure 19–1 shows base current and base emitter voltage versus collect emitter voltage with emitter current held constant. Notice that nothing spectacular happens at LVceo. This is simply the point where base current is equal to zero. A transistor with LVceo = 50V and BVcbo = 90V can often be operated at 60V to 70V if the design will tolerate a low value of negative hFE (reverse base current). Above 50V, some means must be provided to absorb the reverse base current, but this is often just a high-value resistor across the base emitter junction. At voltages close to BVcbo, reverse base current climbs rapidly, and active reverse drive may be needed.

I have had many designs in production for years, operating well above LVceo, with no loss of performance or reliability. There is one caveat though: if a transistor is operated at high power levels above LVceo, there

Figure 19–1.
Operation above
LVceo is safe when
provision is made
for reverse base
current.

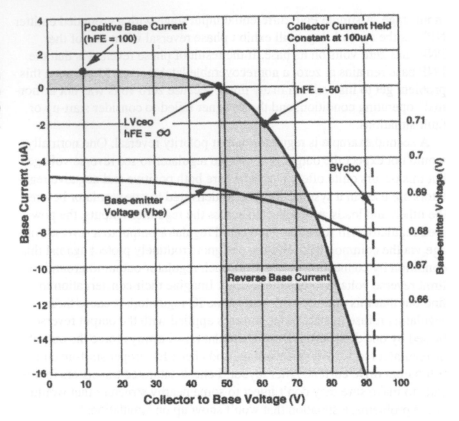

is a danger of forward-biased secondary breakdown, a phenomenon where current crowds to one tiny area of the transistor and breakdown plummets to half its normal value. This is normally only a problem in power transistors subjected to simultaneous high voltage and high current, but caution should be used in lower-power designs where the transistor could be subjected to a transient overload condition. Secondary breakdown can occur in less than a microsecond, and unless the voltage across the transistor is quickly reduced to well below LVceo, it will be permanently damaged.

20. Analog Design—Thought Process, Bag of Tricks, Trial and Error, or Dumb Luck?

Allow Me to Introduce Myself

Rather than leave the reader wondering where I got the weird ideas to be presented here, and maybe whether I should be allowed to run loose, I think it best to tell about my past: I spent my entire money-making career doing research and development for the U.S. Government ("the Gov"); the Department of Defense, to be exact. None of the authors of the first book of this series were in this category, and I will be surprised if any in the second are. However, this background does give one a different perspective, which can be useful.

DOD gets a lot of bad press these days. Most of the accusations have some basis in fact, and some are absolutely correct. But the more experience I have with industry and academia, the more I see the same problems. People are people wherever they are. The laws of physics apply indiscriminately to both military and civilian arenas. An idea that does not work in one can often be adapted to not work in the other. It increasingly seems that when I buy something for home use, I had better be prepared to fix it, or even re-engineer it! I am thinking primarily of mechanical and electro-mechanical gadgets, for example my daily battles with the car and the drink machine (I am not talking about the mornings I sleepily try to insert my Exxon card in the Coke machine). Mechanics aside, however, the electronics industry is not without fault. I have a car radio that sometimes emits sounds that are truly awful. There is room for improvement all around.

Given my employment, my experience has been in the design of relatively simple systems, produced in relatively small quantities, often with inadequate development time. I will necessarily emphasize these aspects in my philosophy of analog circuit design. My type of work is not as glorious as designing an integrated circuit (IC) that will be produced by the zillions, but it is just as necessary, and applies more often than one might think. Examples are: in-house lab equipment that will not be sold or even replicated, a jerry-rigged solution to a problem holding up an expensive field test, a quick demonstration that a proposed project has a chance of working (or doesn't)!

The military often makes headlines using a $100 part in place of a $1 part. (That's 20dB or 40dB, depending on whether you use 10 log or 20 log. I say use 10 log because money is power.) However, if it would take $10,000 worth of testing to ensure that the $1 part is indeed adequate and only 100 units will be built, it is a toss-up as to which part is really cheaper. Given the horrendous cost of field failure, pick the one that is most likely to work.

Philosophical question #1: Is an inexpensive widget that does not work better than an expensive one that does not work? You can buy more of them, but so what?

Philosophical question #2: If you were going into battle and your life depended on your equipment, which you didn't have to pay for, would you pick military or commercial?

The military (and NASA!) are extremely concerned about reliability; failures may be spectacular. So is industry; a design failure could easily mean a recall of 100,000 cars for General Motors. There is an ongoing discussion (argument, really) of how to achieve reliability. It is not likely to be settled soon, especially given that we have not agreed on exactly what constitutes failure!

Problem #1: Supplier A's widget meets all specs, but just barely in every case. Supplier B's widget is right on target in all cases except one, where it is unfortunately slightly out of spec. Which would you pick? Hint: the Gov picks A.

Problem #2: As you get farther from the transmitter, FM radio sounds great out to a point then drops out rather suddenly, while AM just gets noisier and noisier. Which is better? Hint: good music stations are on FM; emergency broadcast information is on AM.

A couple other items: I taught a course on Applications of Analog Integrated Circuits for ten years, mostly to students who weren't terribly interested. I know that some people don't get excited when they see an analog circuit, even a beautiful one. I learned which concepts were easy to pick up, and which were difficult. After it was all over, I realized I had never specifically mentioned one of the most important aspects of analog design—it is FUN! Too many digital projects consist of taking an arbitrary bunch of numbers and performing some questionable calculations on them in order to produce something I am not really interested in. I liken it to that marvelous invention, the kitchen compactor, that takes 20 pounds of garbage and transforms it into 20 pounds of garbage. I get the feeling the only time the bit flippers get any excitement is when the system crashes. I am not totally against computers; I enjoy playing back my phone messages and hearing my answering machine having a discussion with some store's computer. I don't know about artificial intelligence, but they definitely have artificial stupidity! Pages of ones and zeros just don't excite me. (An exception is my checking account; that's close enough to reality to get my attention.) Digital design will soon be just computers designing more computers, if it isn't already.

Analog, on the other hand, does not seem to be amenable to automatic design. And it usually has to connect to the real world. You hook up your new amplifier and get the joy of observing sounds coming out of your speaker; or smoke, depending on your level of expertise. Pushing a button on a transmitter you've designed, hearing the acoustic pulse go out, then feeling the earth shake under your feet as 50 lbs of explosive go off is an experience unmatched by anything I've seen in amusement parks.

Computer designers don't know what to do with a good op amp; in fact, there is nothing they can do. We analog people get to play with all sorts of neat stuff, including digital circuits! In reviewing 20 or so systems I've designed, I found that not one was free of digital circuits! In fact, half the time it was not clear whether the system was predominantly analog or digital. But we get to count these as analog!

If you read Bob Pease, you know that some of the world's most sophisticated measuring equipment (his) relies on such high-tech items as cardboard boxes, spray paint, dishwashing soap, plastic scraps, and RTV silicone glue (use electrical grade; some of the regular type contains acid!). To that I would add: Reynolds Wrap, duct tape, paper clips, refrigerator magnets, and Coke cans.

Lastly, I claim to be an expert on mistakes, for the simple reason that I've made most of them already, and am working on the rest. When I advise against something, it's usually because I've already tried it, with disastrous results.

I have never really been able to explain how I go about designing something, and doubt that I ever will. Nevertheless, Table 20–1 gives some aspects that are involved. These are not steps in the sense of finishing one, then going on to the next. They overlap, and one should try to keep all of them in mind at all times. I will ramble through these; you will see that many items could have been placed in more than one section.

Table 20–1 Six "Steps" to Analog Design

1. You want me to what?
2. A better mousetrap—because the mice are getting better.
3. Breadboard—the controlled disaster.
4. If it doesn't work, take two capacitors and call me in the morning.
5. Look, Mom, no smoke!
6. The job's not over till the paperwork is done.

Note: I will not attempt to distinguish between small systems and large circuits; with ICs there is a lot of overlap. A switched-capacitor filter may be listed as a circuit, but you better be aware of Nyquist's theorem, which is really system theory. Do not attach undue significance to whether "circuit" or "system" is used in any given place.

You Want Me to What?

First, make sure the problem is clearly defined in your head. This is so obvious it often gets overlooked. Did you understand clearly what your supervisor wanted? Did he understand what the customers wanted? Did they understand what was really needed? You will not likely get many brownie points for doing exactly as told if what you were told was idiotic. In the Gov, engineers are not allowed to talk to prospective contractors to answer questions during negotiations. I understand the legal reason—to prevent favoritism—but technically it's exactly backwards. One of three things usually happens:

1. We talk with likely contractors before the bidding starts.
2. We talk during the bidding anyway, with the warning that, "I am not allowed to talk to you; you are only imagining that I am; if asked later I will not remember any of this."
3. There are monster misunderstandings.

It is sort of like designing an op amp circuit without feedback; i.e., impossible. It is my view that engineering implies getting something done, and if that requires bending the rules into a triple granny knot with a half hitch, so be it.

Once you understand the goal, don't lose sight of it. I once fiddled with a circuit until I had a very efficient form, and gleefully presented it to my supervisor. He agreed that it was very efficient, but pointed out that it performed the wrong function. I had gotten so engrossed in the details that I had lost the big picture.

I do not mean to exclude pursuing a tangent, or even idle dreaming on your own; that has led to several of my inventions. But once a tangent becomes promising, make it a secondary clearly defined goal. Ants accomplish quite a lot with their Brownian motion, but they haven't designed an analog circuit yet, not even a digital one!

A Better Mousetrap—Because the Mice Are Getting Better

This used to be a joke, until I read that the Gov is trying to breed better mice. Just what we need, right? My cat can't catch the ones we have now. . . . Anyway, the next step is to get started toward your now clearly defined goal. Getting started right is important; speed is not terribly relevant if you're headed in the wrong direction. False starts are inevitable, but admit them early. Maybe you have trouble getting started; I do. Selecting the best idea from all the ideas in the world, thought of and not thought of yet, overwhelms me. But fear not:

AXIOM:
There may be an optimum system, but you don't want it.

A system can be optimized for one, maybe two, variables only, at the expense of all others; maybe serious expense. Furthermore, maxima are usually fairly broad and flat-topped, so normally you can move a ways off the peak without losing much, possibly gaining a lot on another variable where you were way down the slope.

Hypothetical problem: You want to maximize two functions, one proportional to cos q and the other to sin q. You shouldn't need higher mathematics to tell you it's impossible. One method of attack is to decide which is more important, let's say the cos one, and maximize that. At q = 0 cos q = 1, 100%, but sin q = 0, zip, nada, -∞dB. Oops. But by moving out to q = 0.3 rad, you can have sin q = 0.3 and still have cos q = 0.95; or to q = 0.5 rad and still get sin q = 0.5 and cos q = 0.9. Not bad, huh?

Similar problem, different subject: When adjusting a tuned filter, don't try to "peak" it. The response changes very little around the peak. Adjusting for zero phase shift is a far more sensitive method. If you can't do that, it is also more accurate to adjust so the 3dB down points straddle the desired center frequency.

OTHER AXIOM:
If you've done the job, it's done. Sort of.

There isn't much reward for reinventing the wheel. However, a guy named Rader invented a new type of wheel, and if he got a patent, he should have a lot more money than I do. The obvious starting point is: has the job been done before? If not, is there something close? Table 20–2 gives my favorite sources for ideas.

Table 20–2   Sources for Ideas

Source	Comment
Personal memory	The mind is pretty good at remembering and correlating patterns.
Others' memory	Two heads are better than one, if they're on different people.
Mfrs' spec sheets and application notes	Usually work, use available devices, assistance available.
Magazine articles	I clip any that might be useful and keep them in a notebook.
Textbooks	Optional; good ideas usually show up in above items.
Patents	May be necessary anyway to avoid paying royalties or fines.

Note: decreasing order of importance

Be aware of conventional wisdom, but don't be limited by it. An inventor, whose name nobody remembers, worked on a telephone before Alexander Graham Bell, but was advised that the telegraph was perfectly adequate. Things are done the way they are for a reason, but it may not be a very good reason. Feel free to find out. Wear safety goggles, or at least some kind of glasses. Ordinary plastic lenses will stop most types of electronic shrapnel. Life is dull if you follow the instructions.

Back in 1966 we needed a sample-and-hold with a very long hold time. This implied a buffer with a very high input impedance. (Capacitors are only available so big, especially ones that have low self-leakage.) MOS transistors had become available, but "everybody knew" they were unstable, noisy, and susceptible to damage from static electricity. Howsomever, they were so cute I couldn't resist. I figured out how to make a reasonably accurate buffer. Temperature stability wasn't good; in fact, if you got the device too hot the characteristics changed permanently! But the circuit was to be used in a controlled environment. It was a DC application, so I could beat down the noise with capacitance on the output. I did lose a few MOSFETS through careless handling, but once in the circuit with a microfarad on the gate, they were safe. I don't recall the exact hold time, but I know I measured droop by sampling a voltage one day and measuring it the next! At first it looked like the hold time was infinite, at least until I realized it drifted toward max voltage, not zero. . . . The reader should wonder what switching device was good enough; it was a relay!

I published the circuit[1] and there must have been considerable interest because I received a dozen or so inquiries. Later RCA succeeded in making IC op amps with MOS transistors. These were pretty much poohpoohed because the input specs weren't good, but look at the variety of CMOS devices available now!

Adapting an old idea has the advantage that you are starting with something that presumably worked, but be aware of: Pitfall #1: A good idea applied to the wrong situation is a bad idea. Pitfall #2: Murphy's Law, applied to drugs, adapted to circuits: Any modification which produces a good effect will also produce numerous bad side effects.

Seldom are two applications identical. Some subtlety may trip you up. The Band-Aid approach has its limits. Exception: politics. A few years back our laboratory got no money at all for new construction, but a sizable pot for alterations. They took a tool shed, added three wings and an upper story, and made a respectable building out of it. The original building became the foyer. It had to retain its "T" number, "T" meaning Temporary (since 1945), but who cares? I have designed new equipment with very strange nomenclature borrowed from other equipment to avoid running afoul of some rule. Use your imagination.

Look at It Another Way

Very often the solution to a problem appears immediately upon formulating the problem differently. I like to recall a story I read of mountain climbers who attacked a lesser but still-unclimbed peak. They reached a huge chasm and had to turn back. They related the information to another party who tried a different route and went right to the top. If they instead had tried to best the chasm, the mountain might still be unclimbed.

Example: The standard way to measure phase difference is to set a flip-flop on the zero crossing of one signal and reset it on the zero crossing of the other. The fraction of the time that the flip-flop is set gives the fraction of a cycle the second signal lags the first; averaging and scaling gives a DC readout of 0 degrees to 360 degrees. This gives an ambiguity at 0 = 360. Phase jitter around zero gives an average readout of 180 degrees, exactly wrong! This is normally solved by adding 180 degrees by inverting one signal, moving the ambiguity to 180. But we had to build a phasemeter into a hands-off system, where the necessary automatic switching would have added considerably to the complexity. The solution was to measure the angle in sign-magnitude format (0 to 180 degrees, plus or minus), which has no ambiguity. The circuitry for this method turned out to be fairly simple, also,[2] and had an additional advantage for unattended operation: a modest amount of noise caused only a modest error; extra zero crossings can drive a set-reset phasemeter crazy.

Sometimes you have to reverse your thinking entirely. The standard way of protecting against reverse battery connection is a diode, but the voltage drop is sometimes unacceptable. I, and probably many others, tried unsuccessfully to do it with a power MOSFET. The obvious way doesn't work because the inherent back diode conducts when reverse voltage is applied. Bob Pease got a patent by realizing all you have to do is turn the transistor around backwards! The FET doesn't really mind, and the back diode is working for you!

Bag of Tricks

Certain concepts appear over and over again. I like to think of them as a bag of tricks, in the sense that a magician's "tricks" are really scientific principles, skillfully applied, with special attention to how the human brain works (and doesn't work). Here are a few of my favorites:

PLLs and FLLs

Phase-Lock-Loops (PLLs) are cute devices, widely used, even where they shouldn't be. A similar device, the Frequency-Lock-Loop (FLL)[3] has some features the PLL does not, at the expense of giving up some you may not need in a given application. Possible advantages are: no out-of-lock state and hence no lock transition; insensitivity to phase inversions or even arbitrary phase jumps; frequency can be offset in a linear, continuous manner. The two devices together cover a wide range of applications. For an example, read on:

Frequency Synthesizers Many systems need one or more accurate frequencies. Even the crystal manufacturers themselves don't stock all possible frequencies; it's prohibitive. They will cut any frequency for you, which will necessarily cost you more and take considerable time. And what if you need to switch the frequency? An indirect frequency synthesizer takes a reference frequency (e.g., from a standard crystal oscillator) and multiplies it by one arbitrary integer and divides it by one or two others.[4] It uses a feedback loop (a PLL) and some counters. Thus you can take one accurate frequency source and create a host of others semi-digitally. Often there is an accurate clock around; even microprocessors have crystals attached these days! There are some design techniques you need to know and some limitations, but they are not bad. I have these in half a dozen systems.

Tone Detectors What if instead you have to *detect* a signal of known frequency? Generate the expected frequency with a synthesizer, then compare the input signal with it in a simple circuit (see also Note 4). The center frequency and effective bandwidth, and also the shape, of the effective bandpass filter can be precisely controlled. Frequency hops can be programmed. I have used this in several systems, too.

Pseudo-Noise Pseudo-random noise (PRN or simply PN) generators[5,6] generate a neat signal that looks like noise, but is actually deterministic, and hence has precisely defined properties. They are made from a few shift registers and gates, possibly followed by filtering. Why generate more noise, when we are plagued with enough of it already? Well, noise testing for one thing. Secure communications for another. And how else do you generate a reasonable broadband signal?

Modulation/Demodulation When I say "modulation," you probably think radio or TV. But it is useful in a surprising number of other applications. Chopper op amps use modulation. It can be used to do some fancy filtering tricks; how about a 60,000dB/octave filter?[7] Need narrowband noise? Use a PN generator, filter the output to the exact shape and (half) bandwidth you want, then modulate it up to the desired center frequency!

Sine and Triangle Generators Generating a sine wave is one of the classic problems of our discipline. Some really terrible ways of doing it have been devised. You can take a microprocessor and a D-A converter and in less than a year generate a stairsteppy thing that looks like a sine wave if you stand across the room. Unless you really need a low-distortion sine wave, just generate a square wave and remove the harmonics with a low-pass or bandpass filter. Triangle wave? Just run the square wave through a pseudo-integrator. If a square wave isn't already available, you can get it from the triangle wave itself with a hysteresis clipper (Schmitt Trigger). (One of the two circuits has to invert.) This makes a loop and is the basic function generator circuit.

Thevenin and Norton Equivalents; Frequency and Impedance Transformations

These "tricks" can simplify a lot of problems and allow you to juggle circuits into more desirable forms. They should be in any good circuit or filter book; if they're not in yours, trash it and I'll send you mine.[8] From time to time an article appears on how to build gain into a filter stage, usually using a computer program. It is not necessary.[9] The filters of Figures 20–1A and 20–1B have the same characteristic; only the gain is different. In both cases the open-circuit voltage (mentally break the loop) at e1 is equal to e2 but comes through an impedance of $C\sqrt{2}$. (For any G, the two capacitors to the right of the dotted line in Figure 20–1B sum to $C\sqrt{2}$.) The circuit to the left of the dotted line does not know what is on the right side (unless it peeked). Therefore, for any input the voltages at e1 and e2 will be the same in either case. The output is simply e2 multiplied by whatever gain the op amp is set for by the negative feedback divider. As a quick check, let G go to zero; the circuit of Figure 20–1B reduces to that of Figure 20–1A (with an extraneous load resistor).

If all capacitors in the circuit of Figure 20–1A are increased by a factor X (Figure 20–1C), it should be obvious that the time response to an impulse will have the same shape, but will be expanded X times (slower). Since the frequency response is the Fourier transform of the impulse response, the frequency characteristic retains the same shape but is compressed by a factor X in frequency. This also should tell you that all capacitors in a filter should be of the same type so they will drift together. The cutoff frequency will necessarily drift, but at least the filter shape will not change. In fact, when building the circuit of Figure 20–1A, instead of looking for two similar capacitors whose values differ by exactly a factor of two (which seldom happens), I get three of the same value, hopefully from the same lot, and parallel or series two of them.

I once had to design a sinusoidal oscillator of frequency 0.004Hz. That's a period of about four minutes. And it took at least 10 cycles to settle after power-up. After running a few strip-chart records I realized I might not live long enough to complete the design. I got smart and reduced the capacitors by a factor of 1000. Using a 'scope, I got the bugs

Figure 20–1.
Gain, frequency, impedance manipulations on a Butterworth filter.

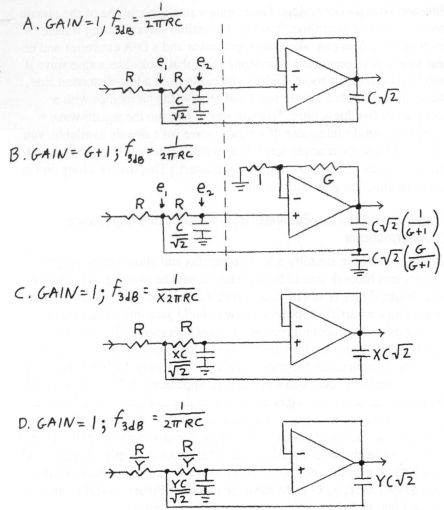

A. $GAIN = 1, \; f_{3dB} = \dfrac{1}{2\pi RC}$

B. $GAIN = G+1; \; f_{3dB} = \dfrac{1}{2\pi RC}$

C. $GAIN = 1; \; f_{3dB} = \dfrac{1}{X2\pi RC}$

D. $GAIN = 1; \; f_{3dB} = \dfrac{1}{2\pi RC}$

out of the design in about the same time it previously took to make one adjustment and check it. Then I reduced the capacitors by factors of ten, making sure no side problems cropped up. This works for high-frequency filters, too. Get the circuit working correctly at a frequency where the op amps are nearly ideal, then start reducing the capacitors and watch the effects of finite gain-bandwidth (and stray capacitance) show up!

If all the impedances in the circuit of Figure 20–1A are reduced by a factor Y (Figure 20–1D), the voltage transfer ratio is unchanged, since voltage transfers are determined by ratios of impedances. The input impedance is indeed Y times lower, but remember, I said *voltage* transfer ratio. This allows the three capacitors in my version to be juggled to a power of ten; oddball precision resistors are easier to find. There are other things that can be done, too, but they take a little more math.

Starting from Scratch

How does one generate an honest-to-goodness, brand-new, out of-the-blue idea? I can see steps leading up to it and numerous alternatives discarded, but I can't explain the spark, the actual jump from the old to the new. Let me walk you through some of my favorite creations:

I was working with elliptic filters, which require zeros. I could not find a single op amp filter section having zeros in any of my books, so I invented my own. (As far as I know; I have since run across two others, but both are more complicated than mine.) Elliptics are relatively easy with passive filters—the impedances of a capacitor and an inductor are equal but opposite at some frequency; cancellation produces a zero. (Skip ahead to Figure 20–4 if necessary.) I reasoned that the differential inputs of an op amp could do the differencing, or subtraction. If the input had two paths to the output, via the two op amp inputs, which had the same voltage divider ratio at some frequency, the output should be zero at that frequency. It would have to be in order to maintain zero voltage across the op amp inputs. The one path could provide the negative feedback required by the op amp, and the other could provide the positive feedback required for filter peaking.

In reviewing frequency-selective circuits, I noticed that the Wein bridge, used as a voltage divider, had phase lead at low frequency and phase lag at high frequency (or vice versa, depending on which end you look at). Somewhere in between, phase shift had to be zero. I did the equations, and, sure enough, at one frequency it looks like a ⅓–⅔ voltage divider with no phase shift. Now I was excited.

This would give me a pure notch, with equal amplitude on either side. This was not exactly what I needed, but I could probably fudge one end or the other to get different amplitudes. I thought of several possibilities; the most promising was that I could split off part of one capacitor or resistor in the Wein bridge, using Thevenin equivalents, and not alter the fundamental properties of the bridge. The end result is shown in Figure 20–2. It seems pretty minimal for all it has to do. There are no obvious nasty requirements on the op amp. But hold on; there is more!

I was fascinated that at one frequency the op amp output did exactly nothing. It was a true zero; there was no approximation in my calculations. Did I really need an op amp, or would any old differential amplifier do? I would still need positive feedback, but why couldn't that work, too? It did work (Figure 20–3)! Heady with success, I pushed on. One by one I took the standard op amp circuits and converted them to "diff-amp" circuits.[10] Who needs gobs of gain? Who needs op amps?

My revelation to the world generated a tidal wave of apathy. Overnight I was propelled from obscurity to oblivion.

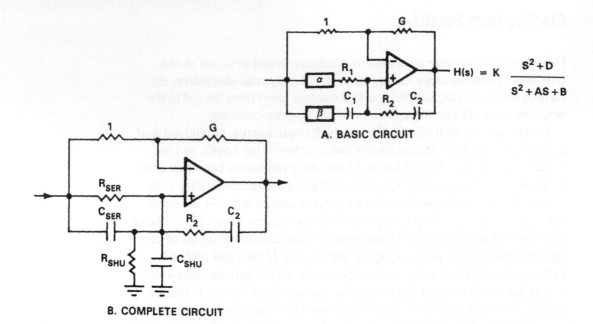

COMPONENT VALUES

HIGHPASS	LOWPASS
LET $\beta = 1; C_1 = C_2 = 1$	LET $\alpha = 1; R_1 = R_2 = 1$
$R_1 = \dfrac{(1 - \dfrac{D}{B})}{A}$	$C_2 = \dfrac{(1 - \dfrac{B}{D})}{A}$
$R_2 = \dfrac{1}{BR_1}$	$C_1 = \dfrac{1}{BC_2}$
$G = \dfrac{(R_1 + R_2 - \dfrac{A}{B})}{R_1}$	$G = \dfrac{(C_1 + C_2 - \dfrac{A}{B})}{C_2}$
$\alpha = \dfrac{G + \dfrac{D}{B}}{G + 1}$	$\beta = \dfrac{G + \dfrac{B}{D}}{G + 1}$
$R_{SERIES} = \dfrac{R_1}{\alpha}$	$C_{SERIES} = \beta C_1$
$R_{SHUNT} = \dfrac{R_1}{(1 - \alpha)}$	$C_{SHUNT} = (1 - \beta)C_1$

Figure 20–2.
Single op
amp resonator
with zeros.
(Appeared in *EDN*,
24 January 1985.)

POSTULATE:
Ideas, although having no mass, do have inertia. They are hard to get
going, but once moving they are hard to stop. This applies to both good
and bad ideas.

Although probably ancient history by now, here's another example of
what can be done with a little cleverness: I needed a fairly sharp 5KHz

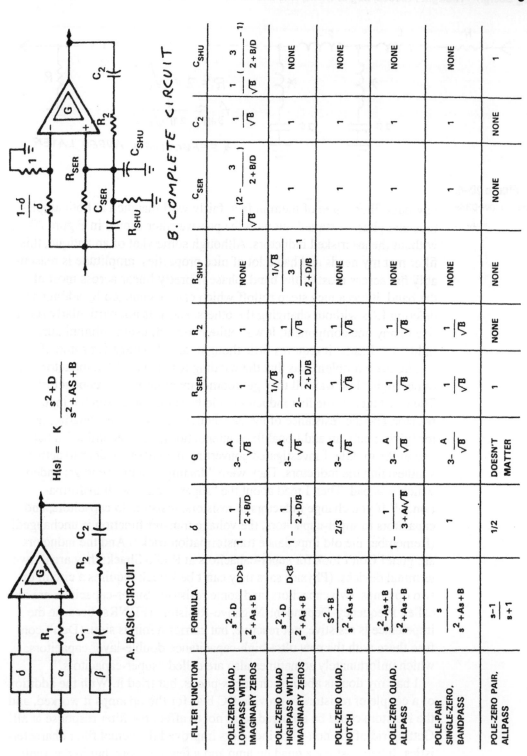

A. BASIC CIRCUIT

$$H(s) = K \frac{s^2 + D}{s^2 + As + B}$$

B. COMPLETE CIRCUIT

FILTER FUNCTION	FORMULA	δ	G	R_{SER}	R_2	R_{SHU}	C_{SER}	C_2	C_{SHU}
POLE-ZERO QUAD. LOWPASS WITH IMAGINARY ZEROS	$\dfrac{s^2+D}{s^2-As+B}$ D>B	$1 - \dfrac{1}{2+B/D}$	$3 - \dfrac{A}{\sqrt{B}}$	1	1	NONE	$\dfrac{1}{\sqrt{B}}\left(2 - \dfrac{3}{2+B/D}\right)$	$\dfrac{1}{\sqrt{B}}$	$\dfrac{1}{\sqrt{B}}\left(\dfrac{3}{2+B/D}-1\right)$
POLE-ZERO QUAD HIGHPASS WITH IMAGINARY ZEROS	$\dfrac{s^2+D}{s^2+As+B}$ D<B	$1 - \dfrac{1}{2+D/B}$	$3 - \dfrac{A}{\sqrt{B}}$	$2 - \dfrac{3}{2+D/B}$ $1/\sqrt{B}$	1	$\dfrac{1/\sqrt{B}}{\dfrac{3}{2+B/D}-1}$	1	1	NONE
POLE-ZERO QUAD. NOTCH	$\dfrac{s^2+B}{s^2+As+B}$	2/3	$3 - \dfrac{A}{\sqrt{B}}$	$\dfrac{1}{\sqrt{B}}$	$\dfrac{1}{\sqrt{B}}$	NONE	1	1	NONE
POLE-ZERO QUAD. ALLPASS	$\dfrac{s^2-As+B}{s^2+As+B}$	$1 - \dfrac{1}{3+A/\sqrt{B}}$	$3 - \dfrac{A}{\sqrt{B}}$	$\dfrac{1}{\sqrt{B}}$	$\dfrac{1}{\sqrt{B}}$	NONE	1	1	NONE
POLE-PAIR SINGLE-ZERO, BANDPASS	$\dfrac{s}{s^2+As+B}$	1	$3 - \dfrac{A}{\sqrt{B}}$	$\dfrac{1}{\sqrt{B}}$	$\dfrac{1}{\sqrt{B}}$	NONE	1	1	NONE
POLE-ZERO PAIR, ALLPASS	$\dfrac{s-1}{s+1}$	1/2	DOESN'T MATTER	1	NONE	NONE	NONE	NONE	1

Figure 20–3. Resonator with zeros with no op amp. (Appeared in *EDN*, 20 February 1986.)

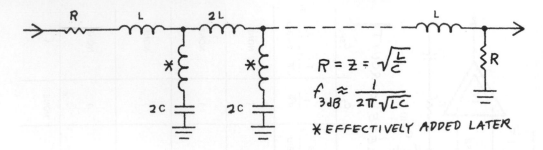

$$R = Z = \sqrt{\frac{L}{C}}$$

$$f_{3dB} \approx \frac{1}{2\pi\sqrt{LC}}$$

* EFFECTIVELY ADDED LATER

Figure 20–4.
Passive low-pass
ladder.

low-pass filter; three of them in fact, fairly well matched in both amplitude and phase. I started out with the passive filter shown in Figure 20–4,
without the asterisked inductors. Although somewhat of an antique, this
filter met my needs and had a lot of nice properties: amplitude is reasonably flat across most of the band, phase is pretty linear across most of
the band, it has a nice steep rolloff which can be changed by adding or
deleting LCs without changing the others, and it is not particularly sensitive to any one component. It was quite compact, using subminiature
inductors. Lastly, it requires little thought, an advantage for some of us.

The main problem was that the winding resistance of the inductors was
rather high. (Just wait till they get room-temperature superconductors!)
The resistance of the first inductor could be subtracted from the input
resistor, and the resistance of the last inductor from the terminating resistor (giving only an additional fixed attenuation), but that still left a bad
one in the middle. I investigated converting the passive ladder to active by
synthesizing the inductors. They were "floating" (neither end grounded),
which was bad. Then I read about the "super-capacitor" transformation.[11,12] If you change inductors to resistors, resistors to capacitors, and
capacitors to super-capacitors, the voltage transfer function is unchanged!
(Remember the old impedance transformation trick?) And the inductors
are gone! Don't look for super-capacitors at Radio Shack; they aren't two
terminal devices. (Physics says they can't be.) Each requires a circuit of
two op amps, two capacitors, and some resistors. Super-capacitors are also
called Frequency-Dependent-Negative-Resistors (FDNRs) because the
impedance is resistive, not reactive, but carries a minus sign. (Don't confuse these with the new ultra-high-capacitance double-layer capacitors,
which unfortunately sometimes also are called "super-capacitors.")

I had my doubts about such hocus-pocus, but tried it. With the addition
of a couple of resistors to provide DC bias for the op amps it worked, and
the resistors could be arranged so as not to affect the filter response at all!
Getting rid of the non-ideal inductors improved the actual filter characteristics. It had cost me a quad op amp and a few resistors, but in the application it was a good trade.

I found a couple more tricks. I had discovered that varying the terminating resistors (in the passive version) would improve one part of the
frequency response curve at the expense of some other part. The resistors
obviously should be frequency dependent. That sounded vaguely familiar.

Sure enough, what I needed was a pair of super-inductors; worse yet, one floating. But that was in the passive version; in the active version they reverted to ordinary inductors! After all that trouble to get rid of inductors, should I put two back in? Yes indeed! It reduced the droop at the bandedge noticeably. Since they added just a minor correction, they were not critical; and the winding resistances could be subtracted from the adjacent resistors anyway!

There was still some "fuzz" on the output signal, as the system used tones at 7.5KHz and 15KHz. Making the filter an elliptic-like would be easy in the passive ladder; you just add inductors in the shunt legs to create the transmission zeros (refer back to Figure 20–4). And in the active version it meant adding resistors, a virtual freebie! (Note that this is not a true elliptic; if you place the zeros at specific places, the humps in the reject band will be unequal.) The zeros did increase the sag at the edge of the passband, but I could minimize this by toying with the two terminating inductors some more.

The overall circuit is shown in Figure 20–5 and the response in Figure 20–6. It has proven quite satisfactory. Note that the precision capacitors are all equal and have been juggled to a nice value using impedance transformation. Passive-derived filters can be hard to troubleshoot, as they cannot be split into independent sections. I had one that met spec, but definitely looked different from the adjacent two. I found an op amp shorted to ground; the sensitivity was so low it worked with a part missing! In the

Figure 20–5.
Active version of low-pass ladder.

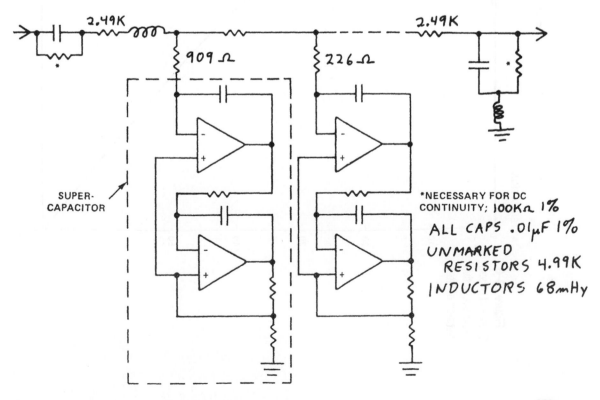

SUPER-
CAPACITOR

*NECESSARY FOR DC
CONTINUITY; 100KΩ 1%
ALL CAPS .01μF 1%
UNMARKED
RESISTORS 4.99K
INDUCTORS 68mHy

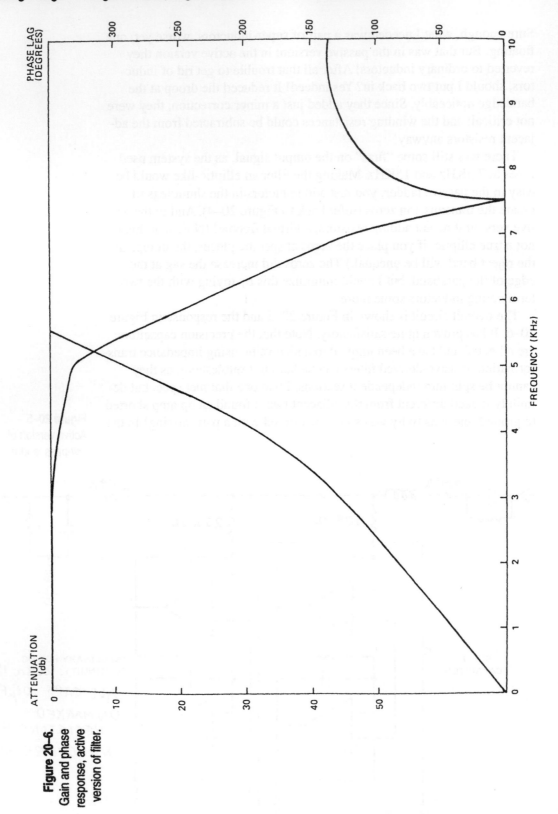

Figure 20-6.
Gain and phase response, active version of filter.

elliptic-like, however, the shunt legs can be easily checked. The voltage across each leg should drop to zero at the frequency of the zero it creates. Once these are working properly, there isn't much left to check.

Would I do it again? Probably not. Today you can get elliptic filters in mini-DIPs, thanks to switched-capacitor technology, and these would probably do the job and have better matching. Engineering consists mostly of trade-offs; you usually don't get something for nothing. However, there are some "freebies." Be on the lookout for them; they are pearls of great price. We are lucky to be in one of the few businesses where new devices not only work better than the old ones, but are likely to be less expensive, too!

Breadboard—The Controlled Disaster

If there is one point that is central to my design method, the focus, the peak, it is the breadboard. I mostly design by making mistakes and then correcting them. I don't particularly recommend this method, but it works for me. I don't think I would have succeeded in a discipline where I couldn't test my ideas. I think on paper; I don't even like to answer the phone without paper and pen in front of me. After the basic design, I think directly on the workbench. That hairy rat's-nest with a bunch of leads connected to it is very important.

Exception: When I work with the explosives people I am a lot more careful. Aside from the possibility of drastically reducing local real estate values and putting one's self into low earth orbit, a single accident can mean the end of a project. Even if no one gets hurt, it is obvious someone could have. There are other areas that require extra caution: high-voltage or high-power systems, radiation, medical electronics, equipment destined for Mars . . .

I recently had to design a system involving magnetics, a subject I had been able to avoid since college. I came up with a system that worked great—in my head. I wound one coil on a Coke can and another on a piece of roofing flashing. They didn't work worth beans. I scratched my head until I remembered why coils aren't wound on aluminum forms. Although aluminum is not magnetic, it is conductive, and those cylinders looked like one-turn secondary windings, shorted. I tried it again with a plastic wash bottle and a glass beaker, and it worked 1000% better. If I had made drawings and waited for cylinders to be machined, I would have wasted a lot of time and money.

When I had to get a signal off a rotating drum, I wondered why I couldn't just insulate the ball bearings on the shaft and run the signals through them. I dug a couple out of my junk bin, rigged them up, and immediately found out why people use slip rings. The bearings generated almost a volt of noise!

Push-in breadboarding strips have been much maligned. I agree they are not the way to go for state-of-the-art design, but for mundane work they are great. I have some laboratory boxes where, if you take the cover off, you find a push-in strip inside! Just use common (engineering) sense. You don't leave inch-and-a-half leads on the components on a circuit board; why should you expect to get away with it on a push-in strip? Ground unused strips. Put an old metal panel underneath as a ground plane (connected to ground, of course). Where possible, connect strips adjacent to sensitive points to guard terminals (low-impedance points that are nearly the same potential). Clip off unused IC terminals; don't count on them being unconnected inside. I recently published an article on a fairly fast circuit.[13] The 'scope trace was unfortunately left out; it is shown here as Figure 20–7. Note that the output rise time is 20ns. It was done on a push-in strip.

At least nine times out of ten the circuit card will work better, which is nice. But watch out:

WARNING: The layout is part of the circuit.

Figure 20–7.
Response of wideband transconductance amplifier differentiator.

I designed a crystal oscillator on a push-in strip. I ran it through a wide range of temperature and supply voltage with no problems. But when we put it in a fancy hybrid circuit, some units intermittently oscillated at a

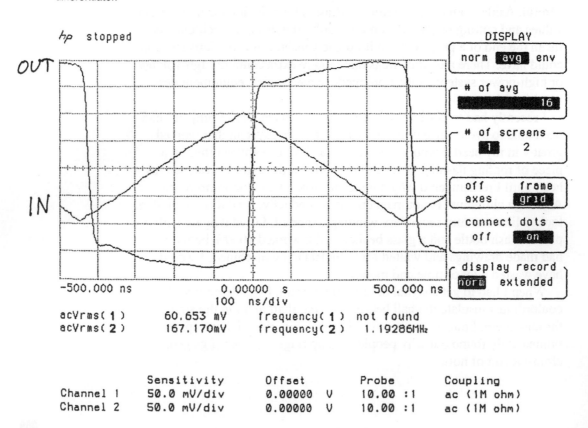

	Sensitivity	Offset		Probe		Coupling
Channel 1	50.0 mV/div	0.00000	V	10.00	:1	ac (1M ohm)
Channel 2	50.0 mV/div	0.00000	V	10.00	:1	ac (1M ohm)

much higher frequency. I couldn't make the breadboard do it. I guessed that the circuit liked the extra stray capacitance of the breadboard. This seemed consistent with the breadboard version refusing to go high. I estimated the stray capacitance on the output, experimentally found the maximum the circuit would tolerate, and picked a value in between. Adding this to the hybrids fixed them all.

Being able to breadboard has a number of career-enhancing advantages. When a question comes up, I can go to my bench and get the answer. And it's the *real* answer, not what I *think* is the answer, or what some *computer* thinks is the answer. The projects that get done are the projects that get funded. The projects that get funded are the projects that get approved, normally done at a managers' meeting held in a room with no windows to the real world, literally or figuratively. Computers these days can make some pretty fancy vu-graphs, but when I pull a breadboard out of one pocket and a battery out of the other and hook them up and put on a demonstration, it's no contest.

WARNING:
Don't put the battery in the same pocket as the circuit.

I did this once, and as I pulled the circuit out of my pocket, the flashbulb went off. The circuit had somehow made contact with the battery and powered up. Bad demonstration.

ALTERNATE WARNING:
Don't put the battery in the same pocket as the car keys, either.

Again, against all odds, the battery made contact. This gave a whole new meaning to the term "hot pants."

On an acoustic link we developed, I had the project manager take the receiver downtown to the sponsor, lay it on his desk next to the speakerphone, call me back at the lab, and tell me what code he was setting it on. I set the transmitter to that code, laid the phone beside it, and sent the tones. A flashbulb went off in the sponsor's face. Now, granted, the phone company does this sort of thing all the time, but it still makes for an impressive demonstration, showing that your idea really works.

In the Gov we are not supposed to work on anything until the money arrives, which, due mostly to Congress, can be nearly at the end of the year. But the deadline for completion never shifts with the delay in funding. Also, it seems that THE ANSWER is always needed by COB (Governmentese for Close-Of-Business). Whether it be the value for a resistor or the meaning of life, it is necessary for a meeting the next morning. Therefore, my systems usually have to be designed with parts on hand; there isn't time to order some. The best I can hope for is to upgrade later. This makes the following item important:

My Private Stock

Table 20–3 gives a summary of what I try to keep on hand, and why. Do *not* try to buy everything in the world, especially all at once. My rules for getting parts are

1. When ordering a part, order extra so I'll have some next time.
2. Order a different part, too; one I think I might need in the future.

Table 20–3 Stock of Parts

Parts	Quantity	Comment
Analog ICs	14 trays	Plus drawer of less-used ones
Digital ICs	8 trays	Basic CMOS
Vacuum tubes	Junk bin	To impress junior engineers
Transistors	2 trays	Plus bin of power transistors, bipolar and MOSFET
Diodes	1 tray	Plus bin of assorted and power types
Resistors, 5%	All values	Kit
Resistors, 1%	5 trays	Semi-sorted; get a kit if you can
Capacitors, 10%	2 trays	All values ceramic
Capacitors, 1%	2 trays	All multiples of 10 plus bin
Capacitors, electrolytic	Same bin	Keep 'em small; avoid if possible
Inductors, xfmrs	Junk bin	Except 1 tray subminiature shielded, multiples of 10
Zener diodes	1 tray	All low-V low-I values; some high-V high-I snubbers
Current-limiting diodes	1 tray	All values I can get my hands on
Crystals	1 tray	Smattering of frequencies; mostly low-frequency TO-5 cased
Pots	1 tray	TO-5 cased, single and multi-turn, good cermet, most values
Other	Small quantities	Low-power SCRs, TO-5 relays, LEDs, opto-isolators, networks of matched resistors and capacitors, flashbulbs, DIP switches, subminiature fuses, Sonalerts, "black blob" miniature power supplies ±5V, ±6V, ±12V, ±15V; aspirin

Note: Trays are plastic 18-compartment 7" × 11" × 2".

3. Save old parts in junk bins; clean them out only when the bin overflows or the parts become unrecognizable.
4. Save old breadboards in a drawer. If I have ever used a part before, it's in there somewhere.
5. Take advantage of free samples, within reason.

If It Doesn't Work, Take Two Capacitors and Call Me in the Morning

When students bring me circuits that don't work, they are usually surprised that I am not surprised. With all the little details that need attention, which I am not good at, I don't expect a circuit to work the first time. In fact, I plan on it. I put in terminals for observing critical points, and jumpers for separating stages and opening feedback loops. A circuit board always gets a revision, so you can take them out later. Plus, it gives you spaces that can be commandeered for those bypass capacitors and protection diodes you just found out you needed.

ASSERTION:
Circuits don't just fail; they fail in a certain manner, in a certain spot.

I grill the student: What did it do or not do? Was there an AC signal on the output? A DC level? What were the power supply readings?

Pass the Pease, Please

Bob Pease has written a complete book on troubleshooting.[14] I will mention only a few things I have had the misfortune to become acquainted with.

The most common problem is that the power supply is wrong, not hooked up, or simply not turned on. CMOS will often power up from the input signals via the protection diodes and work to a certain extent. The symptom is that every signal exhibits remnants of every other, since the power supply depends on the signals. Check the supply voltages, on the card, right at the trouble spot, with a voltmeter *and* a 'scope.

The next most common problem is that the circuit is not wired according to the diagram! Connections and/or parts are wrong or missing altogether. Do not expect the circuit to work with even one mistake; mother nature is unforgiving.

When I had checked out the encoder signal generator and power amp driver for the magnetic system mentioned earlier, I hooked them together and threw the power switch. Red lights flashed on the power supply. I had visions of exotic nonlinear oscillations resulting from the high-powered output signal getting back to the sensitive crystal oscillator, and how I was going to apply my superior expertise to cure them. The problem? In my jerry-rigged setup the power supply leads had gotten smashed together, and the insulation eventually gave up. (Teflon creeps badly with time.) I felt rather sheepish when I separated them and everything worked.

And the list goes on. 90% of the time the problem is a stupid mistake. Assume you have made one. But, oh-oh, I just said a dirty word.

Assumptions

EXPERIENCE:
Assumption is the mother of [unprintable].

The first rule on making assumptions is: Don't. Find out for sure if you can. If you can't, proceed, never forgetting that your work is based on something that may be wrong. If things just aren't working out, it may be that the assumption you made is invalid. While your circuit is doing nothing is a good time to review your assumptions, and also your

Approximations

Approximations are the lifeblood of engineering, but they can also be the death of a system. As above, Don't, unless you have to. $\frac{22}{7}$ is a cute approximation for pi, but punching a button on any scientific calculator will get you the actual value to a disgusting number of decimal places. Pi = 3 is a poor approximation, for emergency use only. However, should you get into trouble using it, the appendixes give proofs that pi = 2 and pi = 4. Pi = 3 may be obtained by averaging the two proofs. This will distract your supervisor long enough to forget about writing up your deficiency report. These proofs should also teach you two valuable lessons:

1. Don't believe everything you read.
2. Don't deal with disreputable persons.

When you do have to approximate, keep the fact not too far down in your memory. Are the approximations cumulative—piling up on you? There is a tendency to make an approximation that is in itself reasonable, but then to proceed as if it was absolute truth.

In my PhD thesis I calculated the signal and noise frequency spectra based on the best models I could come up with, and derived the optimum filter. It came out a very narrow spike, infinitely steep on the upper side. I traced the causes to an approximation I had made in the noise calculation to make the math doable, which made the noise spectrum fall off extremely rapidly with increasing frequency; and an assumption about the target that gave a spectrum less steep, but with a precisely defined maximum frequency. If the real target spectrum was actually a little bit lower than I had estimated, the "optimum" filter would miss it completely. I settled for a flat-topped bandpass, which worked fairly well. It did pay to make the lower cutoff as steep as practical, as the noise spectrum was indeed quite steep (but not as steep as my model indicated); keep the cutoff frequency as low as possible without admitting a horrendous amount of noise; and settle for what signal was left in the resulting passband. The inaccurate analysis did offer a possibility for improvement—

rather than try to make the model more accurate to reflect the poorer results, try to alter the system to be more like the original inaccurate model and actually achieve the optimistic results!

Ground: As Solid as the San Andreas Fault

Here is everybody's favorite approximation. Ground is one of the most useful concepts we have, but it is only a concept. You can define one infinitesimal point on the card as zero voltage, but all others are at least slightly different, possibly seriously different. Entire chapters have been written on grounding; probably books. Suffice it to say that there are two popular methods, which paradoxically are virtually opposite! One is to use only the one point as ground. Each circuit must have its own individual ground lead to that point so ground current from no circuit flows in the ground lead of any other, inducing an undesirable voltage. This is generally impractical, but useful in some special cases. This is the idea behind "sense" leads on a power supply, "four-terminal" measurements on an impedance bridge, and separate analog and digital grounds.

I prefer the brute-force approach—the ground plane. One side of my cards will be near-solid copper. Power supply buses may be integrated; a well-bypassed supply looks like ground to AC signals. Short leads may be integrated; long leads should be run around the edge. A copper sheet is about as low an impedance as you can get, at least at any temperature you would care to work in. Plus, there are a number of side benefits. Ground, the most common (no pun) connection, only requires a feedthrough. Leads mostly have capacitance to ground rather than to each other, the latter generally being harder to deal with. The cards are basically self-shielding; electromagnetic interference isn't going to get any further than the surface of the next card.

Clean Thoughts

Just as two adjacent leads on a circuit board make a dandy capacitor, two adjacent leads on a *dirty* circuit board also make a resistor. Even a flux that is initially non-conducting may carbonize after repeated overheating, and one can make resistors out of carbon. I hereby lay claim to having invented the light-emitting circuit board. Also the smoke-emitting circuit board. Not a component, mind you, the board itself. I figured the grunge accumulating from numerous changes didn't matter because it was from power supply to ground, but apparently even that has its limits.

REMEMBER:
Smoke is one of the seven warning signs of circuit trouble.

There is a lot of argument about cleaning boards. They are working on new "no-clean" fluxes. I hope they work better than the old ones. I built a Heathkit depth finder which had specific instructions *not* to clean the board, which I thought rather optimistic for electronics that had to operate in saltwater atmosphere. It worked for a day and a half. I took it apart and cleaned the board. It then worked until something mechanical failed years later.

There is a possibility of solvents leaching contaminants into non-hermetically-sealed packages, such as epoxy DIPs. I have never experienced this, but I do not submerge the cards, just brush/spray them off. I *have* experienced the problem with switches and pots, even the "sealed" types. Keep fluids away from them, or add them after cleaning.

My personal favorite cleaning method is acetone followed by ethyl alcohol. In spite of the dire warnings on the label, acetone is pretty innocuous. At the dispensary (that's Navy talk for first-aid station) they clean adhesive tape goo off with acetone. And if things are going really badly, you can drink the alcohol instead of wasting it on the board.

Covering a mess with plastic spray doesn't get you off the hook. Water molecules do get through plastic coatings. If the board is clean, it will be distilled water and probably not hurt; but if it is dirty, you just get plastic-coated slime.

Instrumentation—Your Electronic Eyes

Of utmost importance in troubleshooting is proper test equipment. Table 20–4 gives a list of items I would not want to be without. Herewith some further comments: a friend of mine was actually told he could have only one piece of test equipment. (He quit.) If I had only one choice, it would be a high-speed variable-persistence (memory) analog 'scope. It is your best shot at seeing what's really going on. Digital 'scopes have some excellent features, but keep in mind that you are only seeing a processed version of part of what happened some time ago. If there is any doubt, connect both analog and digital 'scopes to the point in question. If they don't agree, at least one is lying. If the trace on one changes significantly

Table 20–4 Stock of Equipment

Instrument/Equipment	Comment
Analog 'scopes	#1 measuring instrument; fast variable-persistence is best
Digital 'scope	Pretty pictures, but rely on #1
Spectrum analyzer	Mine does filter responses in one sweep—nice
Printer/plotter	If it's digital, it should give you a printout
Voltmeter, DC, digital	Good accuracy, but remember, it's an average
Voltmeter, AC, true-RMS	Digital plus analog meter, which is great
Function generators	AM, FM, sweep, noise, pulse, synthesized, variable-phase
Filters	Butterworth, Bessel, Elliptic
Counter/timer, LCR meter	See text
Attenuator box	1dB calibrated steps; stop fiddling with pots
Power supplies	Constant-current and regular
Temperature chamber	See text
Microscope, binocular	Amazing, the crud you see with 8× magnification
Calculator, scientific	Cheap
Slide rule	Backup for above

when the other is disconnected, it was influencing the circuit unduly. I had a circuit that appeared to have a low-level 25KHz oscillation; it disappeared when I turned off the digital 'scope. If you have a glitch that appears at the beginning of the sweep on an analog 'scope no matter what point you probe, suspect that it belongs to the 'scope.

A spectrum analyzer is handy for a lot of jobs, but know that it does not really compute a Fourier transform, or even a FFT, but a DDFT—a Doubly Discrete Fourier Transform, which has some limitations. Digital meters give you so much apparent accuracy they can be misleading. They can define only one parameter, and have to average that one. Is a 1.000V DC signal meaningful if it has 1V AC noise on it?

AXIOM:
The neater the display, the more likely it is hiding something.

I like synthesized function generators, with dial option if possible. I know the frequency is right where I set it. I have four, and have trouble keeping one in the office. Pulse generators should have variable rise and fall times to reproduce the real signal accurately. Laboratory filters are indispensable; again, I keep both continuously variable and precisely settable. Any old-timer who had to fiddle with an impedance bridge appreciates modern LCR meters. Read the manual, which should point out that it is simply a tool using a particular method to determine a parameter which is only a definition. Mine will measure inductors two ways, and the numbers are usually quite different. Parts do get damaged, or even mislabeled, once in a while. A capacitor labeled "100" can be either 100pf or 10 (followed by "0" zeros). Also, the only thing you can be sure of about a 0.01-microfarad capacitor is that it is not exactly 0.01 microfarad, or at least not for long. Put some heat on it and watch it change. Which brings up the most controversial item:

I keep a small temperature chamber right in my office, and do not consider a prototype circuit design finished until I have used it. Temperature is generally the best way to test the sensitivity of your new circuit. If you don't do it, mother nature or the air conditioning man is going to do it for you. Spray-freeze and soldering iron tips are good for isolating an offending part, but too crude for anything else. After all, most parts will fail if you melt them. If nothing else, put your circuits in the refrigerator, bring in your blow drier (or your wife's if you have no hair left). Here on the East Coast, where the temperature is usually disagreeable, I used to hang circuits out the window.

All the equipment in my room adds up to less than half of my yearly salary plus overhead. Do try to explain to management that good equipment will more than pay for itself by increasing your productivity, and I hope you have better luck than I did. When I finally got a spectrum analyzer after years on a project, I took one look at the system output and threw away all my test data. The inductance of a transformer winding was resonating with a coupling capacitor, and my spectrum that should have been flat had a huge hump in it. Of course, had I suspected I would have borrowed an instrument or checked it another way, but that's the point: without the analyzer I never suspected.

Classic case of false economy: In developing a system, the one potential problem we were unable to check was hermetically sealing the special hybrid package. Management wouldn't approve the purchase of a $10,000 sealing machine. Guess what gave us the most trouble, being the last problem solved before successful production—achieving a hermetic seal. The hidden costs of the delays involved are hard to quantify, but I figure it cost us over a million.

On Disproving the Laws of Physics

True story: I designed a system that worked from a battery, 28V @ 50mA. Years later we wanted to adapt it to another system whose battery was 14V @ 2.5mA, a voltage reduction of half and a current reduction of 20. (The battery was special, and hence a given; a last resort was twin batteries.) I thought I could do it with minor improvements rather than a complete redesign. Not redesigning would have several advantages: a lot of retesting would not be necessary; we could be sure it would fit in the special hybrid packages; the layouts could be reused, at least as a "mule" for demonstration. New low-power op amps, comparators, and voltage regulators had become available in the decade and a half it took DOD to get the original system into production, which were of some help.

PREDICTION:
If an ideal op amp is ever produced, it will inexplicably be unavailable in a quad.

I went through each separate circuit, looking at every part, to minimize power drain. I discarded two of the three regulators, reducing current drain and saving voltage headroom. Some adverse interactions occurred, but were cured with better design and lots of capacitance here and there (my mythical aerosol can of "spray capacitance"), neither of which cost current. A lot of impedances were unnecessarily low. Savings snowballed; a lower-power circuit had a lower input current, which could use a larger biasing resistor, which put less load on the previous stage, which could then be lower power, etc. I thought I had it solved, but had some discrete boards made to be sure. The first board exceeded 2.5mA considerably. I rechecked it section by section. I made another board, which was no better. Finally I realized the system required significantly more current than the sum of its parts! That pointed to an interface problem, and I soon found it.

The oscillator (the one mentioned earlier that needed stray capacitance) worked fine, but the rise time was slow. It was driving CMOS, which draws no current in either digital state, but a lot in the time spent in the linear region in between. The obvious solution: insert a Schmitt trigger. The not-so-obvious non-solution: prefab Schmitt triggers don't do the job. The input impedance is infinite and the output switches cleanly, but something in between is still conducting. I devised my own (Figure 20–8) out of the only CMOS logic circuit I could find where I could get at the individ-

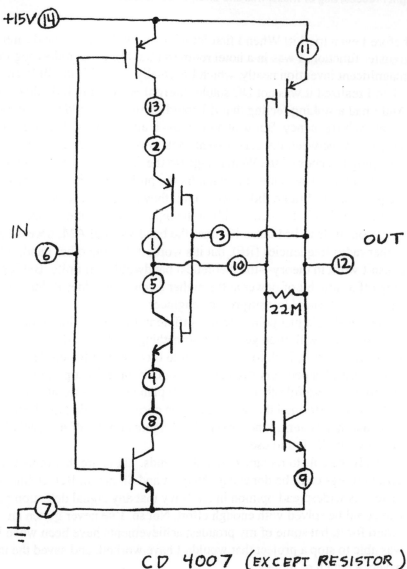

Figure 20–8.
"No current"
Schmitt trigger.

CD 4007 (EXCEPT RESISTOR)

ual transistors, the 4007, and one resistor. The output inverter is unfortunately hard-wired between the supplies, and the resistance must be chosen according to the frequency, but it fixed the problem. Adding an IC is easier in a hybrid than a circuit board; the resistor chip was as big as the IC!

The new circuit did introduce another problem. It switched close to the rails, and the op amp driving it only got within about a volt of the rails. This was fixed by adding forward diodes in series with the power supply and ground, effectively reducing the supply voltage of the 4007. There was still plenty of output swing to drive the rest of the CMOS.

Having Achieved True Failure

OK, so now you've checked everything, and your circuit definitely is not going to work. Don't give up just yet. (There's always tomorrow.) Is there some spot where you approached something the wrong way, maybe even backwards? My first version of the filter of Figure 20–2 failed,

369

before I even tried it! When I first found a circuit that gave the desired transfer function, I was in a hotel room (on travel). I was redrawing my magnificent invention neatly, which I normally wouldn't do till later, when I realized it was not DC stable. Aarrgh! How could it do this to me? And I had a sinking feeling that if I could fix that, it would then be unstable at high frequency. But wait a minute—it already was! That rang a bell somewhere between my ears. I went back to work on it, and sure enough, swapping the ends of the Wein bridge (to the form of Figure 20–2) fixed *both* problems. Can some alteration fix the problem without destroying the purpose? FM radio didn't work until they realized it took more bandwidth than simply the bandwidth of the input signal or the frequency deviation. In fact, FM takes 10 times the bandwidth of AM, necessitating higher radio frequencies (RF), but it's worth it. On the other hand, it still doesn't work in theory—the theoretical bandwidth is infinite. But lopping off a little bit of power at the higher frequencies doesn't hurt appreciably, which shouldn't surprise an engineer.

Failure should just point you in a different direction. This is an iterative process which will often get you to something workable. If you wind up where you started, look for some point to break out of the circle.

It doesn't happen very often, but the circuit, or at least part of it, may do something useful other than what you planned. The circuit I spoke of earlier that performed the wrong function correctly could have been useful in another system. It never was, but I figure about half my oddball ideas eventually found use.

Lastly, be able to recognize real dead ends. There are theorems that say certain things can't be done; e.g., Nyquist and Shannon. Before Shannon, there was widespread opinion in the Navy that any signal detection problem could be solved with enough effort. Not so. I've never gotten an award for it, but some of my proudest achievements have been when I was able to stop a project that wouldn't have worked, and saved the taxpayers a bunch of money.

If it can't be done, what is the nearest thing you *can* do, and is it useful? When the Ground-Fault Interrupter (GFI) first appeared, I couldn't figure out how something electromechanical could open the circuit fast enough to prevent electrocution. The answer is, it can't, if you provide the initial path to ground. It instead hopes to detect a prior leakage to ground and open the circuit *before* you touch something that should be ground but has become electrified. It's a lot better than nothing!

Failure Analysis

Don't automatically throw deceased parts in the trash can. Failure analysis laboratories can do some amazing detective work. There is always a slight chance the problem is not your fault! When power MOSFETS became available, we had a rash of failures, even though we were not exceeding the ratings. Our Failure Analysis Lab detected something going wrong in the substrate, and found a publication detailing the problem. The inherent reverse diode was actually part of a transistor which self-destructed at high current. The manufacturer cured the problem, but the part number didn't change; one had to look at the date code. Note: It is very difficult to remove an epoxy case without destroying the device itself. You may have to use units in a ceramic package or metal can just so the lab can get them apart.

Look, Mom, No Smoke!

Now your system (apparently) works. The next step and the one following are important, but one or both often get neglected or omitted entirely in the rush of things. In the Gov it's the end of the fiscal year when the money expires; in industry I gather it's the time-to-market goal.

How *well* does it work? Sure, it meets the specs, but that's not the whole story. I like to say, "Play with it." That turns some people off; rephrase it if you like. What I mean is exercise it, use it, misuse it, abuse it, duplicate it (even if you only need one). A lot of bad results can show up, and better you find them than someone else.

1. It can't be reproduced.
2. It only works sometimes.
3. It works for a while, then quits.
4. It exhibits peculiarities under certain conditions.
5. It fails when the temperature or supply voltage varies.

Believe bad-looking data points, unless you have a very good reason not to. I had a circuit that worked, but required more battery voltage than I thought it should. I checked voltage drops, and found momentary peaks of 4V (no decimal point) across a Schottky diode, which I had used to minimize voltage drop!, paralleled it with an ordinary silicon diode, and gained 3V on the allowable battery voltage range. Why not just use a heftier Schottky? Reverse leakage was a problem.

Another time I was plotting a filter response which looked OK, but I noticed the amplitude was way down. In reaching around to the back of the signal analyzer, I had connected to "Source Sync" rather than "Source Out"! The former was a short pulse, having a flat spectrum like the expected pseudo-noise, and naturally in sync with it. The result was correct in this case, but in turning the drive amplitude all the way up to get more output, I could have overloaded the circuit. Just looking at the output may not be good enough, due to the "hidden node" problem. This most commonly occurs in multi-stage low-pass or bandpass filters. The output looks clean, but back along the line some stage is overloading. (I call this "going digital.") You will get signals showing up in parts of the spectrum where they don't belong. I put the highest-Q (peakiest) stage last. It is the most likely to overload, and I should see it. The problem is particularly insidious in filters like that of Figure 20–5, where some op amps are not in-line, but off to the side. Check every op amp output and input to be sure its range is not being exceeded.

Don't assume it will get better in production; it usually goes the other way.

RULE:
If it only happens once, it might be a mirage. If it happens twice, it's real.

Intermittent failures are the nastiest to locate. First, duplicate the conditions under which it happened *exactly*, including variables you don't think should matter. I can cite instances where the time of day had an effect. If duplicating conditions doesn't cause the problem to reappear, start varying things, *everything*. The circuit may be marginal with respect to some parameter.

THEOREM:
Zero is the reciprocal of infinity. Infinity does not exist; therefore neither does zero.

This has some practical ramifications: in our world a voltage typically decays exponentially. After a few time constants it's pretty far down, but it is not zero. If you started with 10kV, you better wait a lot of time constants, or you may get some do-it yourself shock therapy. Secondly, once you get below about half a volt, semiconductor junctions cease to conduct and capacitors may stop discharging, especially electrolytics, which have a tendency to recharge some all by their lonesome. One result is a circuit which always works right the first time it is turned on, but sporadically if power is turned on and off. It may be getting preset into a wrong state, requiring some bleed resistors across capacitors.

A related problem is that we usually trust to luck what happens when the power is turned on or off. That is, until we experience an unignorable number of failures. If a circuit works once or twice and then fails, the problem may be large capacitors charging or discharging into a sensitive node. Most recent devices will tolerate rail-to-rail swings, but observe that with the power turned off that is zero volts! Connecting low-impedance sources with the power supply turned off can damage ICs, even those with protection diodes.

I have seen so many "impossible" occurrences I long ago lost count. Once we had a receiver apparently trigger on the wrong code, a serious problem, one that "couldn't" happen. We kept pinging, and after awhile it happened again. We tried some other codes with no problem, then returned to the original code, and it happened again. We finally realized it only happened when the receiver had an "8" in the code where the transmitter was (allegedly) transmitting a "9," and then only sometimes. It turned out the shipboard generator wasn't quite up to the task. The line voltage would drop below the spec on the power supply during transmit, its output would drop out of regulation, the VCO could not achieve its maximum frequency, and the transmitted signal was somewhere in between an "8" and a "9"! It was fortunately a temporary generator and the problem was solved by taking some other equipment off the line, but we did add a note to the operating procedures to make sure the line voltage was up to par.

Be aware of three realities which are similar, but different:

1. What you want to see.
2. What you actually see.
3. What is actually there.

Try to keep toward the bottom of the list.

Increasingly I get failures in devices I have purchased, look closely at them, and spot an obvious flaw that would have been caught in a reasonable testing program. Testing is expensive, so do it as efficiently as you can, but don't skip it. Also, be aware that we engineers have an inherent problem with testing in that we naturally handle our products with respect, not abuse. Loan it to a college student; send it through the U.S. mail.

HERESY #1:
I do not use ESD protection when breadboarding.

If I am designing a part that is sensitive, I want to know it as soon as possible. Actually, in 50-year-old buildings in the Washington, DC climate I have never had a problem show up. Production? Different story. I use as much protection as possible for equipment going out to a customer. I may or may not run ESD tests on the product, depending on the application. Many of my devices don't get handled after assembly.

CONUNDRUM:
Is a good device useful if it can't be tested to show that it is indeed good?

The military generally says no, but there are obvious exceptions. Very limited testing can be done on explosive devices. I am sure Chrysler doesn't test each air bag. NASA cannot completely duplicate the lunar environment.

HERESY #2:
I have a dislike for self-test indicators.

It's a great idea, but they often inspire false confidence. Many don't check much more than the battery. Often it's impossible. The only way to really test a smoke detector is with smoke. Pushing the button tells you the battery can sound the buzzer, which is nice, but it should be labeled "battery test."

A particular problem I deal with continually is this: a fuse (electrical type; the ones that set off explosives are spelled "fuze") makes a dandy compact, inexpensive, one-bit, non-volatile write-once one-way memory

(WOOWM?). It is great for "sterilizing" explosive devices, performing basically the same function they do in civilian life. The problem is how to test it. The sterilize function must be tested on each unit, somehow; this is a requirement for all safety features. To do this without actually blowing the fuse and dudding the device, we test the cards using a "constant-current" power supply. These limit at a precise current; set it below the fuse rating and it will not blow. (Most fuses blow around 100% overload—twice the rating.) Note: An ordinary "current-limiting" power supply will not do. It limits only after the monster output capacitors have discharged, by which time your fuse is probably blown, or worse yet, damaged. For the same reason you have to be careful about how much capacitance is in the circuit.

Two additional cautions: The material inside the fuse is very similar to the solder you are using on the outside. Overheat it and you change its electrical and/or mechanical properties. Also, since it takes a certain amount of power to melt the fuse, the low-valued ones can require as much as 8V across them to blow. They are not meant for 5V supply circuits!

I like constant-current power supplies for testing ordinary circuits, too. If you make a mistake, they are far less likely to damage parts. Also, you find out exactly how much capacitance you need on the power supply bus. A marginal circuit may work on an ordinary power supply, work on a fresh battery, then fail as the battery discharges and its internal resistance goes up. Turn the current setting down until the voltage starts to drop. Does the system oscillate or latch up? Turn it off and on and see whether the system will power up on a marginal battery. I use the Hewlett-Packard 6177 constant-current power supply. Keithley also makes some. I have four, and they are often all out on loan. In an emergency you can usually fake it using a constant-current diode, which is another device I use a lot.

One Last Look

Stand back from your design and evaluate it objectively, as if it were someone else's. What did you set out to do? How did the objectives change along the way? Is there now anything that needs reevaluating? It is embarrassing to have someone point out parts that are no longer needed, and it's happened to me. Can that which you have accomplished be applied to something else? Or extended further to create something new? Half my patents were side issues, "bootlegged" off my assigned work.

The Job's Not Over Till the Paperwork's Done

Engineers are often so enraptured with their creations they don't bother to advertise. I speak as a guilty party.

OBSERVATION:

If you build a better mousetrap, the world may beat a path to your
door, but it will be to demand a contribution for fatherless mice.

At the very least, document your work to the extent that someone else
can figure it out if you lose an argument with a semi on the way home. If
you have trouble writing it down, do you really understand it?

HERESY #3:

I do not write everything down in bound notebooks (or computers).

80–90% of my ideas are worthless; why let them pile up and make it
more difficult to find the good ones? On the latter, I keep the first sketch
(for patent purposes), the most recent (for obvious reasons), and just
enough in between so I can retrace the evolution of a design if necessary.

I use a vertical, time-dependent filing system, otherwise known as
letting it pile up. How long ago I referred to it determines how deep it is
in the pile. If the pile gets too deep, it will avalanche of its own accord.
Then I sort it: unnecessary and outdated stuff (most of it) into the trash;
important stuff into loose-leaf notebooks and a couple of alphabetical
files.

PROBLEM #1:

Books must be put on a shelf. They make the pile grow too fast,
and they hurt when they land on you.

PROBLEM #2:

For a while I was afraid to touch the pile because occasional noises
indicated something was living under it.

I also use a secondary filing system for administrivia: I paper my walls
with organization charts, purchase requests, time sheets, etc. I have received complaints that this made the room uglier, but that is a weak argument for steel walls painted battleship gray.

Beware the Neatniks

When documentation becomes an end unto itself, it becomes self-
defeating. The following people will be among the last to board the
lifeboat if I am in charge:

People who are more concerned with how pretty the diagram looks
than whether it is understandable. In the past few months I have wasted
time both because: (1) a dot on a four-way connection was almost invisible on the Xerox and the leads didn't get connected, and (2) where there
was only a crossover with no dot, the leads got connected anyway. If

there is any doubt, I put a hump in a crossover and use only "tee" connections plus dots anyway.

Drafting types have no appreciation of how a circuit works; they can't be expected to. Don't allow them to show the input resistor next to the output with lines running clear across the page because there was a little more room there. The circuit has to be arranged logically; the diagram should be arranged similarly. It recently took me a long time to figure out a diagram for a simple system, which was particularly aggravating because I had designed it. The lines to a resistor crossed, like it was twisted. Another lead crossed itself; it did a loop-the-loop. Computer drafting seems to have made this worse, but I'm not sure why. Insist on good drafting. A few people can see how a circuit works no matter how badly it's drawn, but most of us need all the help we can get.

I had seen a particular narrowband filter circuit in several books, but it didn't appeal to me. In fact, it wasn't obvious to me how it worked. Eventually I got around to analyzing it and found out it was a circuit I was already using! I draw it as shown in Figure 20–9. If you are used to the other way, you may not recognize this one. I prefer this way because I think it makes what is going on more obvious, at least if you are familiar with the properties of the bridged-tee. The circuit has unity gain at resonance only by virtue of the input being brought in through a large resistance to a low-impedance point. (Observe the note that the circuit is intended for high-Q applications.) The true circuit gain as far as the op amp is concerned is greater than Q squared! I was getting poor performance using a 741 and didn't understand why until I appreciated this fact. 7.5KHz and a Q of 50 might sound like 741 stuff, and *separately* could be, but here it means I needed a gain-bandwidth significantly greater than 35MHz! I got noticeable error with the fastest op amp I had. The op amp still has to be unity-gain compensated because at high frequency (where oscillations will occur) the circuit has 100% feedback. These properties were not obvious to me with the circuit drawn the other way.

Similarly, layout personnel are going to simplify their job, not yours.

OBSERVATION:
If there are two leads on a board, the layout person will want a two-layer board.

Insist on a ground-plane board for serious analog work. One of the times my request was ignored the board proved absolutely hopeless. We had to scrap it and start all over. The opposite may happen if the drafting room needs work. I got back a board with three identical channels laid out three different ways. So much for matching . . .

Next is technicians who tie-wrap all the leads together tight as a banjo string. One single lead will tighten first, and together with the solid mass it makes a high-Q system, and if the system is hit with a vibration at its natural frequency, that lead is a goner. Loose leads vibrate individually, usually not much, and bang against each other if they do, damping the

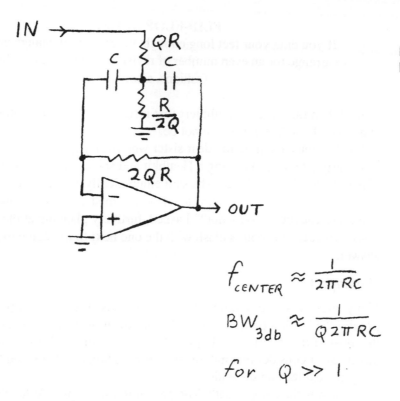

Figure 20–9.
Narrowband filter.

$$f_{CENTER} \approx \frac{1}{2\pi RC}$$

$$BW_{3db} \approx \frac{1}{Q 2\pi RC}$$

$$for \quad Q \gg 1$$

oscillations. They also have a lot less coupling capacitance. I once designed drivers for a high-voltage electroluminescent display which worked until all the leads were neatly lashed together. Then when one segment lit, they *all* lit. It was decreed by the powers that be that leaving them loose was unacceptable, so I finally had to use ribbon cable with every other lead grounded.

My branch once built a computer, an entire 6-foot rack back then, using only black wire. It looked very tidy, but it was nearly impossible to trace anything. We almost gave up and rewired it before we got it debugged.

Next come supervisors, the ones who think your desk and lab bench should be cleaned off at the end of the day. If I took all the piles on my desk, combined them, and squared them up, I would have trouble finding anything. But I usually remember which pile I put things in, and if a corner of a page is sticking out with section D of an LM339 on it, I know what that drawing is. For some reason I can remember *that*, although I never seem to remember that the pin numbers go *counterclockwise* looking at the top of a board . . .

Particular emphasis for those who "correct" my reports which are already correct. Computer spelling checkers and secretaries come to mind. My "baseband" signal became "basement." Once a secretary switched something in a way I didn't like, but she was insistent and it wasn't worth fighting. Apparently the delay to the next revision was longer than her memory, because she then switched it back.

FLIP-FLOP RULE:
If you drag your feet long enough, the rules will change back,
so arrange for an even number of revisions. Applies to clothing
fashions, also.

And then there are the quibblers that insist that "low-pass" must be hyphenated but "bandpass" cannot be.

Lastly I must mention my dear sister who nearly gave me heart failure at an early age. I was building a Heathkit and she came over to inspect. "That doesn't belong there!" she said emphatically, pointing to a resistor I was about to solder. I was mortified. My first mistake on a kit, and my kid sister had caught it. "Why not?" I asked, furtively glancing at the instructions. "Because the colors clash with the one next to it!" came the logical answer.

The Report

Many engineers hate writing reports, but I enjoy it, mostly. Beginning is usually the hardest part, so I don't. What I mean is, I start in the middle, the meat of the report. It is simply what I have done, so I just write it down. But I consider that merely an outline. Then I go back and fill in the gaps and tack on the ends.

What was the assignment? Who gave it to me? When? Where? (The four Ws) I try to forget that I am at the end of the project looking backward with hindsight, and go all the way back to the beginning, when I first began to think about the project. That is where the reader is. It can be difficult for experts to teach because they know the subject reflexively. Remember, you *are* the expert on this particular thingamabob, possibly the only one in the world, because you just invented it! This part constitutes your introduction.

How did you do it? (The big H) How does it work? How do you *know* it works? Also, what was rejected? What didn't work? Analyze it. (A as in Aardvark) This expands the body of the report.

Then Terminate it. (Gimme a T) By now, if the report is well written, the reader has reached the same conclusions you did, but list them anyway. Managers often read only the introduction and conclusion. I had one line manager who sent every report back with red marks all over the first three pages, but none thereafter. That was obviously all he read. This may be why some editors want a summary right at the beginning. Conclusions should include recommendations and plans for future work. The end may not be the end! Put all the letters together and you get W-W-W-W-H-A-T?—which is not a bad description of what the report should answer.

Bureaucracy

Engineers should not be content with theories about how nice things should be, but should apply their talents to making things work in the real world. Bureaucracy *is* part of the real world. Plan on it, just as you would keep in mind that your system is probably going to have to fit in some sort of package.

Horror story #1: I needed some shielding between the transmitter card, which was generating 100V, and the receiver card, which was detecting 1 microvolt, a 160dB difference. Not surprising. It was fairly easy: put them at opposite ends of the rack, and leave unused card slots adjacent to each, inserting empty cards. Any old card would do, since I used all ground plane cards. What was *not* easy was convincing the documentation department. They could not handle an undefined card. They made up a drawing for cutting and drilling a blank piece of board material. This shorted all the pins on the connector together, including the power supplies, which were bussed to every slot. So they relief-drilled all the holes, but then the ground plane wasn't connected. They fixed that, but I'm sure that had not ground been the middle pin, some would have gotten mounted backwards and failed. I would have saved money in the long run by laying out and fabricating a card with nothing on it. Using defunct cards would have worked, too, but it did occur to me that every card has to have a corresponding testing spec, and I could envision having to write one that said, "This card shall *not* function properly in any of the following ways: . . ."

Horror story #2: The same system used only ¼W 5% resistors. In one spot I needed ½W, so I simply paralleled two. This was not acceptable to the powers that be; I was obviously wasting a resistor. So another spec was called out, the parts list changed, and the circuit board redone. Then progress stopped half a dozen times while someone located me to ask if one callout or one pad spacing was really supposed to be different from all the rest. Worse yet, some testing should have been redone since the value had changed slightly (for some reason values have been carefully arranged so half a standard value is almost never another standard value), but they were too busy doing the paperwork to notice.

And then there are those who insist on assigning arbitrary numbers instead of codes. Rev B could have been yesterday or a decade ago; a date tells me for sure. The military takes a perfectly readable part number for a capacitor which already contains the necessary information and replaces it with a meaningless number that you have to look up in a table that nobody has. And what does the table tell? It gives the cross-reference to the original part number! How to do it right: you may have heard of the "miracle memory metal" NiTiNOL. Its name tells you it is an alloy of NIckel and TIn and it was developed at the Naval Ordnance Laboratory, so you have not only an idea of what's in it, but where to go for information! Much better than "Alloy X-1B."

Fight these bad people. Engineers are usually not argumentative and just look for ways around roadblocks. I think we are the only group other than maybe the left-handed Albanian guitar players not protesting for our rights. Snarl as you go back to your cage. It keeps them on their toes, and may make it a little easier for the next engineer.

Intellectual Honesty

This is not a separate section, nor does it fit into another section. It belongs in all of them. You may at times get ahead by fooling someone else, but you will not make much progress fooling yourself. If you have read my diatribes before, you know I feel the computer side of our profession has at times oversold its product. But we analoggers are not in good stone-throwing position, either. Don't make the truth shortage any worse; it's bad enough already. Consider how far we have slipped:

I have several catalogs labeled "digital." There are few digital circuits in them; most are binary. I worked with a real digital computer back in 1957, probably one of the last. It used vacuum tubes, was the size of a furnace, and in fact had a stovepipe on it to get the heat out. The only memory was punched cards, its input and output.

I have a large number of catalogs labeled "linear." Half the devices in them are nonlinear. "Analog" is better, although we seldom still compute analogues of anything, other than an occasional inductor. Every "sample-and-hold" in my book is really a track-and-hold. "Differential amplifier" refers to anything from an op amp to a matched pair of transistors; the term has become useless. I sent for info on a "hex op amp" in a 14-pin DIP, an impossibility. I figured they were prewired as followers or inverting-only, which might be useful. But it wasn't even that: it was six CMOS inverters. I guess "op amp" meant they wouldn't oscillate with feedback, which wasn't too surprising with only 20dB gain. Most zener diodes are really avalanche diodes.

We talk about voltage and phase, forgetting that both exist only as differences. An unspecified voltage is presumed to be referenced to ground, but what ground? Phase reference is often very obscure, leading to a lot of errors, for example in FFT systems. In modulation systems the DC component may vanish *either* because its amplitude is zero or because its phase is zero (oops; relative to sine).

Does it really matter? I'm afraid so. In the next section I will cite a downfall from misuse of the term "integrator." One of my favorite gripes: op amps have high gain and wide bandwidth. WRONG. Op amps have high gain *or* wide bandwidth. That is the meaning of gain-bandwidth product, inherent in op amps. This carelessness can lead to op amps being used where another device would work better.

Example: A large number of precision full-wave rectifier circuits using op amps have been published. Usually frequency response is not mentioned. This is a difficult application for an op amp; the frequency response can be surprisingly terrible. A sine wave when rectified has significant (–40dB) components out to ten times the fundamental frequency, and the time it takes the op amp to slew across diode drops and other nonlinearities can be disastrous. There are ways of doing it without op amps. One way is to feed the signal to both linear and clipped inputs of a balanced modulator so the signal gets multiplied by its own polarity (sneak a look at Figure 20–10A if you like). The old 796/1596 balanced modulators were fast, although gain and DC stability were poor.

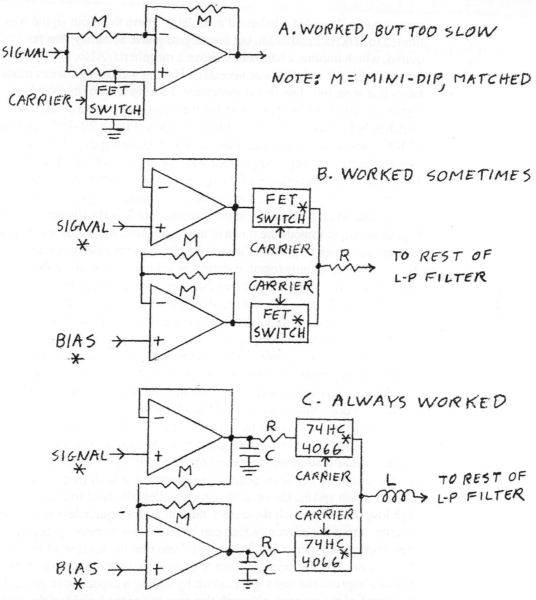

A. WORKED, BUT TOO SLOW

NOTE: M = MINI-DIP, MATCHED

B. WORKED SOMETIMES

TO REST OF
L-P FILTER

C. ALWAYS WORKED

TO REST OF
L-P FILTER

✱ IF FET SWITCHES ARE RUN SINGLE-SUPPLY,
BIAS = $+\frac{V}{2}$ AND SIGNAL MUST BE REFERENCED TO IT
ALTERNATELY, IF BIAS = 0 SIGNAL IS REFERENCED TO GND,
BUT FET SWITCHES MUST BE RUN SPLIT-SUPPLY

Figure 20–10.
DC–accurate
balanced
modulators.

Next gripe: Balanced modulator circuits having an op amp on the output. The modulator output, by definition, is a switching waveform having infinite slopes. An op amp cannot accommodate these for at least two inherent reasons—bandwidth and slew rate limits. Although a form of low-pass filtering, slew-rate limiting is nonlinear, and hence generally unacceptable. I needed a balanced modulator where the input signal was limited to 20KHz bandwidth, but one-degree phase accuracy was required, which implies a bandwidth above a megahertz. Also, better than one-millivolt DC stability was necessary. Having a choice between modulators that were too slow or too inaccurate, I had to devise my own (see Figure 20–10). I buffered and inverted the signal with op amps and then switched between the two with CMOS switches (Figure 20–10B). (Digital CMOS transmission gates can transmit analog signals quite nicely, and they are fast!) The output impedance was that of the switches, but above 1MHz the output impedance of the op amps was no better. Note that with this arrangement I could put considerable capacitance to ground on the op amp outputs since they only had to accommodate 20KHz (linear), which kept their output impedance down at the higher frequencies. It was necessary to find a trick to prevent the switches from momentarily connecting the two op amp outputs together during switching, which drove them crazy. The balanced modulator was followed by a low-pass filter, as is often the case. Putting its input resistor *before* the switches (Figure 20–10C) prevents the outputs of the op amps from being tied directly together if one switch closes before the other opens. (What if one switch opens before the other closes? The filter momentarily gets no signal, which is a truly minimal glitch, since the signal is in the process of switching to its opposite anyway!) Note that the resistance does not double, as only one resistor is connected at a time.

Important note: If a filter like that of Figure 20–1A is used, you may find surprisingly large switching glitches on the output, exactly what a low-pass filter is supposed to get rid of! The path that is supposed to be positive feedback becomes positive feed-forward at high frequencies because: for fast spikes, the op amp output is allegedly held to zero by (1) high loop gain (of which there isn't much at high frequencies) and (2) low op amp output impedance (which can be 100 ohms or more, going up even higher with increasing frequency). Note that the version of Figure 20–1B is much better because there is a capacitor to ground in the path. You can improve the unity-gain circuit by adding a capacitor to ground at the output of the op amp, although this may lower the height of the spikes by making them wider. The capacitor does *not* change the filter characteristic as long as the op amp can drive it at any frequency in the passband, and does not oscillate. In this case the filter of Figure 20–5 was used. Here the inductor blocked the current spikes (Figure 20–10C).

Incidentally, I was surprised to find the digital driving circuitry slowing my analog circuit down! I found that one side of a 4000-series CMOS flip-flop lagged the other by a significant amount; I had to switch to high-speed CMOS. This shouldn't have surprised me because a linear circuit need only pass the signal frequency accurately, but logic needs orders of magnitude more bandwidth. For analog work, a 1MHz digital switch may not be of any more use at 1MHz than an op amp with a GBW of 1MHz, namely none.

Simplify, Simplify, Simplify

Simplify as much as possible (but no more)! I have read countless windy dialogues which claimed to reveal some new truth but in reality only obscured an old one. A real classic:

Years ago I read an article in a magazine where the author claimed that by adding an inverting transistor in the loop after an op amp and switching the feedback to the noninverting input he had (1) increased the time constant of an integrator by a factor of beta, (2) made a noninverting integrator, (3) achieved a high-input impedance integrator; none of which was true. I wrote to the magazine editor, who forwarded my letter to the author. I received back a 6-page "proof" of his claims. (The circuit only had 1 op amp, 1 transistor, 1 capacitor, and 3 or 4 resistors.) I plowed through his convoluted analysis. His math was correct; it gave the standard result when his complicated answer was simplified properly. Instead, he associated some terms with things "everybody knows" and ignored others. He fooled himself and the magazine, but he didn't fool me or the circuit. I breadboarded it to be absolutely sure. The author claimed he had tested it successfully, but gave no details.

MURPHY'S LAW, APPLIED TO OBFUSCATION:
There may or may not be a simple way of looking at a problem,
but there is always a complicated way.

Part of the problem came from the common practice of referring to an RC low-pass as an "integrator." It may be technically correct to model an integrator as a low-pass filter having a DC gain of 1,000,000 and a breakpoint of 0.0001Hz, but it just distracts from the true use. In the old days op amps were sometimes described as having an inverting gain and a noninverting gain, both with loose tolerances. One might not have appreciated that the two were almost perfectly matched, the basis for a lot of good circuits.

I was work supervisor for a thesis student whose analysis of his system produced a result we both knew was incorrect, but neither of us could see where he had gone wrong. I suggested a simpler analysis, but he was determined to find out why his didn't work, which I didn't want to discourage. His school supervisor found the error: he had canceled a complicated expression from both sides of an equation. It turned out to be equal to zero, so he had unknowingly divided by zero, leaving nonsense.

Humility

Most people, particularly engineers, like to think they are in full control of the situation at all times. This is not so. A graphic demonstration is a Washington, DC ice storm. It's amazing how the world changes when the coefficient of friction changes an order of magnitude. We had a particularly bad one where, after all the cars piled into each other at the local intersection, the drivers got out—and all fell down! It was so slick you could not walk; some people literally crawled home. Some loons flying south (I am speaking of the birds, not the congressmen) iced up and crashed into suburbanites' yards! Recalling this scene reminds me how close to the ragged edge we really are.

THOUGHT PROBLEM:
What would happen if the value of pi suddenly changed?

I recommend an occasional dose of humility. If you are overconfident, it usually happens automatically. True, you need some self-confidence to achieve anything, but you are not likely to learn if you believe you already know it all. If you need help, recall that before the days of transistors, not to mention ICs, the vacuum tube crew had 'scopes, oscillators, voltmeters, counters, regulated power supplies, radios, TVs, radars, sonars, i.e., most everything we have now. True, usually not as nice, although many audiophiles are still hanging onto their tube amplifiers. If all the time and money put into semiconductors had been put into tubes instead, they would probably be pretty good by now. We would surely have integrated circuits, and maybe heaterless versions and complementary devices!

I had almost forgotten that as a graduate student I designed an op amp. At the time (1962), tubes worked better than transistors. The op amp had only 62dB gain, but that was flat to 10KHz, giving it a gain-bandwidth close to 10MHz, an order of magnitude better than a 741. (Did you ever see a 741 with a warning label—"USE OF THIS PRODUCT ABOVE 1KHZ MAY LEAD TO LOSS OF ACCURACY"? It's true!) It would tolerate a 4000pF load. Settling time to 0.1% of final value was less than 10 microseconds, including a 40V step, with the 4000pF load. (Output current was 100mA.) Other parameters weren't as good as a 741, except input impedance, which was reported as infinite. (Infinity was smaller back then, perhaps because the universe hadn't expanded as far.) It had a true balanced input, but the only use envisioned for the noninverting input was a handy place to connect the chopper amplifier. How shortsighted! Supply current was significantly higher, especially if you count the heater current. It covered an entire card, and we won't talk about cost.

Have we progressed all that far? Well, yes and no. I sometimes miss the warm, cheery glow of vacuum tubes on a cold winter's night and the thrill of seeing blue lightning bolts inside them when I exceeded the voltage rating. They don't have transistors that glow purple like the old regulator tubes, or even LEDs for that matter.

Luck of the Irish and Non-Irish

More than once I have improved a circuit by accident. Typical is: moving a 'scope probe, and hence its ground lead, and finding out I had too many grounds (ground loop) or too few (no ground connection). Or dropping the probe on the circuit and seeing oscillations disappear. (The shielded cable acts as a shield for whatever it falls between.) Or sticking a capacitor in the wrong hole. (So *that's* the point that needs more capacitance!)

Jim Williams tells of getting the answer to a circuit problem by observing the monkeys at the zoo. I would have thought that more applicable to management, but the point is that ideas seem to be held up in the brain until some trigger springs them loose. If you seem to have a block you can't get through, try to get around it, using whatever sources happen along, no matter how unlikely.

I once had a balky circuit card that would always work for me, but never for the person I made it for, which made troubleshooting difficult. But it also gave me a clue. The problem was an unconnected CMOS input. These have such a high impedance they can be switched by static fields. One of us was apparently charged positive and the other negative. I was lucky it showed up before it got any farther.

Engineering Ethics

In the uniformed military, when an accident happens it is, by definition, somebody's fault. Somebody has to be responsible, and you don't want to be that somebody. This system has its shortcomings, but it works better than the civilian government, where nothing is anybody's fault, no one has the responsibility, and mistakes happen over and over again.

UNANSWERED (LEGAL) QUESTION:
Where does an engineer's responsibility stop?

I suspect the average engineer would say a widget is just a widget, and how it gets used or misused is beyond our responsibility. It may be logical, but others say different, and they have been winning some in court. Companies have been held responsible for damage their products did, even when warnings were ignored or the equipment totally misused. A friend of mine faced a million-dollar lawsuit when the plaintiff named the private contractor involved, the highway department, which let the contract which set the rules, and the head of the department, who delegated the authority. *Design News* carries a legal column every issue. Spec sheets carry disclaimers on use in life-support equipment. I recently received a shipment of capacitors which included a sheet telling me not only not to eat them, but what remedies to take if I did! I am not making this up.

Two good books have recently been published on disasters involving engineering.[15,16] Most were mechanical problems, but several involved computers, two concerned electrical power, and one was an effective failure of an analog system. Losses and/or lawsuits in most cases involved millions of dollars. Many involved injuries or fatalities.

What would you do if one of your products accidentally hurt somebody? I hope you would feel genuinely sorry, but I also hope you would not give up your profession. But, first of all, try to prevent it. Design your products as if you had to plead your case to a judge, who can't operate the controls on his VCR, against a lawyer, who is out of jail only because she is a lawyer. It could happen.

Go to It!

Engineers are needed, now more than ever. We old-timers are wearing out one by one. Yet I have mixed feelings about encouraging young engineers these days. Our paychecks don't reflect our contribution toward the incredible improvement in our standard of living that has been achieved in the last century. And we seem to get blamed for everything that goes wrong, and indeed for everything we haven't been able to fix yet. The best I can offer is the satisfaction of knowing that you have accomplished something worth doing, and that's worth more than money.

Analog is not a curiosity. It is out there, both on its own and helping computers interface to the real world. I hope I have helped you in some small way in our small corner of the profession. Returning to the question posed by the title of this chapter, if I have presented my case well, you know the answer is ALL OF THE ABOVE. Use any tool available to you. Touch a computer if you have to; just wear rubber gloves (mentally, at least). And don't forget, HAVE FUN!

References

1. A. Delagrange, "Amplifier Provides 10 to the 15 ohm Input Impedance," *Electronics* (August 22 1966).

2. A. Delagrange and C.N. Pryor, "Waveform Comparing Phasemeter," U.S. Patent 4025848 (May 24 1977).

3. A. Delagrange, "Lock onto Frequency with Frequency-Lock Loops," *Electronic Design* (June 21 1977).

4. A. Delagrange, "Need a Precise Tone? Synthesize Your Own," *EDN* (October 5 1980).

5. M. Damashek, "Shift Register with Feedback Generates White Noise," *Electronics* (May 27 1976).

6. A. Delagrange, "Simple Circuit Stops Latching," letter to the editor, *Electronic Design* (May 28 1981).

7. A. Delagrange, "It Could Be the Ideal Filter," *Electronic Design* (February 16 1976).

8. A. Delagrange, "An Active Filter Primer, MOD 2," Naval Surface Warfare Center Technical Report (September 1 1987): 87–174.

9. A. Delagrange, "Op Amp in Active Filter can also Provide Gain," *EDN* (February 5 1973).

10. A. Delagrange, "High Speed Electronic Analog Computers Using Low-Gain Amplifiers," U.S. Patent 5237526 (August 17 1993).

11. Bruton and Trelevan, "Active Filter Design using Generalized Impedance Converters," *EDN* (February 5 1973).

12. A. Delagrange, "Design Active Elliptic Filters with a 4-Function Calculator," *EDN* (March 3 1982).

13. A. Delagrange, "Feedback-Free Amp makes Stable Differentiator," *EDN* (September 16 1993).

14. R. Pease, *Troubleshooting Analog Circuits,* Butterworth–Heinemann (1991).

15. Steven Casey, *Set Phasers on Stun,* Aegean Publishing Co. (1993).

16. Henry Petroski, *To Engineer is Human,* Vintage Books (1992).

Appendix A

Proof That PI = 2

Circumscribe a sphere with an equatorial circle C (see Figure 20–A1). Draw a line R from the equator to the pole; this is the radius of the circle. Obviously C = 4R. Therefore the diameter is ½ of the circumference, or pi = 2. "No fair!" you will undoubtedly say, "You used spherical geometry!" But that is precisely the point. It has been known for half a millennium that we live on a sphere; plane geometry is the wrong method to use.

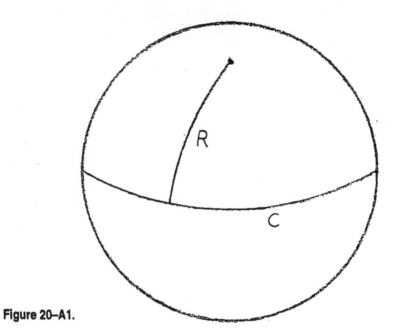

Figure 20–A1.

Appendix B

Proof That PI = 4

Circumscribe a circle of diameter D with a square (see Figure 20–B1). The perimeter of the square is obviously 4D. Now, preserving right angles, "flip" the corners of the square in until they touch the circle, as indicated by the dashed line. The perimeter of the new shape is obviously still 4D. Now, "flip" the eight new corners in (dotted lines). The perimeter is still unchanged. Keep "flipping" the corners in until the shape becomes a circle. The perimeter, still 4D, becomes the circumference, so pi = 4.

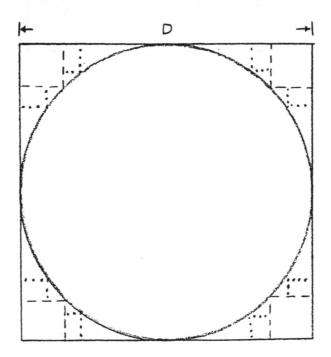

Figure 20–B1.

Index

Printed and bound by CPI Group (UK) Ltd, Croydon, CR0 4YY

03/10/2024

01040336-0014